Green Energy and Technology

More information about this series at http://www.springer.com/series/8059

Kwame Awuah-Offei
Editor

Energy Efficiency in the Minerals Industry

Best Practices and Research Directions

Springer

Editor
Kwame Awuah-Offei
Department of Mining and Nuclear Engineering
Missouri University Science and Technology
Rolla, MO
USA

ISSN 1865-3529 ISSN 1865-3537 (electronic)
Green Energy and Technology
ISBN 978-3-319-54198-3 ISBN 978-3-319-54199-0 (eBook)
https://doi.org/10.1007/978-3-319-54199-0

Library of Congress Control Number: 2017955235

Printed on acid-free paper

This Springer imprint is published by Springer Nature
The registered company is Springer International Publishing AG
The registered company address is: Gewerbestrasse 11, 6330 Cham, Switzerland

Contents

Chapter 1
Introduction

Kwame Awuah-Offei

Abstract The goal of this book is to present the current knowledge regarding energy efficiency implications of mining processes and future research directions. This introductory chapter explains the purpose and motivation for this book, provides highlights of the book, provides strategies that a reader can use to read the book, and identifies the key unanswered questions that require further research. It is my hope that this book will be a valuable resource for industry professionals and researchers and stimulate further discussions on energy efficiency in mining.

Keywords Energy efficiency · Energy · Mining · Minerals industry

1.1 Introduction

Total energy costs are high for most mines because mining is an energy intensive activity. Energy cost is a key consideration for mining professionals and researchers that drive decisions on research and initiatives on energy efficiency. As pointed out by Levesque and co [1], the prevalence of energy efficiency initiatives in the minerals sector is closely correlated with energy prices. In recent years, however, concerns about climate change, the carbon footprint of products, and the related internalizing of associated costs are also driving decisions related to energy consumption and efficiency in the minerals sector. Finally, energy costs are important to mining professionals because the overall energy efficiency of mining is affected by the efficiency of all parts of a mineral project. Energy is consumed by all the processes in the mineral life cycle.

A volume like this one that discusses best practices and provides research directions for the future is long overdue. As energy consumption, the associated climate change impacts, and costs have become increasingly important, mineral

K. Awuah-Offei (✉)
Mining and Nuclear Engineering Department, Missouri University of Science and Technology, 226 McNutt Hall, Rolla, MO 65401, USA
e-mail: kwamea@mst.edu

K. Awuah-Offei (ed.), *Energy Efficiency in the Minerals Industry*, Green Energy and Technology, https://doi.org/10.1007/978-3-319-54199-0_1

industry professionals and researchers have had to look at a variety of sources to gather information about best practices and ongoing and future research initiatives. To the best my knowledge, no volume like this exists in the English literature that collates contributions from various experts into one resource for industry professionals and researchers.

This volume presents the current state of the art regarding energy efficiency implications of mining processes. The book is divided into four main sections: ground fragmentation; material handling; mineral processing and extractive metallurgy; and miscellaneous topics. The main sections follow, to a large extent, the unit processes of mining so that the reader can instinctively know where to find things. Besides attempting to explain the purpose and motivation for this volume, this introductory chapter tries to summarize the highlights of the chapters contained in this book, provide strategies that a researcher or an industry professional can use to read the book and identify the key unanswered questions that require further research.

1.2 Highlights

The first section of this book includes five chapters that discuss energy efficiency implications of blasting and ground fragmentation in mining [2–6]. The section starts with an overview of the energy distribution during ground fragmentation by blasting [2]. This chapter provides a basic introduction to the basic theories on the energy content of explosives, how that energy is released to the surrounding rock mass during detonation, and the different forms that energy is transformed into during explosives. By itself, the discussions in this chapter can inform a mining engineer's decisions on explosive selection, blast design, and execution. However, the chapter also serves as a useful introduction to the next four chapters in this section.

The last of these chapters experimentally examines the energy efficiency of rock fragmentation using blasting [6]. The researchers determined the proportion of the explosive energy transformed into seismic wave energy, kinetic energy, and fracture energy transferred during the blasting process. They conducted experiments at two quarries to determine the energy proportions using the seismic field from seismograph records, initial velocity of the blasted rock face obtained from high-speed video footage, and fragment size distributions from image analysis of the muckpile material, respectively. Their work shows that the maximum total energy measured, in these experiments, accounts for at most 26% of the available explosive energy, indicating that the energy efficiency of blasting is rather low.

One of the remaining three chapters deals with the energy efficiency of drilling, which is the method for creating a means of loading explosives into rock for fragmentation [5]. The remaining two chapters deal with the effect of hole stemming and detonation wave collision on fragmentation results and energy efficiency of blasting [3, 4].

The second section of the book, which covers material handling, contains four chapters that deal with the energy efficiency of material handling operations [7–10]. The section opens with an overview of energy efficiency implications of loading and hauling equipment [7]. The next two chapters deal, respectively, with shovels and trucks, which together constitute the most common loading and hauling method [8, 9]. The first of the two provides a review of the current literature on cable shovel energy efficiency while the second deals with approaches for benchmarking energy efficiency of trucks.

The final chapter in this section presents a framework for assessing dragline energy efficiency using equipment monitoring data [10]. The authors present a three-step approach involving: (1) assess energy efficiency using data from dragline monitoring systems to estimate an overall performance indicator; (2) quantify the relationship between different operating parameters and the energy efficiency indicator; and (3) improve the energy efficiency performance of operators by using the results to optimize operator training.

The third section deals with energy efficiency implications of mineral processing and extractive metallurgy and contains four chapters [11–14]. Given the significance of comminution energy in any discussion of energy efficiency in mining, it is perhaps befitting that this section begins with a chapter on the best practices and future research needs for energy efficient comminution [11]. The next chapter discusses electrical energy consumption in electrowinning of metals from solution [12]. The chapter shows that to achieve energy efficient electrowinning a plant has to maximize current efficiency and optimize electrolysis parameters. The chapter also concludes that significant energy savings can only be achieved by changing one of the underlying electrochemical reactions or reducing the anode overpotential. The next chapter deals with plant process control and real-time optimization approaches that are used to achieve lower specific energy requirements by lowering variability in key process variables and determining more appropriate operating points [13]. The chapter also presents case studies to illustrate the current state of the art in process plant automation for energy efficiency. The final chapter in this section presents the case of the Atlantic Copper in Huelva, Spain, which was the first copper smelter in the world to receive ISO 50001 energy management certification [14]. The chapter presents Atlantic Copper's experience, process, and results in energy management.

The final section, which includes three chapters, deals with renewable energy in mining and other miscellaneous topics [15–17]. The use of renewable energy in mining is one of the major energy innovations in the last decade. The first chapter in this section discusses the integration of solar energy in mines' energy supply to enable them address energy and sustainability challenges [15]. The chapter discusses recent developments in solar energy in the mining industry and presents case studies where this framework has been successfully applied to incorporate solar energy into the mining energy supply mix. The second chapter in this section deals with energy efficient practices in mine ventilation [16]. Mine ventilation is an ancillary operation in underground mining that can have significant energy efficiency implications. The second chapter of this section discusses the energy efficiency implications of this important aspect of underground mining systems.

The final chapter of the section investigates the technical and economic feasibility of installing an energy recovery system (ERS) on diesel electric drive mine haul trucks [17]. On a mine haul truck, an ERS saves energy by recovering energy when the truck brakes during descent into the pit and puts that energy back into the system during ascent out of the pit. The chapter evaluates the technical and economic viability of various ERS technology using simulation. The work shows that lithium-ion batteries are infeasible because of poor charging rate while electrolytic double-layer capacitors are infeasible because of its low cycle life. Electromechanical flywheels are judged the most cost-effective option.

1.3 How to Use This Book

This book intended to be a resource for mine managers and engineers who want to improve the energy efficiency of their operations and, thereby, increase production efficiency and sustainability. It is also intended to be a resource for researchers looking for a comprehensive review of the literature on energy efficiency in the minerals industry. Each chapter is written by subject-matter experts who have contributed to the literature on the topics they have written in this book and are familiar with the current knowledge and outstanding questions that need further research. I anticipate that there will be two kinds of users for this book: industry practitioners and researchers.

For industry professionals, I suggest they start with this introduction. They can take note of the description of the various sections of the book as outlined in Sect. 1.2. The professional can then refer to the particular section of the book or the particular chapter that is of interest. It is my hope that each chapter provides an adequate overview of the energy efficiency consideration in the particular area for an industry professional. However, in case further reading on the subject is required, the list of references in the chapters is a good starting point for any professional.

For researchers, this volume is a good starting place for research on various energy efficiency topics in mining. Each of these chapters is a good review paper that summarizes the state of the art and provides citations to the relevant literature in the area. For the beginning graduate student or the seasoned researcher, the chapters in this book represent a valuable resource for energy efficiency research in the minerals sector. In some cases, complimentary chapters (especially, those in the same section) can provide additional resources that will be useful for any energy efficiency research endeavor.

1.4 Future Research Directions

This volume provides many suggestions for industrial best practices that managers and engineers across the mine life cycle can use to improve the energy efficiency of the mines. Each of these suggestions is backed by sound research. However, many other areas still require further research so we can bridge the gap between theoretical benchmarks and actual energy efficiency performance.

Comminution and material handling are still the main areas that show the most potential for energy efficiency improvement. The gap between theoretical benchmarks and current best practice is still wide. For example, as I have pointed out earlier, Sanchidrián and colleagues estimate that the maximum total energy measured during a blast does not exceed 26% of the available explosive energy [6]. We will need to reimagine how we remove in situ material and reduce its fragment sizes to the sizes required to liberate valuable minerals from gangue in order to bridge this gap.

Certainly, research that uses holistic, systems-based approaches (e.g., mine-to-mill techniques) that lead to globally optimal systems should be encouraged over approaches that optimize subsystems alone [4, 11, 18, 19]. At a minimum, any energy efficiency improvement initiative should evaluate the effect of improving a subsystem along the mining energy chain on the global energy efficiency of the mine. Management should not pursue any energy improvement initiative that improves a subsystem but does not result in overall energy improvement. Hence, future research that aims to improve mining energy efficiency should always consider the system-wide effects of any efforts to optimize energy efficiency.

Specific areas that can provide the necessary improvement in energy efficiency of comminution and material handling include research that helps us improve our understanding of energy transformations during blasting so that we can direct more of the explosives energy toward useful work. Also, we need better understanding of how operators affect energy efficiency of material handling machines. This is necessary to clarify the relationship between specific operator practices and energy efficiency. Finally, to facilitate more energy efficient comminution, we need to better understand fracture mechanics in rocks.

In addition to comminution and material handling, especially in light of climate change and its implications for mining, there is a need for research that facilitates optimal integration of renewable energy sources into mines. We need more research that provides means for determining optimal hybrid systems for different operating and economic conditions [20]. This is particularly important for mines in remote areas where there is little to no energy infrastructure.

Finally, as I have argued elsewhere [21], we need research that properly articulates the return on investment for public policy that facilitates energy efficiency in the minerals industry. Typically, manufacturing and residential energy consumers get the bulk of the policy attention when it comes to energy efficiency. However, mining is very energy intensive and a significant energy consumer in many economies. In particular, electricity consumption by the mining industry makes up a significant portion of the energy use in developing countries with significant extractive industries [22]. Hence, it will be of public interest to increase the energy efficiency of mines in order to improve the energy efficiency of the entire economy.

1.5 Conclusions

It is my hope that this volume will be a valuable resource for industry professionals and researchers. The work in this volume represents the state of the art regarding key topics in energy efficiency in the minerals sector. The breadth of coverage and the depth in each of the chapters make it a useful resource for all managers and engineers interested in energy consumption and efficiency at the mine site. Above all, I hope that this volume will spur on further discussions on all aspects of energy efficiency in mining.

References

1. Levesque M, Millar D, Paraszczak J (2014) Energy and mining—the home truths. J Clean Prod 84:233–255. doi:10.1016/j.jclepro.2013.12.088
2. Lusk B, Silva JJ (2018) Energy distribution in the blast fragmentation process. Energy Effi Miner Ind Best Pract Resour Dir
3. Konya CJ, Konya A (2018) Effect of hole stemming practices on energy efficiency of comminution. Energy Effi Miner Ind Best Pract Resour Dir
4. Johnson C (2018) Effect of wave collision on fragmentation, throw and energy efficiency of mining and comminution. Energy Effi Miner Ind Best Pract Resour Dir
5. Karpuz C (2018) Energy efficiency of drilling operations. Energy Effi Miner Ind Best Pract Resour Dir
6. Sanchidrián JA, Segarra P, López LM (2018) Energy efficiency in rock blasting. Energy Effi Miner Ind Best Pract Resour Dir
7. Soofastaei A, Karimpour E, Knights P, Kizil M (2018) Energy efficient loading and hauling operations. Energy Effi Miner Ind Best Pract Resour Dir
8. Awuah-Offei K (2018) Energy efficiency in cable shovel operations. Energy Effi Miner Ind Best Pract Resour Dir
9. Sahoo LK, Bandyopadhyay S, Banerjee R (2018) Benchmarking energy consumption of truck haulage. Energy Effi Miner Ind Best Pract Resour Dir
10. Abdi-Oskouei M, Awuah-Offei K (2018) Role of the operator in dragline energy efficiency. Energy Effi Miner Ind Best Pract Resour Dir
11. Klein B, Wang C, Nadolski S (2018) Energy efficient comminution: best practices and future research needs. Energy Effi Miner Ind Best Pract Resour Dir
12. Moats MS (2018) Energy efficiency of electrowinning. Energy Effi Miner Ind Best Pract Resour Dir
13. Bouchard J, Sbarbaro D, Desbiens A (2018) Plant automation for energy efficient mineral processing. Energy Effi Miner Ind Best Pract Resour Dir
14. Tejera JM, Rios G, Martínez T, Palacios M (2018) Energy management systems in copper smelting: the atlantic copper case study. Energy Effi Miner Ind Best Pract Resour Dir

15. Sanchez JPP (2018) Solar energy applications in mining: a case study. Energy Effi Miner Ind Best Pract Resour Dir
16. Demirel N (2018) Energy efficient mine ventilation practicesjosé pablo paredes sánchez. Energy Effi Miner Ind Best Pract Resour Dir
17. Terblanche PJ, Kearney MP, Hearn CS, Knights PF (2018) Technology selection and sizing of on-board energy recovery systems to reduce fuel consumption of diesel electric mine haul trucks. Energy Effi Miner Ind Best Pract Resour Dir
18. Schofield N, Moore J, Carswell J (2013) Mine to mill reconciliation. AusIMM Bull 38–42
19. Nadolski S, Klein B, Elmo D, Scoble M (2015) Cave-to-Mill: a Mine-to-Mill approach for block cave mines. Min Technol 124:47–55. doi:10.1179/1743286315Y.0000000001
20. Carvalho M, Romero A, Shields G, Millar DL (2014) Optimal synthesis of energy supply systems for remote open pit mines. Appl Therm Eng 64:315–330. doi:10.1016/j.applthermaleng.2013.12.040
21. Awuah-Offei K (2016) Energy efficiency in mining: a review with emphasis on the role of operators in loading and hauling operations. J Clean Prod. doi:10.1016/j.jclepro.2016.01.035
22. Oladiran MT, Meyer JP (2007) Energy and energy analyses of energy consumptions in the industrial sector in South Africa. Appl Energy 84:1056–1067. doi:10.1016/j.apenergy.2007.02.004

Part I
Ground Fragmentation

Chapter 2
Energy Distribution in the Blast Fragmentation Process

Braden Lusk and Jhon J. Silva

Abstract The study of energy distribution in a blast fragmentation process is the subject of active research. The complexity of the phenomena and the high intensity and speed of some of the physical processes occurring during an explosion such as high pressures and temperatures make measuring of the energy distribution a very difficult task. Because of the limitation of current technologies to measure the actual energy released in an explosion, the assessment of energy distribution is done considering the balance between the ideal energy stored in the explosive and the effects of the released energy in the surrounding media. To study the ideal amount of energy in the explosive, it is necessary to use thermophysics and thermodynamic principles, while the effects in the surrounding media are explained using materials deformation theories, material fracture models, and dynamics. This chapter will review the basic principles behind the assessment of the ideal energy in the explosives and discuss the most accepted theories about the distribution of the energy in the surrounding media when an explosion takes place.

Keywords Explosive energy · Fracture · Vibration · Detonation
Blasting · Fragmentation

2.1 Introduction

In the rock blasting process, there are two main considerations: the explosive and the confining rock media. The effects of the energy released by the explosives in the rock media and measured as fragmentation, ground movement, and vibration among others, are dependent on the characteristics of the explosive products.

B. Lusk
Missouri S&T, 226 McNutt Hall, 1400 N Bishop Avenue, Rolla, MO 65401, USA
e-mail: blusk@mst.edu

J. J. Silva (✉)
University of Kentucky, 504 Rose Street, 234H M&MRB, Lexington, KY 40506, USA
e-mail: jhon.silva@uky.edu

K. Awuah-Offei (ed.), *Energy Efficiency in the Minerals Industry*, Green Energy and Technology, https://doi.org/10.1007/978-3-319-54199-0_2

It is also a function of the initial conditions and the properties of the rock media. The varying levels of initial conditions and properties of the rock media are often characterized simply as confinement. When the energy is released from the explosive, a chemical reaction involving a fast oxidation reaction occurs. As a consequence of the oxidation, a significant amount of heat is generated. The heat increases the temperature of the gasses, expanding the gasses into a volume much larger than the original volume. As a result of the rapid expansion of the gasses, a high pressure is developed. The high pressure in the gasses will perform different tasks in the confining media in the form of plastic and elastic deformation, movement, rock damage and fracturing, among others. The release of energy in the explosive involves changes of the matter between different states from solid to liquid and gaseous and also will produce changes in its chemical state. Engineers and scientists use *thermophysics* principles to assess the temperatures and amounts of energy required to change the matter (the explosive) between the different states. On the other hand, they use *thermochemistry* to study variations in the chemical state. Both disciplines allow us to quantify the changes in internal energy (U) and enthalpy (H) [3]. A low input energy is required to start the process of releasing the internal energy stored in the explosive. In the initial stage, the bonds in the molecules are broken in an endothermic process. Once the process has started, the changes in the chemical state of the explosives will allow the formation of products resulting in the liberation of the energy of the explosive in an exothermic reaction [5, 7]. To better understand the balance of energy stored in the explosive, this chapter will review the basic physical concepts of heat, temperature, and work.

Because the explosive matter changes between different states, the chapter will also review the heat capacity, latent heat of fusion, heat of vaporization, and heat of transitions. Changes in the chemistry of the explosives will produce heat of reaction, heat of formation, heat of combustion heat of detonation, and heat after burn. Chemistry and state changes are usually measured and quantified by assessing the internal energy and the enthalpy of the explosive.

To study the tasks or effects of the energy released by the explosives in the surrounding media, it is also necessary to review the concepts of rock fracture and rock fragmentation. Rock fracture deals with the study of generation and propagation of one or more cracks while rock fragmentation not only studies generation and propagation of many cracks but also the total energy consumed by the surrounding media and the final size distribution of the fragments [14]. Fragmentation also includes rotation and movement of individual rock particles, which produces grinding during movement. To better understand the rock fragmentation process, this chapter will review the Griffith energy balance concept, the strain energy released rate, fracture surface energy, and the fracture surface energy consumed. It is commonly accepted that the release of energy from the explosives in the rock media will be distributed in different functions such as: (1) the energy that produces plastic deformation in the crushed zone adjacent to the blasthole and is used in the borehole expansion, (2) the energy to create new cracks and new surfaces in the fractured zone, (3) the energy used to extend and open the current cracks in the fractured zone, (4) the energy to mobilize the rock fragments and provide an initial velocity to flying rock fragments, (5) the energy that provides initial angular velocity to rotating rock fragments, (6) the energy consumed in heating the rock mass, (7) the seismic energy carried by the stress waves out of the fractured zone, and (8) other forms of energy including radiant energy in form of electromagnetic radiation, sound, light, and others.

All of the functions described briefly above consume portions of the overall energy contained in the explosive product. The distribution of this energy is dictated by many factors. Each function is critical in the analysis of the fragmentation process. As a finite quantity of energy exists in a column of explosive, the sum of energy consumed by all functions must be equal to that available. Many of these functions are well defined in the literature and practice; however, the overall energy balance and some specific functions are not clearly understood and thus are debated widely in the literature. This chapter discusses some of the most accepted and proposed distributions found in recent research.

2.2 Stored Energy in the Explosives and Release Balance

The energy produced when explosives react is an oxidation process. A similar process occurs in a combustion reaction. The main difference between the two chemical reactions (detonation and combustion) is the high velocity of the reaction in an explosion. In the simplest definition, an explosion can be defined as a chemical process in which a substance reacts with oxygen and releases heat. Under this concept, the explosive can be called the reactant, and the source of oxygen is called the oxidizer. During the process, new chemical substances or products are created from the reactant and the oxidizer. In chemistry, the difference between the internal energy of the reactant and the products is called heat of reaction. In the case where the reactant is an explosive such heat is known as heat of explosion. The amount of heat generated in an explosion is indirectly measured by the energy content in an explosive. It is related to the characteristics of the explosive material and the available quantity of oxidizer to burn all the reactant to its most highly oxidized products. The resulting products will have the lowest internal energy. The relevant characteristics of the explosive are quantities often determined experimentally. Because of the speed of the reaction and the complex process involved in an explosion, some of the quantities can be measured, but others need to be estimated from secondary information. Among the most significant characteristics of an explosive used to estimate the level of energy stored in the explosive are (1) density, (2) detonation velocity, and (3) detonation pressure. Based on these parameters, it is possible to assess the theoretical energy available in the explosive and the energy available to do useful work. The amount of oxidizer required can be determined using oxygen balance analysis of the explosion phenomena.

The chemical reaction of the explosion needs to generate expanding product gasses. If no gasses are generated in the process, the released energy of the explosive will remain in the reactant as heat and no mechanical work will be transferred from the explosive to the surrounding rock. In this, we explain the oxygen balance analysis and include a discussion of the most significant explosive properties relevant to the stored energy.

2.2.1 *Reaction Product Hierarchy*

Most of the industrial explosives consist of carbon, hydrogen, nitrogen, and oxygen and are called CHNO explosives [3]. The order of products formed during an explosion is known as the reaction product hierarchy. The "rules of thumb" state that:

1. All nitrogen becomes N_2,
2. All available oxygen goes first to convert hydrogen to water H_2O,
3. Any oxygen left after H_2O formation burns carbon to CO,
4. Leftover oxygen from step (3) converts CO to CO_2,
5. Leftover oxygen from step (4) forms O_2 and is available for use in secondary reactions,
6. Traces of NOx (mixed oxides of nitrogen) are always formed, and
7. Any leftover carbon becomes solid residue.

A practical example is the reaction of cyclotrimethylenetrinitramine $(O_2NNCH_2)_3$, best known by the Research Department Formula X (RDX). Its oxidizing reaction is as follows:

$C_3H_6N_6O_6 \rightarrow$ 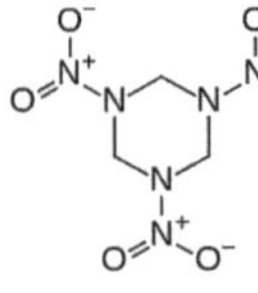

1. All nitrogen becomes N_2:

$$6N \rightarrow 3N_2$$

2. All available oxygen goes first to convert hydrogen to water H_2O

$$6H + 3O \rightarrow 3H_2O\,(3O\ \text{remaining})$$

3. Any oxygen left after H_2O formation burns carbon to CO

$$3C + 3O \rightarrow 3CO\,(\text{all O was consumedat this point})$$

4. Leftover oxygen from step (3) converts CO to CO_2.
   There is no O, so no CO_2 is formed

So the overall reaction of RDX is:

$$C_3H_6N_6O_6 \rightarrow 3N_2 + 3H_2O + 3CO$$

This is the initial reaction sequence of RDX, but some products remain reactive after the initial reaction is complete and they will undergo to further reactions [3].

2.2.2 *Oxygen Balance (OB)*

The amount of energy stored in the explosive and the amount of energy that can be released is related to the capacity to bring the reactant to its most highly oxidized products. If there is enough oxygen in the process, the heat of explosion will be maximized, and the production of gaseous products will be optimum. Knowing the required amount of oxygen is fundamental to knowing the optimum performance conditions for the explosion. The calculation of the oxygen balance is straightforward and can be done assuming that the general formula for the explosive is:

$$C_xH_yN_wO_z \tag{2.1}$$

Assuming that all the carbon could be oxidized to carbon dioxide (CO_2) and all hydrogen is transformed into water (H_2O), then Eq. 2.2 gives the number of oxygen atoms required for exact balance after explosion.

$$\text{OB} = z - 2x - \frac{y}{2} \tag{2.2}$$

If OB in Eq. 2.2 is zero or positive, then there is enough oxygen for the explosion. However, a negative value will indicate a lack of oxygen in the reaction, meaning not all the energy is released in the process. If the OB is expressed in terms of the weight of oxygen compared to the weight of explosive (considering that the atomic weight of the oxygen is 16), Eq. 2.2 becomes:

$$\text{OB}(\%) = \frac{1600}{\text{MW}_{\text{expl}}}\left(z - 2x - \frac{y}{2}\right), \tag{2.3}$$

where MW_{expl} is the molecular weight of the explosive.

In the case of RDX, the oxygen balance is given by:

$$C_3H_6N_6O_6 : x = 3; \quad y = 6; \quad w = 6; \quad z = 6$$

and

$$MW_{expl} = 12.01 * 3 + 1.008 * 6 + 14.008 * 6 + 16 * 6 = 222.126,$$

where 12.01, 1.008, 14.008, and 16 are the atomic weight of carbon, hydrogen, nitrogen, and oxygen, respectively. So Eq. 2.3 results for RDX in:

$$OB(\%) = \frac{1600}{222.126}\left(6 - 2 * 3 - \frac{6}{2}\right) = -21.61\%$$

In the case of RDX, the OB indicates that the combustion will be incomplete, and a large amount of toxic gasses such as carbon monoxide will be present. Commercial explosives target the OB close to zero to minimize the production of harmful gasses and optimize the explosive properties. The OB evaluation of explosives is a tool to improve the mixture of explosives (combination of negative with positive oxygen balanced explosives) but the analysis, and the explosion process is much more complex, and the final results cannot be predicted only using the OB concept.

2.2.3 Properties of Explosives Related to Energy Stored in the Explosive

2.2.3.1 Density

The density of the explosive is closely related to the velocity of detonation (VOD) of the explosive. The density is frequently used to estimate the VOD of the explosive. There are different definitions of density depending on the application and conditions including:

- Theoretical maximum density (TMD),
- Bulk density,
- Loaded density.

The TMD is the mass per unit volume of a single crystal of the explosive. It is sometimes referred as the "crystal density of the explosive". Eremenko [6] proposed a relationship between the TMD and the hydrogen content of substitute organic molecules as:

$$\rho_{TMD} = a_i - k_i H, \tag{2.4}$$

where a_i and k_i are constants dependent upon the chemical structural group and H is the percentage by weight of hydrogen in the explosive molecule.

2.2.3.2 Velocity of Detonation (VOD)

The velocity at which the detonation wave travels through an explosive column is called the velocity of detonation. The performance of an explosive is often measured by its VOD. In general, a reduction in the VOD will cause a reduction in the detonation pressure as well as in the availability of the released energy of the explosive.

There are different methodologies to calculate the VOD of an explosive, ranging from empirical calculations based on the molecular structure of the explosive [3], up to direct methodologies such as the Mettegang method, which is based on the breaking of wires at known distances and the interruption of electrical currents while the explosive is detonated. VOD is also measured by monitoring the change in resistance in cables or probes embedded in the explosive during detonation. Among the variables affecting the VOD of explosives are: (a) the relationship between VOD and density; and (b) the diameter of the charge and its relationship to VOD are of particular interest for practical applications.

Although there is a relationship between VOD and density for most explosives, the relationship is not linear. However, for some industrial explosives such as trinitrotoluene (TNT) and pentaerythritol tetranitrate (PETN), intervals of the relationship between ρ and VOD can be considered linear and given by:

$$\text{VOD} = a + b\rho, \tag{2.5}$$

where a and b are empirical constants depending to the type of explosive and ρ is the density of the unreacted explosive. The explosive can have varying density according to the amount of void spaces in the total volume of explosive. If the zero void density is known [the theoretical maximum density (TMD)], the VOD at other densities can be estimated using Eq. 2.6 as:

$$\text{VOD} = \text{VOD}'\left(\frac{\rho}{\rho_{\text{TMD}}}\right) + 1.5\left(1 - \frac{\rho}{\rho_{\text{TMD}}}\right), \tag{2.6}$$

where VOD′ is the VOD at the TMD density.

Experimental testing has shown two important aspects of the diameter of the explosive and its relationship with the VOD. The first is the concept of the ideal VOD and the critical diameter or failure diameter. Figure 2.1 shows a plot of VOD versus diameter for ammonium nitrate fuel oil (ANFO) and two types of emulsions [11]. The ideal VOD, (VOD_i) can be seen as the VOD when the diameter of the explosive is infinite, at this point the, maximum VOD is reached for that specific explosive material. As the diameter becomes smaller, the steady-state detonation velocity decreases until the VOD is lower than sound speed in the unreacted explosive material. The steady state can no longer be maintained and the detonation can fail. This diameter is known as the critical or failure diameter (D_{crit}) or (D_f). The value of D_f depends on the confinement conditions, particle size of the

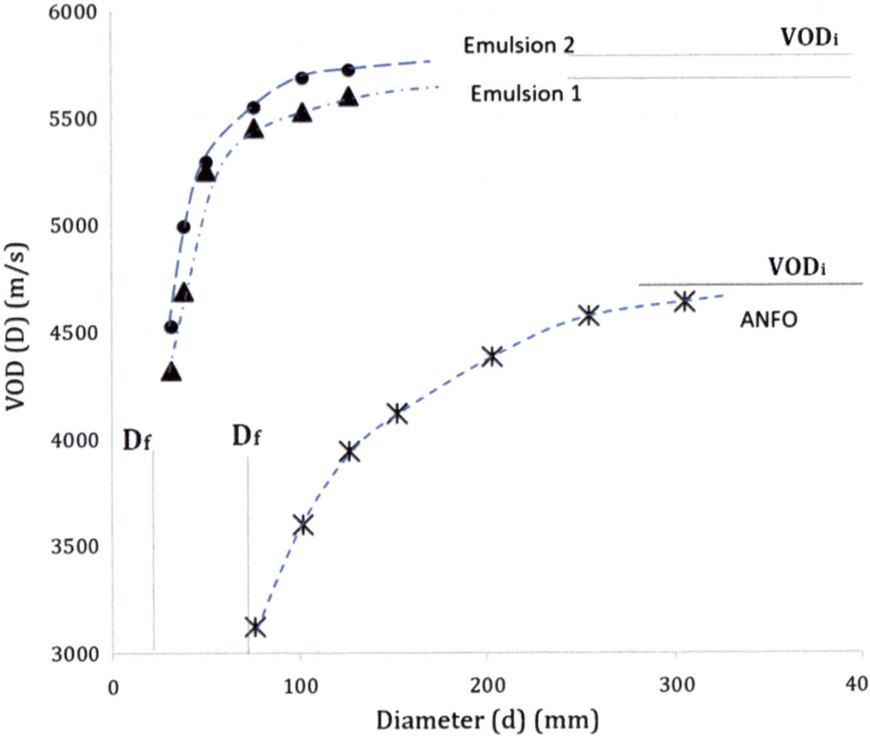

Fig. 2.1 VOD versus charge diameter. (Adapted from [3])

explosive material, initial density, and initial temperature of the unreacted explosive material.

Sun [3] proposed an empirical relationship (Eq. 2.7) between charge diameter and the VOD of an explosive.

$$\mathrm{VOD} = \mathrm{VOD}_{\mathrm{ideal}} * e^{\frac{b}{d^2}}, \tag{2.7}$$

where b is a constant related to the explosive under consideration ($b = -2810$ for ANFO), and d is the diameter of the explosive charge.

2.2.3.3 Detonation Pressure

The detonation phenomena are a very complex process occurring in 3D conditions. Despite its 3D nature, the study of detonation is simpler to describe using 1D theories such as the Chapman–Jouguet (CJ), [8] the Zel'dovich [13], von Neumann [12], and Döring [4] (ZND) theories. The CJ theory assumes that the detonation

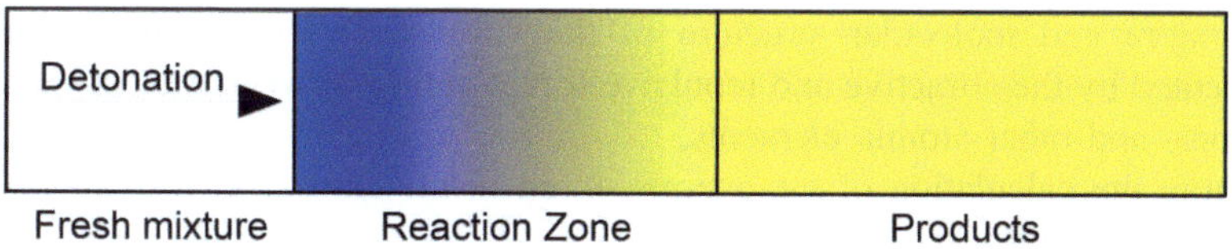

Fig. 2.2 Detonation process

process occurs in three zones called the fresh mixture (unreacted explosives), the reaction zone and the products zone (oxidized explosives) (Fig. 2.2).

Under the Chapman–Jouguet theory, the reaction zone shrinks to zero and the products are assumed to flow at a locally sonic speed relative to the shock, which is called the Chapman–Jouguet condition [8]. Using CJ theory, the detonation pressure can be estimated using Eq. 2.8.

$$P_{CJ} = \frac{\rho * VOD^2}{\gamma + 1}, \quad (2.8)$$

where P_{CJ} is the detonation pressure in gigapascals (GPa); ρ is the density of the unreacted explosive in g/cm^3 and γ is the adiabatic gamma function that can be defined as the ratio of specific heats of the detonation product gases; and VOD is the velocity of detonation in km/s.

The detonation product gases are molecules such as H_2O, CO_2, CO, N_2, etc. Assuming that for most industrial explosives and mixtures, the composition is similar at high temperatures and pressures, at the CJ condition γ can be assumed as 3 ($\gamma = 2.83$ and $\gamma = 2.73$ for PETN and TNT, respectively). Equation 2.8 can be written as:

$$P_{CJ} = \frac{\rho * VOD^2}{4} \quad (2.9)$$

2.2.4 *Energy Stored in the Explosive*

The energy stored in the explosive material can be calculated based on the chemical composition of the explosive, assuming that the chemical reactions have reached equilibrium and that all reaction products have the same temperature. To reach the chemical equilibrium, the OB should be close to zero as in many industrial explosives. The internal energy of a substance is defined as the total quantity of energy that it possesses by virtue of the presence, relative positions, and movements of its components molecules, atoms, and subatomic units. The kinetic energy is represented by the energy of vibration between the atoms of a molecule and the motion of electrons within the atoms. This portion of energy is determined by the temperature and molecular structure of the substance. The potential energy is represented by the attractive and repulsive forces acting between molecules, atoms, electrons, and other atomic elements.

Before the calculation of the energy stored in the explosive, it is necessary to define some key terms.

2.2.4.1 Heat of Reaction

Using thermochemistry concepts, the heat of reaction can be defined as the difference in the internal energy when a chemical substance changes states due to changes in the bond between molecules. The heat of reaction is also known as the enthalpy of reaction. When a chemical reaction occurs at constant pressure and temperature, the heat developed (released or absorbed) is equal to the change in enthalpy.

$$\Delta H = \Delta U + P\Delta V, \tag{2.10}$$

where ΔH is the change in the enthalpy, ΔU is the change in the internal energy, P is pressure, and ΔV is the change in the volume. Using the ideal gas law, Eq. 2.10 can be written in the form of:

$$\Delta H = \Delta U + \Delta nRT, \tag{2.11}$$

where Δn is the difference of gaseous moles for products and reactants, R is the gas constant (8.3143 J/mole), and T is temperature (298.15 K).

For example, if gaseous hydrogen molecules are burned with gaseous oxygen molecules to form water in the gaseous state:

$$2H_2 + O_2 \rightarrow 2H_2O \tag{2.12}$$

In the initial state, there are (H–H) bonds and (O–O) bonds that are destroyed to create in a final state (H–O) bonds. The internal energy stored in the bonds in the initial state are different from those in the final state. The difference is the heat of reaction. Because the heat of reaction is measured regarding the changes in the state, it is necessary to define a "standard state" defined as 25 °C (298 K) and 1 atm of pressure for engineering purposes. The heat of reaction at the standard state can be calculated as the difference between the standard heats of formation of the products and the standards heats of formations of the reactants (Hess's law).

$$\Delta H_r^o = \sum \Delta H_{f(\text{products})}^o - \sum \Delta H_{f(\text{reactants})}^o, \tag{2.13}$$

where ΔH_r^o is the heat of reaction at the standard state and ΔH_f^o is the heat of formation at the standard state. The heat of formation can be considered a special case of heat of reaction and is defined next.

2.2.4.2 Heat of Formation

This is the enthalpy change involved in making a particular compound, or molecule, from its elements where both the elements and the final compound are at standard-state conditions. For example, nitrogen dioxide ($NO_{2(g)}$) is formed from the combination of nitrogen oxide ($NO_{(g)}$) and one molecule of oxygen [2O atoms, ($O_{2(g)}$)] all in gaseous state.

$$2NO_{(g)} + O_{2(g)} \rightarrow 2NO_{2(g)}$$

From calorimetry information, it is known that the values for the standard enthalpy of formation for the compounds involved in the reaction are:

$O_{2(g)}$ 0 kJ/mol, reactant
$NO_{(g)}$ 90.25 kJ/mol, reactant
$NO_{2(g)}$ 33.18 kJ/mol product

So, the heat of formation can be calculated using Eq. 2.13 as:

$$\Delta H_r^o = 2\text{ mol} * \frac{33.18\text{ kJ}}{\text{mol}} - \left\{2\text{mol} * \frac{90.25\text{ kJ}}{\text{mol}} + 1\text{ mol} * 0\frac{\text{kJ}}{\text{mol}}\right\}$$
$$\Delta H_r^o = -114.1\text{ kJ}$$

In this example, ΔH_r^o is negative, meaning that heat is liberated during the reaction and the process is exothermic, on the contrary, positive values mean that it is necessary to supply energy to the reaction and the reaction will be endothermic.

2.2.4.3 Heat of Detonation (Detonation Energy)

The detonation energy is the heat of reaction of the explosive going to explosive products. It does not include any heat generated by secondary reactions of the explosive or its products with air. At this point, it is necessary to highlight that the detonation energy calculated using Eq. 2.11 will be different from the energy in a real detonation. Reviewing Eqs. 2.10 and 2.11, it is apparent that the heat of reaction is also a function of the temperature, volume, and pressure at the moment of the reaction. Some properties of the explosive such as the density and VOD, and some detonation properties such as the detonation pressure along with characteristics of the surrounding media where the detonation take place will affect those variables, thus affecting the detonation energy.

There are different methodologies to measure the detonation energy experimentally in the lab (using a bomb calorimeter device) or in the field under some limitations. The bomb calorimeter device works by measuring the temperature increase of the water surrounding detonated explosives confined in a chamber

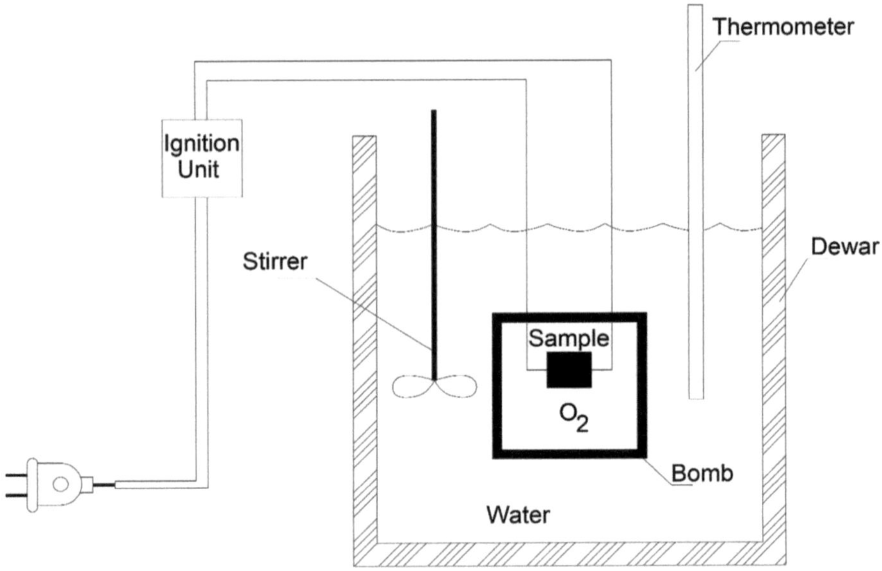

Fig. 2.3 Bomb calorimeter device

(stainless steel bomb). Figure 2.3 shows a basic schematic of a bomb calorimeter device.

For example, the detonation energy of the explosive RDX with the elemental formula given by $C_3H_6N_6O_6$ can be calculated using Eq. 2.13. The overall reaction using the reaction product hierarchy rule of thumb for RDX is:

$$C_3H_6N_6O_6 \rightarrow 3N_2 + 3H_2O + 3CO$$

From calorimetry information, it is known that the values for the standard enthalpy of formation for the compounds involved in the reaction are:

$C_3H_6N_6O_{6(s)}$	14.7 kcal/g mol, reactant
$N_{2(g)}$	0 kcal/g mol, product
$H_2O_{(l)}$	−68.317 kcal/g mol product
$CO_{(g)}$	−26.415 kcal/g mol product

So the heat of formation can be calculated using Eq. 2.11 as:

$$\Delta H_r^o = \left\{3 * \frac{0\,\text{kcal}}{\text{g mol}} + 3 * \frac{(-68.317)\,\text{kcal}}{\text{g mol}} + 3 * \frac{(-26.415)\,\text{kcal}}{\text{g mol}}\right\} - 1 * \frac{14.7\,\text{kcal}}{\text{g mol}}$$

$$\Delta H_r^o = -298.9\ \text{kcal/g mole}$$

If that value is compared to the experimental value of −335.4012 kcal/g mol [3], there is a difference of 36.5012 kcal/g mol or −10.9%. The difference is due to changes in pressure and temperature in a real detonation. Also, when the reaction product hierarchy is changed, for example, assuming that all the hydrogen burns to H_2O and all the remaining oxygen react with the carbon to form CO_2, the theoretical estimated will be different from the previous calculation.

2.2.4.4 Energy as Expansion Work

Ideally, if the reaction products can be expanded all the way down to atmospheric pressure, the expansion work (the work that the released gaseous detonation products exert on the borehole wall) should be almost equal to the detonation energy. If the detonation process is plotted using the pressure–volume (p–v) Hugoniot plane (Fig. 2.4), the expansion work can be defined in Eq. 2.14.

$$E_{\mathrm{w}} = \int_{v_1}^{v} p\,\mathrm{d}v - \frac{u^2}{2}, \tag{2.14}$$

where E_{w} is the expansion work and u is the particle velocity in the detonation process. If the expansion work is expressed using the main variables in the problem, the expansion work can be expressed using:

$$E_{\mathrm{w}} = \frac{P_{\mathrm{CJ}}}{2\rho_{\mathrm{CJ}}}, \tag{2.15}$$

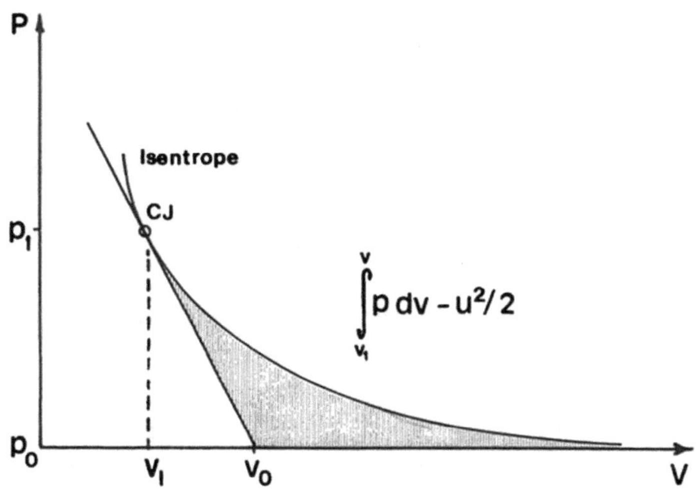

Fig. 2.4 p-v Hugoniot plane

where P_{CJ} is the detonation pressure (Eq. 2.8) and ρ_{CJ} is the density, both variables at the Chapman–Jouguet condition. The density at the CJ condition can be expressed as:

$$\rho_{CJ} = \frac{1+\gamma}{\gamma}\rho_o, \quad (2.16)$$

where ρ_o is the initial density of the explosive (unreacted explosive) and as seen before γ is the adiabatic gamma constant. Combining Eqs. 2.7, 2.8 and 2.15 in Eq. 2.14, the expansion work can be calculated as:

$$E_w = \frac{\gamma}{2(1+\gamma)^2}\mathrm{VOD}_{\mathrm{ideal}}^2 * e^{\frac{2b}{d^2}} \quad (2.17)$$

This expression allows us to examine the incidence of VOD and charge diameter in the expansion work. Higher VOD will result in higher E_w. Also, considering that parameter b is always negative, decreasing the diameter will decrease the expansion work (E_w).

2.2.4.5 Final Remarks

Thermodynamic and thermophysics analysis allows for the calculation of the energy stored in the explosive through the calculation of the heat of formation and heat of reaction; however, whether the assumption of reaction product hierarchy is valid or not would influence such calculations. The expansion work energy can be assessed using Eq. 2.13 or 2.14. Nevertheless, three variables can change the energy attributed to an explosive the: (a) pressure (confinement), (b) temperature, and (c) volume conditions. For example, in a detonation in a rock mass, the generated gasses will begin to escape to the atmosphere through openings and cracks. The temperature, volume, and pressure vary, changing the equilibrium of the reaction between products and reactants, thus changing the expansion work energy. So the actual amount of energy delivered in a blast is unknown but can be estimated from the amount of energy stored in the explosive and the assessment of the amount of energy as expansion work.

2.3 Balance and Use of the Expansion Work Energy Delivered to the Surrounding Rock

Significant research has been performed to determine specific applications of energy produced from the detonation of explosives for rock blasting. The debate is ongoing with regards to the actual quantities of energy expended in different aspects of the blasting process. The release of energy and subsequent fragmentation, movement, vibration, and heat is a very complex process dependent upon many site condition variables and field explosive conditions. The complexity of the conditions and the varying degrees of detonation within the borehole complicate the calculation and partition of the released energy. Many researchers have contributed to the body of knowledge in specific applications of energy; however, the complete process is yet to be quantified accurately. Most analysis has resulted in large percentages of energy remaining unaccounted for. The following sections introduce some common descriptions of energy partitions found in the literature.

The actual energy released by the explosive and manifested as high-pressure and high-temperature gas products will act on the surrounding rock to initially produce expansion work. The energy balance can by expressed by [10]:

$$E_{\mathrm{AEE}} = E_{\mathrm{F}} + E_{\mathrm{S}} + E_{\mathrm{K}} + E_{\mathrm{R}} + E_{\mathrm{NM}}, \tag{2.18}$$

where E_{AEE} is the actual energy of the explosive, E_{F} is the fragmentation energy, E_{S} is the seismic energy, E_{K} is the kinetic energy, E_{R} is the fragments rotation energy, and E_{NM} is the energy in forms that are very difficult to measure. Among the energy difficult to measure are:

- Energy used to expand the borehole,
- Energy to produce cracks in the fragments,
- Energy using to heat the rock mass,
- Energy conveyed as the gasses venting to the atmosphere through stemming and joints,
- Energy of plastic work, radiation, and acoustic energy.

The following sections discuss the measurable form of energy in a mining blast.

2.3.1 Fragmentation Energy (E_{F})

The primary objective of a mining blast is the production of new fragments from the rock mass. If the energy required to create a new unit fracture (γ_{F}) is known, the fragmentation energy can be expressed as [9]:

$$E_{\mathrm{F}} = A_{\mathrm{F}}\gamma_{\mathrm{F}}, \tag{2.19}$$

where A_{F} is the surface area of the fragments generated by the blast. γ_{F} can be estimated using Griffith theory through the fracture toughness (K_{IC}) and the elastic modulus of the intact rock. Such analysis considers only one crack or fracture. Another approach uses the size reduction laws used in comminution. This approach uses Rittinger's law, where the work required to reduce the sizes of fragments from a state a to b is calculated using the average particle diameters before and after crushing, and the Rittinger's coefficient (K_{r}). In a mining blast, the second option is more appropriate because many fractures are involved in the mine blasting process. In that case:

$$\gamma_F = \frac{1}{K_r} \tag{2.20}$$

The value for A_F can be estimated from the muck pile size distribution after blasting, although existing cracks must be identified and their surface area accounted for as preexisting cracks. This process is difficult even with detailed joint mapping.

2.3.2 *Seismic Energy* (E_S)

Assuming that the energy transferred to the rock in the form of seismic waves can be estimated as the integral of the energy flow passing through a surface of control at a given distance, the seismic energy can be estimated as [10]:

$$E_S = 4\pi r^2 \rho C_p \int_0^\infty v^2 \mathrm{d}t, \tag{2.21}$$

where r is the distance from the source, ρ is the rock density, C_p is the wave velocity in the radial direction (assumed as the P-velocity), and $v^2 = v_r^2 + v_t^2 + v_v^2$. v_r, v_t and v_v are the particle velocities recorded in a seismograph for the radial, transverse, and vertical component, respectively. Equation 2.21 is based on many assumptions and hypotheses far from the actual conditions in a mining blast and must be considered as a rough estimation of the seismic energy from a blast [10].

The seismic energy is also difficult to describe due to the attenuation and damping that occurs as the seismic energy travels away from the origin of the blast. The seismic energy is thought to begin as shock in the rock mass very near the borehole. It transitions to a more elastic vibration wave as it travels through the rock media away from the explosive. Some of the initial energy is utilized in elastic deformation in the rock mass, which is difficult to quantify and even more difficult to measure in a dynamic environment.

2.3.3 *Kinetic Energy* (E_K)

In a mining bench blast, if the velocity of the rock face $v(y)$ is known at different heights where y is the vertical position in the face, and assuming that the lateral variations of the velocity are of second order (i.e., the velocity is constant in a horizontal section of the highwall), the kinetic energy can be roughly estimated using Eq. 2.22 [2].

$$E_K = \frac{1}{2} S * B_h \int_0^H \rho(y) * [v(y)]^2 dy, \tag{2.22}$$

where S is the spacing of the blast, B_h is the mean horizontal burden, and H is the total height of the bench. Equation 2.22 assumes that the highwall is composed of various layers of rock with different densities. If the bench is in the same type of rock or the variation of the density is minimum, then Eq. 2.22 can be simplified as:

$$E_K = \frac{1}{2} S * B_h * \rho * H * v_b^2, \tag{2.23}$$

where ρ is the average rock density and v_b is the average rock face velocity. The translational kinetic energy is relatively simple to quantify using the principles of physics because it can be visually verified through video analysis.

2.3.4 Fragments Rotation Energy (E_R)

Once the fragments are ejected from the original position, they not only follow a trajectory but also rotate during the displacement. Assuming all fragments are spherical with different sizes, the rotation energy can be calculated as [2]:

$$E_R = \sum_{i=1}^{N} \frac{4}{5} \pi^2 m_i r_i^2 f_i^2 \tag{2.24}$$

where m_i is the mass, r_i is the radius, and f_i is the rotational frequency of the ith fragment and N is the total number of fragments considered in the analysis. Rotational kinetic energy is somewhat more difficult to quantify because fragments have irregular shapes and rotational angles are not simple to measure even with video analysis.

2.3.5 Energy Nonmeasurable (E_{NM})

As explained earlier the nonmeasurable energy is made up of energy used to expand the borehole, to produce cracks in the fragments, and to heat the rock mass; energy escaping from the borehole through the collar as the gasses vent; and energy used for plastic work, radiation, and acoustic emissions.

If it is assumed that the surrounding rock to a blasthole is homogeneous and there are no joints or cracks, the gases generated as product of the oxidation of the explosive will expand the borehole. In most field applications, this assumption is invalid; however, it is obvious that the initial expansion of the borehole will require expansion energy. The expansion energy can be estimated using expansion cavity theories [1]. If the properties of the surrounding rock are known and the expansion volume of the blasthole is measured, the expansion energy can be estimated.

Some energy is used to generate macro- or microcracks within a fragment of rock as a consequence of the blast. Such type of cracks are different than the fragmentation energy and are fundamental for the downstream mining process of crushing and grinding. This energy is conceptual and there is not a methodology for its assessment. This energy is also sometimes referred to as rock damage.

A blast event is an exothermic process. The high-temperature gases produced by the oxidation of the explosive will heat the surrounding rock in a differential of temperature unknown at this moment. In theory, if the change in the temperature of the surrounding rock is measured, the energy heating the rock mass can be evaluated.

Commonly, stemming is the amount of inert material put in the blasthole, on top of the explosive column, to confine the gases. Having an optimum grade of confinement is important to transfer the energy efficiently from the explosive to the ground. If a large amount of gas escapes from the borehole through the collar, the energy transferred to the ground will be lower than in the case where the gases are properly confined. Stemming is related to the size of the material used, its geotechnical properties, and the diameter of the blasthole. The stemming energy can be estimated by measuring the gas pressure in the borehole close to its collar [1].

2.3.5.1 Energy of Plastic Work, Radiation, and Acoustic

In a mining blast energy is expended to do plastic work, in radiation and as sound. However, there is not a practical way to measure this, so the magnitude is unknown at present.

2.4 Application of Energy Input for Ground Fragmentation

The previous sections have briefly described the quantity of energy available from the detonation of explosives. The simplistic use of this energy during the fragmentation process was also described. The basic concept is simple to understand; however, the complex mechanisms that dictate the specific partitions of energy used for each component of the energy balance are not widely agreed upon in the literature. The calculations are complex, and measuring these energy expenditures in the field proves even more difficult.

In most cases, mining operations can choose to analyze energy utilization on a cost basis. In simple terms, operations that employ blasting will purchase explosives (energy) for the specific purpose of fragmenting rock mass material. The energy used for specific energy balance components is somewhat less important than the resulting fragmentation and remaining rock characteristics. The overall operation must continue to process the material through digging, hauling, crushing, grinding, liberating minerals, and managing waste and byproducts. Obviously, these specific processes are dependent upon the mining method, deposit and mineral type, and final product. These subsequent processes will be more or less efficient based on the success of the blasting phase of the operation. The concept of mine-to-mill optimization considers the total quantity of energy (and thus cost) required to take material from in situ conditions to saleable product.

Many operations are utilizing fragment size distributions to analyze blasting needs with reference to the downstream processes. Mines perform fragmentation analysis utilizing mostly photographic analysis of the muckpile resulting from the blast. Other studies measure the efficiency of the blast according to resulting crusher throughput or energy consumption of primary, secondary, and/or tertiary crushing processes. In each case, very little attention is paid to the fundamental partitioned use of explosive energy. In general, the process is evaluated based on energy in (drill and blast cost) and energy out (efficiency of downstream processes). This optimization exercise generally results in efficient mining processes; however, significant advances could be made with a deeper understanding of the complex processes involved with explosive energy utilized in the fragmentation of a rock mass. For this reason, research continues to quantify the energy partition further.

2.5 Summary

The availability, utilization, and distribution of energy from the detonation of explosives require a detailed understanding of the detonation process. This introductory chapter discussed several concepts and the current state of the art theories. Due to the difficulty in measuring some of the processes that occur during detonation including extremely high pressures and temperatures, the process still remains somewhat ambiguous even to experienced researchers. Nevertheless, many theories add to the understanding of the overall process and distribution of energy from detonation to perform useful work like breaking and moving rock. While some of these studies are based on empirical data and experience, others are based on numerical analysis of the basics of physics and chemistry. The basic information presented in this chapter can be utilized to inform decisions on explosive selection, blast design, and execution.

References

1. Brinkmann JR (1990) An experimental study of the effects of shock and gas penetration in blasting. In: The third international symposium for rock fragmentation by blasting, Brisbane, Australia, pp 55–66, 26–31 Aug 1990
2. Calnan J (2015) Determination of explosive energy partition values in rock blasting through small-scale testing. Dissertation, University of Kentucky, Lexington, KY
3. Cooper P (1996) Explosives engineering, VCH Publishers, Inc.
4. Döring W (1943) On detonation processes in gases. Ann Phys 43:421–436
5. Elements of Armament Engineering, Part One, Sources of Energy (1964) Army material command engineering design handbook series, Washington
6. Eremenko LE (1981) Interrelationship between density and structure in an explosive. Proceedings of the 11th Symp. on Explosives and Pyrotechnics, Philadelphia, Pennsylvania
7. Fickett W, Davis WC (2000) Detonation: theory and experiment. Dovere Publications, New York
8. Jouguet E (1905) On the propagation of chemical reaction in gases. J de Math Pures et Appl 347–425
9. Oucterlony F, Nyberg U, Olsson M, Bergqvist I, Granlund L, Grind H (2004) The energy balance of production blast at Nordkalk's Klinthagen quarry. Bergsprangningskommitten, Stockholm
10. Sanchidrian JA, Segarra P, Lopez LM (2007) Energy components in rock blasting. Int J Rock Mech Min Sci 44:130–147, 154
11. Sun C, Later DW, Chen G (2001) Analysis of the effect of borehole size on explosive energy loss in rock blasting. Int J Blast Frag 5(4):235–246
12. von Neumann J (1942) Theory of detonation waves, OSDR Report 549 (1942), reprinted in Collected Works 1963; vol 6, pp 203–218. Macmillan, New York
13. Zeldovich YB (1940) On the theory of the propagation of detonation waves in gaseous system. Zh Eksp Teor Fiz 10:542–568
14. Zhang Z (2016) Rock fracture and blasting: theory and applications, Elsevier Inc

Chapter 3
Effect of Hole Stemming Practices on Energy Efficiency of Comminution

Calvin J. Konya and Anthony Konya

Abstract In order to increase the efficiency of explosive comminution, the borehole pressure must be maximized and pressure losses minimized. The majority of these pressure losses occur from premature borehole venting and through weak layers intersecting the borehole. With the use of proper stemming material and amount of stemming these pressure losses can be minimized, increasing the efficiency of explosive comminution. This chapter discusses the key considerations in the choice of stemming materials and methods calculate proper stemming size for different borehole sizes. In addition, the pressure models and methods to calculate stemming depth are discussed for both ideal and nonideal stemming material. Following the stemming design section, the chapter presents methods to improve stemming efficiency and reduce total stemming height including airdecks and stemming plugs. The chapter then addresses the issue of minimal fragmentation in the stemming zone. Practical design guidelines are presented for the use of a stem charge, allowing for breakage in the stemming zone.

Keywords Explosives · Rock blasting · Energy efficiency · Stemming Mining

3.1 Introduction

In today's blasting industry, it is understood that the gas pressure in the borehole directly correlates with the degree of fragmentation of the rock and is the only major cause of rock fragmentation [1, 2]. In order to improve the performance of a blast, stemming is used to help maintain this gas pressure over time. Stemming is an inert material that is placed on top of the explosive (powder) column (Fig. 3.1) and was

C. J. Konya (✉) · A. Konya
Precision Blasting Services, Montville, OH, USA
e-mail: Konya@idc-pbs.com

A. Konya
e-mail: Anthony@idc-pbs.com

K. Awuah-Offei (ed.), *Energy Efficiency in the Minerals Industry*, Green Energy and Technology, https://doi.org/10.1007/978-3-319-54199-0_3

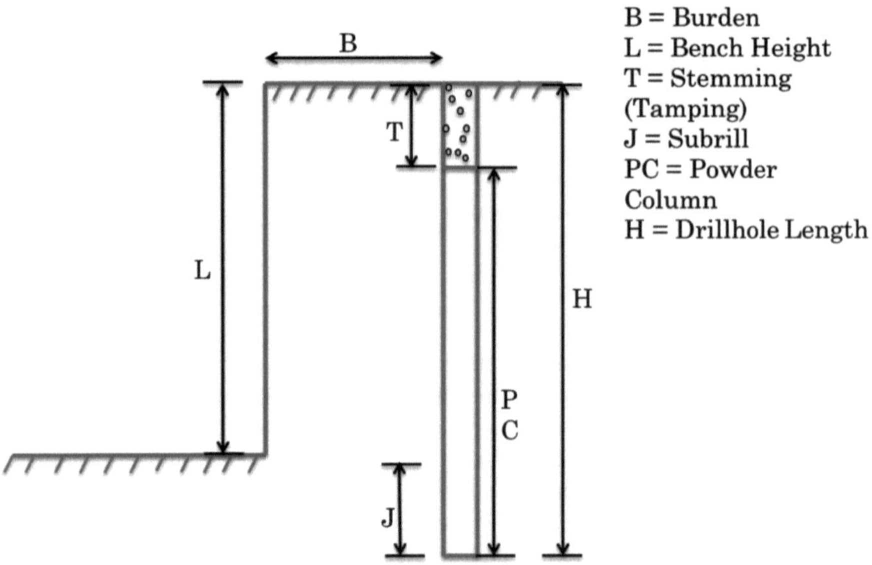

Fig. 3.1 Basic diagram of a bench blast

referred to as "tamping" (T). The goal is to keep it in the blast hole through the entire detonation process. The topic of stemming can only be discussed assuming the rest of the blast is reasonable designed, especially the hole to hole and row to row timing.

Stemming has been shown to increase the efficiency of explosives for many different explosives and blasting situations. Proper stemming was shown to increase the explosive efficiency by over 41% [3]. In addition, stemming is also of extreme importance in overbreak control with proper stemming leading to a 200% increase in the performance of a stemmed presplit hole versus unstemmed presplit holes [4]. An 18.6% reduction in the dig time for shovels and excavators has also been achieved [5] with proper stemming leading to over 3% reduction in P_{80} and a 36% decrease in the variation of P_{80} [6, 7].

Stemming is one of the major components of a proper blast design, with proper stemming leading to up to a 98% decrease in air overpressure [8], a reduction of ground vibration and flyrock, and an increase in rock fragmentation [2].

The use of too much stemming can lower a bench's stiffness ratio, causing it to act in a cratering mechanism [9]. This will lead to an increase in ground vibration and flyrock, as well as changing the fragmentation to a mix of fines and boulders. In addition, any amount of stemming decreases the fragmentation in the stemmed zone and any stemming over the minimum amount needed reduces the fragmentation efficiency.

An example of this is a site using 254 mm (10 inch) diameter borehole with ANFO. The difference between 6.7 m (22 feet) and 4.6 m (15 feet) of stemming would be noticeable in fragmentation with much larger fragments in the first situation. In addition, the second situation allows for an additional 91 kg

(200 pounds) of ANFO per hole, leading to increased breakage throughout the entire shot.

This chapter presents a review of the literature relating to stemming and its effect on energy efficiency of explosive comminution. It discusses the key considerations in the choice of stemming materials and methods to calculate proper stemming size for different borehole sizes, methods to improve stemming efficiency and addresses the issue of minimal fragmentation in the stemming zone. The work reviews seminal papers in the literature including both relevant laboratory and practical field studies that deal with the optimization of the stemming zone in a rock blasting.

3.2 Types of Stemming

The use of a proper stemming material can reduce the total amount of stemming needed by over 40% [10]. This would not only increase the efficiency of the explosive and allow for additional explosive to be used, but it would also drastically increase the fragmentation in the top zone (stemming zone). Most mines have the proper material needed for stemming in their stockpiles and transportation to the bench is the only expense.

The selection of proper stemming can be easily made using the stemming material diagram (Fig. 3.2). The four main categories of stemming that will be discussed here are free to semi-free flowing liquids, free flowing solids, interlocking particles, and solid materials.

Flowing liquids include both free flowing liquids, such as water and semi-free flowing liquids, such as mud. This form of stemming provides the least effective form of stemming, quickly blowing out of the borehole and allowing free flow of the explosive gasses out of the borehole.

Free flowing solids are comprised of both sands and river gravels that generally have high sphericity [10]. This material is only held in place by its weight and viscosity, and does not lock into place. This stemming typically blows out, even with use of large amounts of stemming [11] leading to low borehole pressures and a decrease in explosive efficiency. Solids that are similar in size to sands, such as drill cuttings also fall into this category.

	Low Friction	**High Friction**
Low Viscosity	Flowing Liquids	Solid Materials
High Viscosity	Flowing Solids	Interlocking Particles

Fig. 3.2 Stemming material diagram

Solid materials are materials such as concrete and plaster [12] that are put into a borehole (not to be confused with stemming plugs). The purpose of this is to provide a bridge before explosive detonation, completely sealing the hole. Besides additional cost and time, this stemming is ineffective because it does not have any internal resistance (viscosity) and once it shears from the borehole wall it will eject, releasing gas pressure. Studies have shown solid materials to be less effective than interlocking particles [10]. The combination of interlocking particles and a solid material has been shown to decrease stemming ejection velocity without an increase in fragmentation compared to just interlocking particles [13]. As of this writing, there has been no evidence that using solid material is better than interlocking particles, but improvements to drill cuttings can be achieved when combining a solid stemming method in the middle of drill cuttings [14]. However, this will add significant cost compared to other stemming methods.

Interlocking particles are lower sphericity solids that exhibit high viscosity and can lock into the borehole walls (high friction). These materials, when properly sized, will solidify and "bridge" the blasthole causing a solid plug to form. With the extremely large borehole pressures, the crushing strength of almost any material used as stemming will be exceeded. This means any material, as long as it can lock together and into the borehole wall, and is the proper size for the borehole can be considered an interlocking particle (Fig. 3.3).

Fig. 3.3 Stemming before detonation (right) and the stemming plug after detonation (left)

3.2.1 Sizing of Stemming Material

The selection of proper sized stemming material can decrease the amount of stemming needed on a blast by over 30% and increase blasting efficiency [1]. For example, if 30% more fines are used, compared to optimally sized interlocking particles, the fines blow out, hopefully slightly after rock breakage has began, whereas the interlocking particle will remain locked into the borehole.

The optimal size of stemming is based upon the size of the borehole, in order to maximize the resistance to movement. The optimal stemming size can be calculated with Eq. 3.1 [1], however, a stemming size between 1/25 and 1/15 of the borehole diameter should function similarly [12, 15].

$$T_{\mathrm{s}} = \frac{1}{20} * d_{\mathrm{h}}, \tag{3.1}$$

where T_{s} is the size of stemming (in or mm) and d_{h} is the size of borehole (in or mm).

This relationship has been shown to be most efficient for blasthole diameters from 50.4 mm (2 inches) [11] to 406 mm (16 inches) [13]. Some have tried to quantify the best size of stemming based on the shock velocity through the material [16]; however, shock velocity has a minimal effect on stemming movement and should not be considered an important parameter in stemming size selection. This conclusion that the shock velocity must affect stemming size comes from an attempt to bridge and lock stemming using compressed air [17]. However, the pressures experienced with this loading are less than 1/600 of the pressure in a real borehole situation.

3.2.2 Water Effects on Stemming

In certain cases, such as following screening, the stemming may be moist to wet. This affects the effectiveness of the stemming depending on the type of stemming, with the two main types discussed both affected by water.

The flowing solids, such as sand and clay, can exhibit up to a 50% increase in its ability to hold when large amounts of stemming are used [3]. In such cases, the stemming is minimally wet and should be tamped for packing into the borehole.

Wetting of interlocking particles can greatly reduce the effectiveness of the stemming. When these interlocking particles are wet the pressure transfer from the gas to the particle mass is greater and the plug's compressibility is reduced with water filling small voids between compressed particles. In addition, the contact zone between the compacted particles (stemming plug) is lubricated, lowering the friction and leading to increased ejection velocity.

3.3 Mechanics of Stemming

The way a stemming material functions and how it affects the borehole pressure is a complicated system that has, perhaps, not received as much attention as it should. However, throughout the 1970s and 1980s, there was development done on the mechanics of proper stemming. The majority of this work was done at West Virginia University [10] and the Ohio State University [8, 15] and was aimed at developing a theory on the mechanics of stemming that was applicable to real-field situations. This work consisted of mortar tests and full scale blasts.

3.3.1 Stemming Resistance

The gas pressure in a borehole reaches its maximum, generally in less than 2 ms [18] exerting tens of thousands of pounds per square inch on the stemming. With this nearly immediate pressure, the gas has little time to penetrate deep into the stemming material and the minimal loss of gas into the stemming can be ignored. Just as well, heat transfer to the stemming and borehole walls can be ignored. This is because in the rapid time a minimal amount of heat is transferred; for example, when blasting next to wood, the wood will generally not be scorched from the blast [19].

With these assumptions, the three types of stemming resistance are to be discussed [10] using the stemming resistance diagram (Fig. 3.4). The first way stemming resists movement is through inertial resistance—the weight of the stemming acting downward. All four forms of stemming exhibit this resistance and it can be increased by increasing the total weight of stemming, either through density or quantity. The reason a slightly wet sand will hold better than a dry sand is

Fig. 3.4 Stemming resistance mechanisms

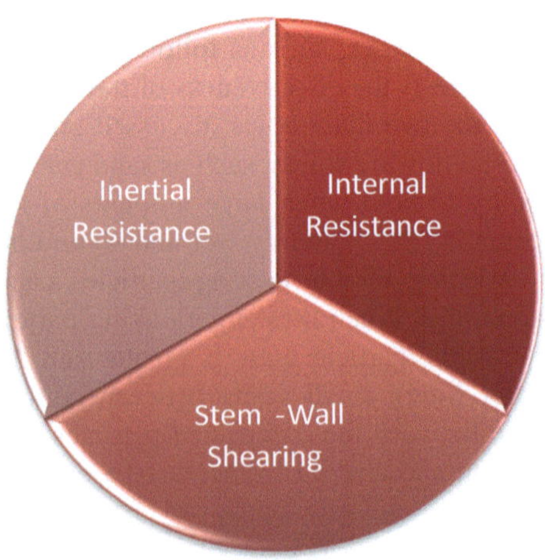

because of the increased density and a greater inertial resistance. This is the least effective method of stemming retention. It affects the retention time and stemming ejection velocity, but will not form a plug and hold the stemming in place.

The second stemming mechanism is the shearing of the stemming material along the borehole wall. This is the frictional force exerted by the stemming as it slides along the rough borehole wall and is the major force pushing down on the stemming. This is minimal in the free flowing solids and is present mainly in solid materials and interlocking particles. As the stemming material shears along the wall, this friction is reduced to almost zero. This is why just a solid material can be ineffective to withhold the gas pressure, after shearing it will act like a single, large mass particle being pushed out of the borehole. The interlocking particles are constantly shearing as each new particle is pushed from the one below, allowing additional time before shearing.

The third stemming mechanism is the internal resistance to flow of the material, which is similar to viscosity. This is the ability of the stemming material to compact together and form a "stemming plug" which crosses the entire borehole. This works as a combination of the gas pressure underneath the material pushing up and the inertial and shearing resistance of the stemming pushing down resulting in the formation of a stemming plug. Previously, it was thought that only interlocking particles could exhibit this mechanism, but the authors have observed possible stemming plugs forming from flowing solids with force balance above and below the material (Fig. 3.5). This is one of the most critical mechanisms of stemming formation and is the major mechanism to prevent stemming blowout (Fig. 3.6).

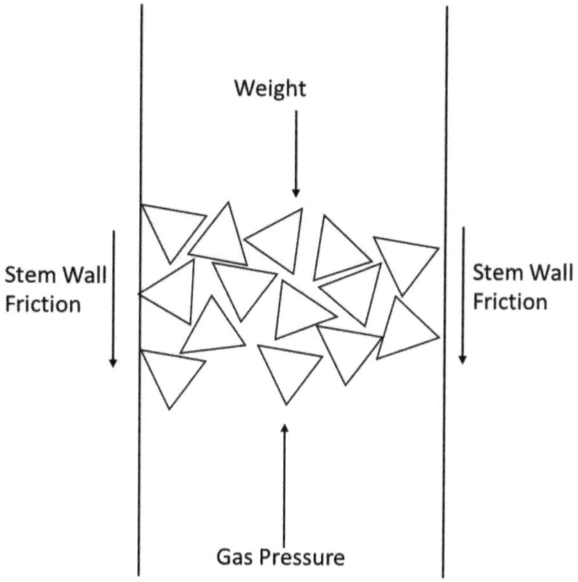

Fig. 3.5 Stemming free body diagram

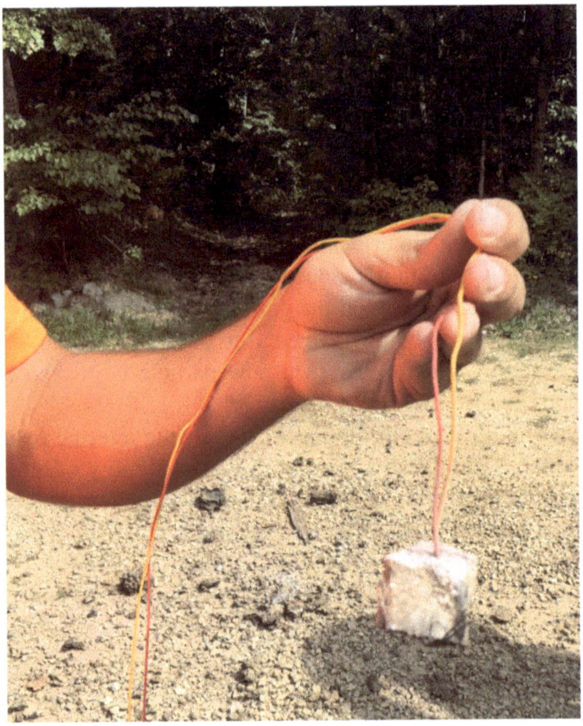

Fig. 3.6 Stemming plug formation with free flowing solids

3.3.2 Borehole Pressure

Borehole pressure must be maximized in order to increase the efficiency of rock fragmentation and movement. Many models exist for determining the pressure within a borehole with most requiring an engineer to calculate the temperature of the explosive detonation. For an in-depth discussion of borehole pressure parameters, one can see Cooper [20]. Two assumptions must be made for this temperature calculation:

- The rapid reaction and pressurization occurs under adiabatic conditions
- The pressure remains constant in the borehole until maximum pressure is produced

With these assumptions, the heat within the borehole can be estimated with the enthalpy heating any water present and the explosively formed gas products of the reaction. These gas products will need to be calculated from proper oxygen balance and the Kistiakowsky–Wilson Rules [20].

$$Q = n_{H_2O}\lambda_{H_2O} + n\int_{T_o}^{T_a} C_p dT, \quad (3.2)$$

where Q is internally generated heat, n_{H_2O} is moles of water, λ_{H_2O} is heat of vaporization of water, n is moles of products, C_p is specific heat of gas at constant pressure, T_o is standard temperature, and T_a is adiabatic flame temperature.

Solving for T_a will yield the temperature to be used in the pressure calculations. C_p is generally a weighted average for all the detonation products. The internally generated heat (Q) can be found by Eq. 3.3.

$$Q = -n\Delta H_d^0, \tag{3.3}$$

where ΔH_d^0 is the heat of detonation.

Models for theoretical calculations of borehole pressures are numerous and are generally accepted if they are ±10% from actual values. The actual pressure in a fully coupled borehole is difficult to determine experimentally, therefore, borehole pressures are generally modeled from decoupled boreholes or from the pressure required to fracture certain materials, such as polymethyl methacrylate (PMMA). This leads to discrepancies and errors. With this in mind, simple models that have been widely used and tested [20] are used by engineers and practitioners.

The first borehole pressure calculation was developed by Cook [21, 22] using a modification of the ideal gas law and hydrodynamic equations of detonation.

$$PV = nRT + \alpha(T, v)P, \tag{3.4}$$

where P is pressure, V is volume, N is moles of material, R is the gas constant, T is temperature, and $\alpha(T, v)$ is the gas co-volume which is approximately $\alpha(v)$.

Allowing for a small amount of error, the gas co-volume can be approximated as the specific volume. A simple calculation for this specific volume was then given [22, 23] as:

$$\alpha(v) = e^{-0.4\rho}, \tag{3.5}$$

where $\alpha(v)$ is specific volume of the gas and ρ is density of the gas.

With the borehole pressure calculated, one can now calculate the pressure losses from the borehole reactions. In the pressure losses of the borehole, a critical concept to understand is that radial cracks are created through the gas pressurization of the borehole causing hoop stresses [1, 24, 25]. This, along with other research [26] shows that the gas does not push the cracks open until the flexural failure of the rock mass occurs. When flexural failure begins, the major radial cracks are pressurized to 60% of their length and at this point, the borehole can be assumed to be depressurized [2]. In calculating pressure losses, one can then assume that the borehole does not lose pressure to fractures until depressurization of the borehole.

Natural fractures, voids, and deformities should be modeled as increases in volume and present a complex nonideal detonation/deflagration of explosives if the critical diameter is reduced. There is currently minimal information in the literature on modeling pressures in such cases. This discussion will not consider such situations.

3.3.3 Gas Flow Models and Optimal Stemming Length

As previously discussed, the importance of stemming is not just to lower air overpressure levels. Instead, proper stemming has major effects on the (energy) efficiency of the explosive used. An explosive has a limited gas volume that it can produce based on its chemistry and detonation conditions. With proper stemming, a "stemming plug" will form which will stop gasses from escaping the borehole and allow them to concentrate their entire pressure to break the rock as a directional charge.

The first case one can consider is when no stemming is present in the borehole. This will result in gas venting out the top of the borehole and resulting in over 10% decrease of maximum burden possible for small holes [27]. As the area of the borehole increases the energy efficiency of the explosive decreases, causing a decrease in the minimal burden. For example, a 406 mm (16 inch) borehole may have the maximum burden reduced to half of what is possible with proper stemming.

The amount of stemming (stemming length) is dependant on the type of stemming used and the proper sizing of the stemming. This stemming amount is also related to the pressure in the borehole and the desired time of gas retention. Stemming retention under 8 ms from firing has been shown to result in bench face velocity of 1/3 that in which stemming was retained for over 8 ms [28].

The most efficient condition would be for the stemming to hold until all breakage has occurred and the gas pressure is released [1]. This is classified as complete retention of the stemming and assumes proper stemming size can be related to the pressure in the borehole (after Otuonye [18]).

In stemming depth design application, using the maximum borehole pressure will ensure the stemming does not blow out. The next major consideration is that of the borehole size immediately after detonation.

Without any stemming in the borehole, the theory of choked gas flow will apply to explosive gasses (Eq. 3.6). This assumes steady-state flow for ideal gasses that have a downstream pressure below 0.587 times the borehole pressure, which will hold in most cases [29]. This theory has not been experimentally confirmed for blasting, but is a good representation and approximation of the effects of not including stemming in the borehole.

$$\dot{m} = C_{\mathrm{d}} A \sqrt{k \rho_o P_o \left(\frac{2}{k+1}\right)^{\frac{k+1}{k-1}}}, \tag{3.6}$$

where $\dot{m}$ is mass flow rate (kg/s), C_{d} is discharge coefficient (dimensionless), A is cross-sectional area of discharge (m^2), k is $C_{\mathrm{p}}/C_{\mathrm{v}}$ of individual gas, C_{p} is specific heat of gas at constant pressure, C_{v} is specific heat of gas at constant volume, ρ_o is gas density at pressure (P_o) and temperature (T_o), P_o is absolute borehole pressure (Pa), and T_o is absolute gas temperature (K). When applying this equation, all the gasses in the borehole can be treated as one gas or individual gasses from oxygen balance. This can be used with one of the explosive gas pressure models [30] to determine the change in pressure of the borehole with time (dP/dt).

The complete opposite of this can also be assumed, where no stemming movement occurs and the borehole pressure is the theoretically calculated pressure of the explosive. This will give the best and worse case scenarios of borehole pressure for design considerations and help one evaluate the feasibility of stemming.

3.3.4 Stemming Pressure Losses

With stemming consisting of sands and gravels containing void space gas can flow through the material with tortuous flow. This can be considered to be turbulent flow that is extremely complex and occurs before stemming movement occurs until a full bridge has formed and the borehole is sealed. This pressure loss is commonly ignored as pressure losses in this flow are minimal, but can be modeled empirically [31].

$$\Delta P_s = AR^B, \tag{3.7}$$

where ΔP_s is pressure drop, R is rate of flow, and A and B are constants.

The constant B is mainly influenced by the particle size, but has minor variations with viscosity and molecular weight of the gas.

$$B = \frac{0.006 + d}{0.036 + 0.51d}, \tag{3.8}$$

where B is a constant and d is the equivalent sphere diameter.

Once the stemming plug has bridged the borehole, the borehole will be sealed and gas will not be able to flow through the material. It is thought that this occurs within microseconds of the detonation. The pressure loss from the stemming is then based on the stemming movement from compression of the material. As the stemming is compressed the volume of the borehole will increase. The movement of the stemming [18] can then be modeled as:

$$M\frac{d^2 l(t)}{dt^2} = P(t)A - R_s, \tag{3.9}$$

where M is the mass of stemming, $l(t)$ = is stemming displacement with time, A is the cross-sectional area of the stemming, $P(t)$ is pressure with time, and R_s is stemming resistance (Fig. 3.7).

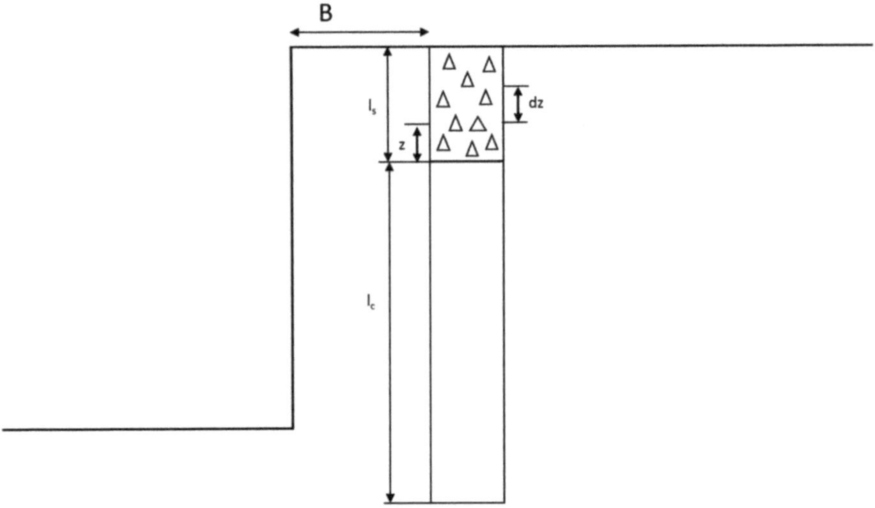

Fig. 3.7 Cross section of stemming for models

3.3.5 Ejection Model

The stemming ejection model is used if one anticipated the ejection of the stemming material or wishes to delay stemming ejection to a particular time. This model assumes adiabatic conditions, with no heat transfer. The pressure relationship is:

$$P_i V_i^N = P_f V_f^N, \tag{3.10}$$

where P_i is initial pressure, V_i is initial volume, P_f is final pressure, V_f is final pressure, and N is C_p/C_v.

Assuming that the blasthole is completely coupled and perfectly cylindrical, the stemming compression model can be combined with the pressure relationship. The pressure changes with time can be replaced with the final pressure to determine the velocity of the stemming material upon ejection. The resistance of the stemming material can be assumed to be negligible when stemming ejection occurs [32]. A brief proof is shown below with additional information available in Otuonye [18].

$$M\frac{d^2z}{dt^2} = P_i A\left(\frac{l_c}{l_c + z}\right)^N, \tag{3.11}$$

where l_c is charge length and Z is distance moved by stemming.

Substituting dimensions for mass and area gives:

$$\frac{\mathrm{d}v}{\mathrm{d}z} * \frac{\mathrm{d}z}{\mathrm{d}t} = \frac{P_\mathrm{i}}{l_\mathrm{s}\gamma_\mathrm{s}\left(1 + \frac{z}{l_\mathrm{c}}\right)^N} \tag{3.12}$$

where l_s is length of stemming and γ_s is density of stemming.

Partial differentiation and boundary conditions with ejection leads to:

$$V_t = \frac{2P_\mathrm{i}l_\mathrm{c}}{l_\mathrm{s}\gamma_\mathrm{s}(N-1)}\left[1 + \frac{1}{\left(1 + \frac{l_\mathrm{s}}{l_\mathrm{c}}\right)^{N-1}}\right]^{0.5}, \tag{3.13}$$

where V_t is the velocity of stemming at ejection.

Following ejection of the stemming material, the theory of choked gas flow would apply until the borehole has been depressurized.

3.3.6 Retention Model

If the goal is to completely retain the stemming material without over stemming a borehole, the retention model can be used to calculate the required length of stemming [33]. This model is based on the theory of particulate flow in bins.

The lateral force acting on the element dy is:

$$F_\mathrm{h} = P_\mathrm{h}\pi d * \mathrm{d}y, \tag{3.14}$$

where P_h is horizontal pressure and d is diameter of borehole.

The total frictional and cohesive forces acting on the sides of the borehole are:

$$F_\mathrm{c} = P_\mathrm{h}\pi d * \mu\mathrm{d}y + C\pi d * \mathrm{d}y, \tag{3.15}$$

where μ is tan $\varnothing$, $\varnothing$ is the angle of friction between stemming and borehole, and C is cohesion (Fig. 3.8).

Since P_v + dP_v is the pressure from the explosive on the bottom of the stemming, then:

$$(P_\mathrm{v} + \mathrm{d}P_\mathrm{v})\left(\frac{\pi d^2}{4}\right) - P_\mathrm{v}\left(\frac{\pi d^2}{4}\right) = \gamma_\mathrm{s}\left(\frac{\pi d^2}{4}\right)\mathrm{d}y + F_\mathrm{c}, \tag{3.16}$$

where γ_s is the density of stemming.

If the ratio of lateral pressure to vertical pressure is set equal to K then the equation can be written as:

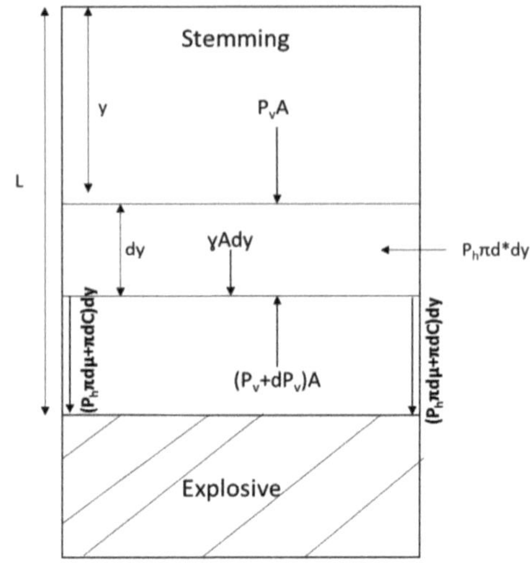

Fig. 3.8 Free body diagram of stemming

$$\left(\frac{d}{4}\right)dP_v = \left(\left(\frac{\gamma_s d}{4}\right) + \mu k P_v + C\right)dy \tag{3.17}$$

After integration and finding the natural log of both sides, the equation can be rearranged to solve for the stemming length required for retention.

$$l = \left(\frac{d}{4K\mu}\right) \ln\left[\frac{4K\mu P_t}{\gamma_s d + 4C} + 1\right], \tag{3.18}$$

where l is length of stemming required and $K = \frac{1-\sin \varnothing}{1+\sin \varnothing}$.

3.3.7 *Borehole Springing*

It has been well documented that following detonation, a borehole will "spring" open to a larger size than that drilled, to a degree based on the rock type [34]. This will create a larger borehole cross section than stemming cross section. Subsequently, when the stemming is pressurized it will be forced to go into this open area reducing its effectiveness. The maximum borehole springing will be approximately 167% of the original borehole [18], therefore three equations are presented to calculate the amount of stemming needed based on borehole springing. Using the retention models, a simplified model has been created for field use. Equation 3.19 assumes no borehole springing, Eq. 3.21 assumes maximum borehole springing (167%), and Eq. 3.20 assumes an average springing of 83.5% [15].

$$T = \frac{0.5}{d}[(12.783 * \ln P) - 107.43] \tag{3.19}$$

$$T = \frac{0.34}{d}[(12.783 * \ln P) - 107.43] \tag{3.20}$$

$$T = \frac{0.18}{d}[(12.783 * \ln P) - 107.43], \tag{3.21}$$

where T is the stemming length (feet), d is the borehole diameter (feet), and P is gas pressure (PSI).

The borehole pressure and the hole diameter are also directly considered in the calculation of the burden for the charge. This can then be simplified where stemming is directly related to the burden of the shot. With proper stemming material for complete retention of stemming, Eq. 3.22 can be used for simple, field calculations of stemming [1, 2]. This simple equation mimics the stemming equation for average borehole springing (Eq. 3.20).

$$T = 0.7 * B, \tag{3.22}$$

where T is stemming length (feet or m) and B is burden (feet or m).

Other research has analyzed stemming retention for other factors where the goal is to prevent stemming from blowing out of the borehole until a certain time has passed. In general, this is due to lack of proper stemming and difficulties in loading such as in underground mine blasting. In underground coal mines, loading horizontal holes with interlocking particles is extremely difficult and often clay dummies are used instead. In such cases, the stemming will most likely blow out, but must be retained until the gasses cool to avoid methane explosions. In general, stemming retention over 13 ms will allow the gasses to cool enough to prevent methane explosions [35].

3.4 Nonideal Stemming

While ideal stemming sizes can be easily calculated, many sites do not have access to ideal stemming. A site may have access to only material that is well graded (poorly sorted) yet wish to obtain high energy efficiency in blasting. Other sites have no access to material except for fines and drill cuttings. While these materials may exhibit higher frequencies of blowout, blasting practices can still be optimized to achieve higher energy efficiency by increasing the stemming retention.

3.4.1 Increasing Stemming Resistance

When a well-graded material is the only option for stemming, the efficiency of the material will be reduced. In addition, many are confused as to what the size of the material is for stemming. In the author's opinion, the P_{50} most accurately represents the actual size of the stemming as long as there are no drastic changes in material sizing. This agrees with previous field studies done on stemming [13, 36]. With proper sizing of the P_{50}, the stem-wall resistance can be optimized, realizing that there will be a decrease in the internal resistance and ability to form plugs.

The use of drill cuttings is a common practice, yet one of the most inefficient forms of stemming. Drill cuttings are commonly used because of the simple loading techniques; however, drill cuttings do not bridge (form a stemming plug) and are very small (generally under 3 mm or 1/8 inch). The only way to increase the effectiveness and length of hold with drill cuttings is to increase the inertial resistance (stemming length). This is a win/lose situation because while the stemming may hold longer allowing for greater face movement the stemming zone is larger and less explosive can be used in the borehole. If drill cuttings are the only option to be used, Eq. 3.23 [1, 2] can be used to estimate the best scenario stemming depth to minimize negative impacts.

$$T = 1.0 * B, \tag{3.23}$$

where T is stemming length (feet or m) and B is burden (feet or m).

3.4.2 Stemming Plugs

Stemming plugs are an effective way to increase the stemming retention time during blasting when using nonideal stemming materials, such as drill cuttings and interlocking particles [36]. These stemming plugs are inserted into the stemming, usually put about 15 cm (6 inch) above the explosive with a layer of material in between the plug and explosive charge to prevent burning. As these stemming plugs are pushed up they redirect the stemming to push into the wall instead of straight up. This increases the stemming to wall shearing and helps to bridge the material (Fig. 3.9).

Many different forms of stemming plugs exist, but all rely on the above principle. Stemming plugs can also be used to plug a hole to place stemming above the explosive charge, such as in presplitting. Stemming plugs have been shown to decrease ejection velocity and improve performance of nonideal stemmings [37]. A comparison of multiple stemming plugs has shown that the best kind of stemming plug is one that will fill exactly 90% of the borehole [38] and has a conical shape. This forces material into the walls, redirecting the energy and increasing the frictional force of the stemming [39]. A plug of the proper orientation can exhibit a

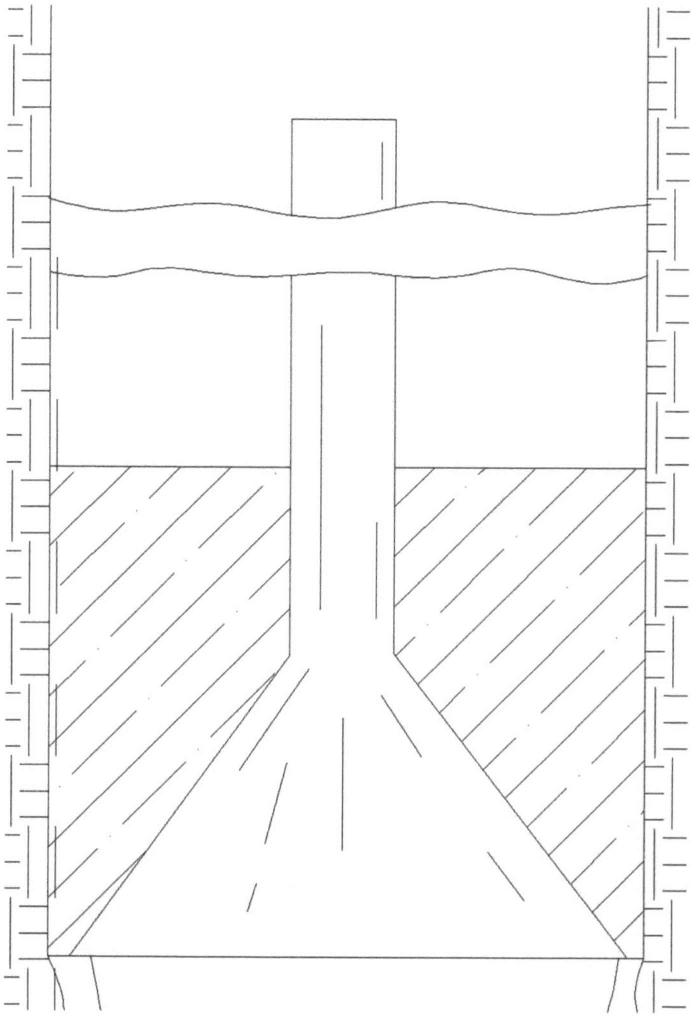

Fig. 3.9 Stemming plug [28]

47 ms longer retention time than that of an improperly sized or shaped plug [40] and increase fragmentation by 22% [41], when switching from the use of improper stemming (drill cuttings) to a stemming plug and the improper stemming.

Stemming plugs also have the potential to improve the effects of interlocking particles, possibly leading to reduced stemming depth to complete stemming retention. In this manner, it may be possible to reduce the total amount of stemming, such as a crushed stone [42]. This would allow for more explosive in the borehole, increasing fragmentation.

In addition, the use of a stemming plug with drill cuttings will generally lead to longer retention times than drill cuttings alone. In some cases, the stemming plug and drill cuttings can perform similar to crushed stone when similar heights of material are used. In certain cases [43], the combination of stemming plug and drill cuttings has led to longer retention times than stemming plug and crushed rock. The major problems with most of the research on stemming plugs are not rigorous enough and tend to not consider all the available options (drill cutting, crushed rock, stemming plugs, and appropriate combinations of these). In many cases, researchers look at the performance of a stemming plug and drill cuttings versus drill cuttings, and do not compare the performance of stemming plugs to crushed rock alone.

The main consideration in the use of a stemming plug will be economics, as each plug has a per unit cost and employee time cost for loading. If the site does not have crushed rock available, then ordering in crushed rock may be more economical than the use of stemming plugs. When considering using stemming plugs a comprehensive study must be completed that considers the cost of the stemming plugs and additional borehole loading time to the additional profit from increased, more uniform fragmentation.

3.4.3 Airdecks

An airdeck is a void space above the explosive charge and below the stemming zone. These air decks allow for some expansion of the explosive gas (depressurization) before pressurization of the stemming zone. This can be done with either the use of a stemming plug, or addition of empty bags. The use of an airdeck instead of stemming material allows for increased fragmentation in the airdeck region and lower ejection velocities of the stemming [37, 43]. Airdecks can also be used to separate layers of stemming and in combination with stemming plugs to increase retention by lowering the pressure on the stemming [44], however no evidence has been presented in the literature to show this to be better than crushed rock stemming.

3.5 Loading Stemming

The way stemming material is loaded is an important factor that affects the energy efficiency of blasting, as improper loading techniques can lead to mixing/moving of explosives, create toxic gasses, and decrease fragmentation. When looking at increasing the efficiency of the stemming loading process, one must carefully consider the explosive type and hole angle. Stemming has been shown to affect over one borehole diameter into the explosive.

The way an explosive is loaded into a borehole has little effect on explosive-stemming interaction, which is mainly formed by how the stemming is loaded. When loading stemming onto emulsions, the blaster has to take extreme care as stemming can sink into the emulsion matrix, causing shrinkage of the explosive diameter and nonideal detonation. Generally, when loading any type of stemming into emulsion, a material with larger surface area should be used first to increase buoyancy (like drill cuttings). After a small amount of this fine material, the regular stemming can be loaded. In addition, a stemming plug or bag can be put into the borehole to act as a barrier between the emulsion and the stemming material. Stemming and ANFO (ammonium nitrate and fuel oil) act in a similar, but the effect is not as drastic as with stemming and emulsion. The mechanism for ANFO disturbance is slightly different, where ANFO is blown upwards into and above the stemming, reducing the total length of ANFO charge. Similar loading practices should be followed to ensure efficient explosive function.

When loading stemming into angle holes, the scaled flow rate (diameter × flow rate) is critical for control of the boundary between the explosive and stemming. If the initial flow rate is too fast, the interface between the explosive and stemming will become an irregular shape, causing a shortening of the explosive diameter in this region [45] (Fig. 3.10).

In addition to this, if the scaled flow rate of the stemming material in an angled hole is too slow, it will leave an air channel out of the hole for gas venting and ineffective stemming confinement. This is because at lower scaled flows, the stemming will not be able to push the previous stemming into the borehole. A higher scaled flow will push the stemming to completely fill the borehole.

The general steps for loading for a proper stemming/explosive boundary are:

1. Load Explosive with as high a scaled flow as possible
2. Load a small amount of stemming at a low scaled flow (with a shovel, allowing for settling)
3. Load additional stemming at fast scaled flow (quicker shoveling or loader)

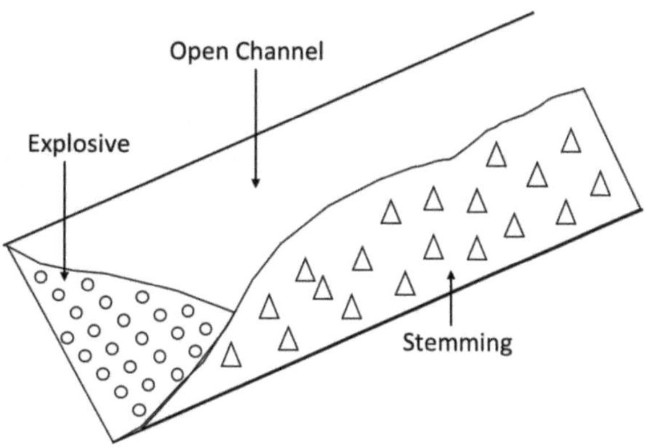

Fig. 3.10 Stemming leaving an open channel for escaping gas

3.6 Stemming Charges

A stemming charge is an explosive charge of reduced diameter put into the stemming to allow for additional fragmentation in the stemming zone of a blast. This has proven to be an extremely effective means of minimizing boulders in a proper blast, where a majority of boulders comes from the stemming zone. This is a more energy efficient approach because explosive energy is not wasted in blasting only for additional energy to be required to break up boulders. Recently, the authors have developed a simple method for calculating the size and depth of burial for stemming charges [46]. This approach is detailed below.

The first step when designing a stem charge is to decide where to maximize the effectiveness of the charge. If the charge is larger in diameter and placed low, it will leave boulders and larger fragments towards the surface. If the charge is small in diameter and placed high it will leave material below unfragmented and can lead to increased air overpressure levels. The starting point for a stem charge diameter is 0.5 times the diameter of the borehole. A maximum stem charge diameter should be 0.75 times borehole diameter. As a starting point, the stem charge length should be approximately four times the stem charge diameter.

The next and most important consideration when designing a stem charge is the proper depth of burial. If a stem charge is buried too low, it will not function at maximum efficiency and if the stem charge is buried too high, it will blow out the stemming increasing flyrock and air overpressure levels. The burial of a stem charge should leave enough stemming above it to prevent blowouts. Equation 3.24 can be used to find the amount of stemming to be left above the stem charge.

$$T_2 = kd_s\left[\left(\frac{2SG_e}{SG_r} + 1.5\right)\right], \qquad (3.24)$$

where T_2 is stemming length above stem charge (m or feet), SG_e is specific gravity of explosive, SG_r is specific gravity of rock, d_s is diameter of stem charge (mm or inch), k is 0.0084 for SI Units, and 0.7 for Imperial units.

The stem charge should then be timed to initiate at the same time as the main charge. This can be done with electronic caps, but is very expensive. Another way to achieve this is by using detonating cord to connect the top of the main charge to the stem charge, allowing for ideal, cheap sequencing of the stem charge.

3.7 Decking

An additional use of "stemming material" is in conditions of weak geologic layers, such as mud, intersecting a borehole. With a reduced resistance in these layers, the gas pressure will quickly blow them out; throwing a large amount of flyrock and causing rapid depressurization of the borehole. This will lead to a significant decrease in fragmentation and muckpile movement, and can increase air overpressure and ground vibration. The amount of pressure lost in these cases will depend on the size of the weak seam. With very minimal research done in the actual pressure loss through these seams, it can be assumed that they would be similar to the ejection model for stemming materials.

To prevent this premature blowout of gas, stemming material is placed into the borehole between the explosives to block this weak area. The area of these weak layers can be determined through accurate drill logs and monitoring of the drill penetration rate, when the penetration rate speeds up a weaker layer has been encountered. Once the weak areas have been identified, the blaster must ensure accurate loading of material, stopping before the weak layer and filling the area with stemming material and proceeding to load additional explosive above that.

While similar to decking to reduce the total weight of charge per 8 ms delay, this should not be confused with equations for decking to prevent sympathetic detonation [2]. Instead, the decks should maintain a set distance above and below the weak seam, and completely cover the weak seam. In the absence of comprehensive research on decking within a borehole, these general rules of thumb for decking can be used:

- Dry holes should have approximately 3× borehole diameter above and below the weak seam
- Wet holes should have approximately 6× borehole diameter above and below the weak seam

The major problem with decking across seams occurs because of initiation of the material. With the inaccuracy of nonel and electric caps [47, 48] random firing of the top or bottom decks could occur. The preferred initiation sequences would have the bottom deck firing just before the top deck when the top deck has not lost confinement before firing. Proper initiation of the top deck has been reported to only occur 27% of the time [49].

This can be dealt with in two separate ways, the first is by using detonating cord between decks to initiate the top immediately following the bottom. This works well if vibration and air overpressure are not of concern. Otherwise, electronic detonators would need to be used for accurate timing, raising the cost of the blasting process at the site.

3.8 Summary

Stemming is a critical component in blasting which can increase the energy efficiency of explosives and increase fragmentation by well over 41% when properly designed and implemented. In rock blasting, this can lead to an increase in burden or in fragmentation, both of which decreases the energy requirements for specific fragmentation. In addition to the efficiency increase, proper stemming reduces air overpressure, ground vibration, and flyrock.

The main components of stemming design and proper use include the type of stemming material, the sizing of the stemming material, and the length of stemming used. The proper stemming material can reduce the total amount of stemming needed by over 40% and for a stemming plug, completely sealing the borehole to retain all gas pressure which maximizes the explosive efficiency.

This chapter also presented techniques for using nonideal stemming and proper loading techniques for increasing efficiency in field conditions. These techniques include the use of commercial stemming plugs, air decks, water, and stem charges.

References

1. Konya C, Walter E (1990) Surface blast design. Englewood Cliffs, New Jersey
2. Konya C (2015) Rock blasting and overbreak control. Intercontinental Development Inc. Montville, Ohio
3. Snelling W, Hall C (1912) The effects of Stemming on the efficiency of explosives. Washington D.C
4. Matthews D (1978) Personnal Communication
5. Sharma S, Rai P (2015) Investigation of crushed aggregate as stemming material in bench blasting: a case study. Geotech Geo Eng 1449–1463
6. Kojovic T (2005) Influence of aggregate stemming in blasting on the sag mill performance. Miner Eng: 1398–1404
7. Cevizci H, Ozkahraman H (2012) The effect of blast hole stemming length to rockpile fragmentation at limestone quarries. Int J Rock Mech Min Sci
8. Konya C, Otounye F, Skidmore D (1982) Airblast reduction from effective blasthole stemming. In: Proceedings of eighth conference on explosives and blasting technique, pp 145–156
9. Konya C (1968) Spacing of explosive charges. Dissertation, University of Missouri, Rolla
10. Konya C, Davis J (1978) The effects of stemming consist on retention in blastholes. In: Proceedings of the fourth conference on explosives and blasting technique, pp 102–112
11. Davis G (1977) The effects of stemming consist on retention in blastholes. Dissertation, University of West Virginia
12. Cevizci H (2013) A new stemming application for Blasting. A case study. Mineracao Min 513–519
13. Eloranta J (1994) Stemming selection for large-diameter blastholes. In: Conference on explosives and blasting technique, pp 255–266
14. Cevizci H (2012) A newly developed plaster stemming method for blasting. J South Afr inst Min Metall 1071–1078
15. Otuonye F, Konya C, Skidmore D (1983) Effects of stemming size distribution on explosive charge confinement: a laboratory study. Min Eng 1205–1208
16. Dobrilovic M, Ester Z, Jankovic B (2005) Measurement in blast hole stem and influence of stemming materials on blasting quality. Rudarsko-geolosko-naftni Zbornik 47–53
17. Boshoff D, Webber-Youngman R (2011) Testing stemming performance, possible or not? J South Afr Inst Min Metall 871–874
18. Otuonye F (1981) Effective blasthole stemming. Dissertation, The Ohio State University
19. Qinan W (1912) High explosives. London, United Kingdom
20. Cooper P (1996) Explosive engineering. New York
21. Cook M (1947) An Equation of State for Gases at Extremely High Pressures and Temperatures from the Hydrodynamic Theory of Detonation. Journal of Chemical Physics: 518-524

22. Cook M (1974) The Science of Industrial Explosives. Albequerque, New Mexico
23. Johansson C, Persson P (1970) Detonics of High Explosives. New York, New York
24. Konya A, Konya C (2017) Precision presplitting—explosive load variations with spacing. In: Proceedings of the international society of explosive engineers, pp 553–562
25. Kutter H, Fairhust C (1971) On the fracture process in blasting. Int J Rock Mech Min Sci 181–202
26. Worsey P (1981) Geotechnical factors affecting the application of pre-split blasting to rock slopes. Dissertation, University of Newcastle upon Tyne
27. Johnsson G, Hofmeister W (1960) The influence of stemming on the efficiency of blasting using 36MM-shot-holes. In: International symposium on mining research, pp 91–102
28. Worsey P (1990) Experimental data indicating a direct link between the rate of stemming ejection and the degree of rock face movement in bench blasting. In: Proceedings of the conference on explosive and blasting procedure, pp 83–95
29. Perry R, Green D (2007) Perry's chemical engineers' handbook. China
30. Cook M (1971) The science of high explosives. Huntington
31. Furnas C (1929) Flow of gases through beds of broken solids. Washington D.C
32. Foldesi J (1974) Nyujtott Toltetekkel Torteno Kozetjovesztes Robbantastechnika Parametereinek Neghatarozasa Kulfejtesben. Doktori Ertekezes Miskolc
33. Outuonye F (1978) Modeling the behavior of blasthole stemming. Dissertation, University of West Virginia
34. Langfors U, Kihlstrom B (1973) The modern technique of rock blasting. New York, New York
35. Knopp J (2000) Stemming ejection and burden movements of small borehole blasts. In: Proceedings of the conference on explosives and blasting technique, pp 1–10
36. Sazid M, Saharan M, Singh T (2012) Effective explosive energy utilization for engineering blasting—initial results of an inventive stemming plug. Harmonising Rock Engineering and the Environment, pp 1265–1268
37. Jenkins S, Floyd J (2000) Stemming enhancement tests. In: Proceedings of the conference on explosive and blasting technique, pp 191–204
38. Worsey P (1990) Stemming ejection comparison of conventional stemming and stemming incorporating blast control plugs for increasing explosion energy use. Frablast 1990:361–367
39. Carr B (2000) New stem plug devices and applications. Stemtite LLC BAI workshop
40. Correa C, Navarrete M (2004) Assessment of stemming plug plastic elements to improve blasting gases confinement in escondida. Explo 2004:95–100
41. Choudhary B, Rai P (2013) Stemming plug and its effect on fragmentation and Muckpile shape parameters. Int J Min Mineral Eng 296–311
42. Varistem (2004) Vari-stem blasthole plugs improve blasting patterns. Min Eng 31
43. Jhanwar J, Jethwa J, Reddy A (2000) Influence of air-deck blasting on fragmentation in jointed rocks in an open-pit manganese mine. Eng Geol 13–29
44. Sazid M, Saharan M, Singh T (2016) Enhancement of the explosive energy utilization with the application of new stemming contrivance. Int J Innovative Sci Mod Eng 1–5
45. Wilkins M, Worsey P (1998) Stemming Technique for loading Angled Hoels charged with ANFO. In: Proceedings of the conference on explosive and blasting technique, pp 159–169
46. Konya A, Konya C (2016) Stemming charge. Acad Blasting Explos Technol
47. Konya C, Walter E (1988) Blasthole timing controls vibration, air blast, and flyrock. Coal Min
48. Konya C (1987) Accurate blasthole timing reduces blasting cost. Mine Quarry
49. Rodgers J, Lee R, Whitaker K (2003) The origins and effects of inter-deck pressure in decked blasts. In: Proceedings of the conference on explosive and blasting technique
50. Akhaven J (2011) The chemistry of explosives. Cambridge

Chapter 4
Effect of Wave Collision on Fragmentation, Throw, and Energy Efficiency of Mining and Comminution

Catherine Johnson

Abstract The chapter presents a review of the current literature on shock and detonation wave collisions in bench blasting and discusses how the resulting fragmentation and throw can influence the energy efficiency of mining and comminution. Wave collisions are now possible in a typical bench blast using programmable delays and the accuracy of electronic detonators. The objectives of this chapter are to: (i) explain how shock and detonation waves are created during a bench blast; (ii) provide a review of current literature on shock and detonation wave collisions with relevance to fragmentation and throw; and (iii) present the results of a recent study at a full-scale mine, which investigated the effect of shock and detonation wave collisions on fragmentation and throw at a granite quarry. This review chapter demonstrates how shockwave collisions can have a negative effect on fragmentation while detonation wave collisions can have a positive effect on fragmentation and throw and, therefore, overall mine efficiency.

Keywords Shock wave · Detonation wave · Throw · Fragmentation
Timing

4.1 Introduction

Fragmentation change is influenced greatly by altering explosive quantity, its distribution and the rock structure. The latter, rock structure, is quite clearly site dependent, but if the energy imparted by the explosive is used to its full potential using the same quantity of explosives through utilizing additional stress wave interactions between consecutive holes or within the same hole, fragmentation could be increased with no cost added.

One of the first processes at many mining operations is blasting, and the purpose of blasting is to fragment the rock mass in the most efficient way for that mine site

C. Johnson (✉)
Missouri University of Science and Technology, 326 McNutt Hall, Rolla, MO 65401, USA
e-mail: johnsonce@mst.edu

K. Awuah-Offei (ed.), *Energy Efficiency in the Minerals Industry*, Green Energy and Technology, https://doi.org/10.1007/978-3-319-54199-0_4

and the material end use. Possibly the greatest improvement to the blasting industry is the introduction of electronic detonators. The improvements related to safety and fragmentation have been invaluable. Several mining operations have identified changes in fragmentation size and distribution with no change to blast design parameters such as burden, spacing and drill hole diameter. Much of this has been attributed to the increased precision and accuracy of electronic detonators compared to traditional pyrotechnic delays. This has opened avenues to look further into timing as a way to alter fragmentation. One such method is the use of ultra-fast timing to introduce shockwave collisions into the medium creating areas of increased shock pressure and preconditioning [1, 2]. Another method is through detonation wave collisions within a blast hole. Reduced overall fragmentation size and increased throw of the muck pile are potential benefits from wave collisions. These fragmentation improvements would be seen throughout the full mine life cycle, resulting in less comminution and therefore lower energy consumption and water necessary in later aspects of the mine cycle.

This chapter presents three main objectives. First, it provides an overview of the theory behind detonation and shock waves and how they relate to bench blasting practice. This overview provides the background for discussing the effect of timing on fragmentation and throw. The second objective is to share a comprehensive discussion of the current literature related to shock wave and detonation wave collisions with relation to fragmentation and throw. The final, third, objective is to present a recent investigation from a granite quarry operation that characterized the effect of shock and detonation wave collisions on fragmentation and throw. Before any of the above can be understood, it is important to understand the basics of rock fragmentation in bench blasts; the following section addresses this.

4.2 Rock Fragmentation in Bench Blasts

Rock fragmentation from blasting is dependent on a number of factors including the properties of the in situ rock, properties of the explosives used, blast pattern design, and shot timing. Rock properties such as compressive strength, porosity, density, Young's modulus, Poisson's Ratio, and rock fracturing and jointing cannot be altered. Thus, any optimization to fragmentation occurs within the limitations placed by the rock mass. This leaves the explosive properties, blast design, timing design, and execution to influence fragmentation. Explosive properties that influence fragmentation include the Chapman–Jouget (C–J) pressure, density of the explosive, and the detonation velocity of the explosive. Blast pattern design elements include burden, spacing, powder factor, stemming length and type, hole depth and diameter, and sub-drill length [3].

4.2.1 Optimized Fragmentation

Traditionally, fragmentation size at a blast site was designed dependent on powder factor: amount of explosive per unit volume of rock moved (kg/m^3). An easy way to increase fragmentation is to increase your powder factor. Powder factor correlates well with blast design parameters including burden, spacing, hole diameter, and face height. Clearly, the closer your blast holes and the greater the weight of explosives per hole, the higher your fragmentation. Increased fragmentation, in this instance, refers to a lower mean fragmentation size. This is, however, not an ideal definition of "increased fragmentation". A bimodal distribution with a large proportion of fines and some large fragments will reduce the mean fragment size significantly but may not be beneficial to the mine. Increased fragmentation should therefore be assessed based solely on the distribution of fragment sizes with the ideal distribution being a normal distribution with majority of particles of the same size around the median fragment size. Some use the term "desirable fragmentation". With increased powder factor, blasting cost also increases so a site must balance these two competing goals.

MacKenzie [4] described optimum fragmentation to be "that blasting practice which gives the degree of fragmentation necessary to obtain the lowest unit cost of the combined operations of drilling, loading, hauling and crushing". Since the introduction of environmental regulations and new developments in explosive and blasting techniques, blasts can no longer be designed simply around monetary cost. An optimized blast design is one that will break or move rock to the required fragment size for secondary equipment to efficiently handle it while minimizing secondary components such as cost and environmental effects including ground vibration and airblast. To that end, the optimization of blasting is not achieved in the same way at every site. This suggests a modern need for optimization tools that allow for adjustments in fragmentation distribution through changes in a number of input variables, including timing.

The degree of fragmentation influences the overall mining process and is the basis of the well-known term "mine to mill", which governs the overall costs to the mine [5]. Optimum fragmentation is documented extensively in literature as referring to a specific site [6–9]. Fragmentation should be optimized to produce the lowest operational cost including all factors of mining, which includes the environmental cost. Often, the fragmented material in mining and construction is feed for a crusher. So anything blasting can do to minimize crusher energy and maximize throughput is important.

4.2.2 Impact of Blast Performance on Fragmentation and Downstream Energy Efficiency and Costs

There are a large number of ways to evaluate the effectiveness of a blast depending on the desired outcomes. Historically, blast effectiveness has been measured based on in-pit results, but given that these results do not fully encompass the areas that blast performance affects, it is necessary to evaluate a blast based on all downstream results. Effective rock fragmentation is key to minimizing downstream costs and efficiency by optimizing crusher and grinder throughput, minimizing wear on equipment, maximizing dig rate and payload, decreasing energy consumption of equipment, and controlling fines production. Some consequences of poor blasting include clogging of comminution equipment due to an excessive amount of wet fines, large boulders often need re-blasting, and oversize can wear digger teeth at a much faster rate. Photographic fragmentation analysis, vibration monitoring, and high-speed video provide quantitative measurements of blast effectiveness and supply data that allows operations to modify blasts to achieve downstream goals [3].

Blasting results affects the overall energy efficiency of mining by affecting the energy efficiency of material handling and comminution. Since blasting is often the first stage in the mining process, changes here can have a large effect on the overall mine cycle. Changing a blast to incorporate wave collisions is achieved with the timing delay you choose, a variable that does not alter your drilling and blasting costs. The following comprehensive discussion sections give an overview of the effect of timing on overall downstream energy efficiency and costs with a specific emphasis on shock and detonation wave collisions.

4.2.3 Rock Fragmentation Analysis Methods

Rock fragmentation distribution can be evaluated in a number of different ways. These methods vary from very simple to perform and qualitative methods to very quantitatively accurate methods that are impractical in production situations. Fragmentation can be evaluated qualitatively on a shot to shot basis by blaster observation and loader operator feedback about sizing and diggability. This method lacks explicit data and is subject to a significant amount of human error and bias. Sieve analysis of shot rock is a very accurate quantitative method of determining fragmentation size, but it is time consuming, impractical, and expensive in active mining operations. Digital image analysis provides a middle ground between the previous methods with a quantitative measure of fragmentation sizing that is minimally disruptive to the mining process, and is, therefore, practical for obtaining fragmentation results of bench blasts. Digital image analysis of shot rock can be performed using images of the muckpile taken with portable cameras, with belt mounted systems, or loader mounted systems [10].

There are various software packages and image capture systems designed to facilitate digital image analysis for fragmentation sizing. These include WipFrag, Split, PortaMetrics, GoldSize, Fragscan, PowerSieve, and BLASTFRAG [10–14]. Many of the image analysis systems operate in a similar manner and most require some type of scaling item to be placed in the photo. Systematic photo acquisition is important both immediately after the shot and throughout the mucking process to eliminate the sampling bias caused by the typically more course fragmentation found on the surface of muck piles [13].

There are a few problems associated with digital image analysis methods that should be understood when utilizing them for fragmentation assessment. These, however, do not negate the usefulness of the analysis. These include the manual editing of rock outlines to ensure correct delineation of fragments, which introduces human error into the analysis, especially when particle sizes are small. Other issues include errors associated with the calculations used to transform rock surface measurements into volumes and the limitations of the resolution of image systems. Additionally, two-dimensional images can sort fragment sizes differently than when using a traditional sieve; for example, the orientation of a thin but long rod shaped fragment that would pass through a finer sieve would be placed in a larger fragment size using image analysis. When utilizing image analysis to do side-by-side comparisons, some of these problems, such as the volume calculations, are irrelevant because any error introduced will apply to all of the images and the difference in size distribution from photo to photo will still be evident. Additionally, despite the issues, when tested, the size distributions found using digital image analysis of muck piles matches those of sieved material well [12].

4.3 Theory of Detonation and Shock Waves

Fragmentation in mine blasting is a result of stress waves and their interaction with the rock mass; the efficiency of such interaction can be lost due to excessive ground or air vibrations. Stress wave propagation is the first step in the fragmentation process, preconditioning the often-strong rock mass with extended fractures in which the gas pressure can expand. Each blast hole creates a complex sequence of stress waves. The two main types of stress waves are detonation and shock waves. The most prominent difference between the two is that a detonation wave can only travel through an explosive mixture, often at a constant rate due to the chemical composition of the explosive until the entire column has detonated. A shock wave results from a detonation wave and travels through the surrounding medium, rock, air or water, dissipating with distance. The detonation wave does have an associated shock wave since it is a pressure wave that travels faster than the speed of sound in that material.

4.3.1 Detonation Waves

Detonation waves travel as energy is released from a chemical reaction, originating from a detonation point. In an ideal detonation, the velocity of detonation (VOD) is the maximum possible velocity when all components in the chemical reaction react fully. Ideal detonation, however, is not a reality in mining explosives. As an oxygen-balanced explosive, such as ANFO, detonates, a large quantity of expanding hot gases is produced. In an ideal situation, only hot steam, carbon dioxide, and nitrogen are produced (Eq. 4.1).

$$3NH_4NO_3 + CH_2 \rightarrow 3N_2 + CO_2 + 7H_2O \tag{4.1}$$

In reality, efficient blasting practice is only possible with sufficient confinement and priming. Furthermore, the explosive products are often contaminated with water and drill cuttings, which affect the explosives quality and chemical kinetics. This results in nonideal explosive reactions, producing toxic gases such as nitrogen dioxide, nitric oxide, and carbon monoxide (NO_2, NO and CO), and reduces the energy imparted onto the rock mass. Explosives are often contaminated with ground or rainwater, leading to increased levels of NOx gases through the breakdown of ammonium nitrate into nitrate and ammonia in solution. The desensitized solution will not fully decompose when detonated, producing nonideal products. Under and over fueling an explosive mixture results in positive and negative oxygen balances, respectively. A positive oxygen balance will produce more NOx gases (Eqs. 4.3 and 4.4), while a negative oxygen balance will produce more CO (Eq. 4.2).

$$2NH_4NO_3 + CH_2 \rightarrow 2N_2 + CO + 5H_2O \tag{4.2}$$

$$5NH_4NO_3 + CH_2 \rightarrow 4N_2 + 2NO + CO_2 + 7H_2O \tag{4.3}$$

Nitric oxide is unstable in air and reacts with oxygen to produce the more toxic nitrogen dioxide (Eq. 4.4), which can be seen after a blast by the generation of an orange/brown cloud [15].

$$2NO + O_2 \rightarrow 2NO_2 \tag{4.4}$$

The degree to which the explosion is nonideal depends on the rate of detonation and degree of confinement. Selecting the correct explosive for a specific site is crucial. Mining explosives, typically ANFO, emulsion, water gels, or blends, have a critical diameter under which they do not fully detonate, reducing the effectiveness of the explosive. If the VOD is lower than its full potential, so is the shockwave velocity reaching the blast hole wall and traveling through the rock mass, affecting the fragmentation.

4.3.2 Shock Waves

Upon detonation of a blast column, a strong shock wave is emitted, which quickly decays into an elastic stress and, in the far field, is responsible for initial ground vibrations [16, 17]. When the shock wave reaches the borehole wall the fragmentation process begins. This shock wave, which starts out at the VOD, decreases quite rapidly once it enters the rock and in a short distance reduces to the sonic velocity of that particular rock. The distance at which this occurs depends on the rock blasted and the VOD of the explosive, as well as other contributing factors such as pore pressure or fractures in the rock. At those areas surrounding the blast column, radial cracks form after a small delay due to the detonation pressure. The outgoing shockwave travels through these cracks radially, expanding them and creating a tangential stress. This pressure is compressive, but since rock is weaker in tension, it is the tensile wave that typically causes the rock to break. When the stress reaches a free face, it reflects back as a tensile wave due to the change in density [18].

When considering a square wave, like that in Fig. 4.1, the front of the shock wave is already in shock. The velocity at the back of the wave is the sum of the particle velocity and the wave speed and appears vertical (1 in Fig. 4.1). The velocity increases at a faster rate at the back of the wave since it is approaching a medium of higher density and pressure as a shock front has already passed through. This velocity eventually catches up with the front (4 in Fig. 4.1), the velocity at this portion, the rarefaction wave, is the greatest velocity the wave will possess. Beyond this point, this rarefaction wave is entering an unshocked medium reducing its velocity until it will eventually slow to the velocity of the rock mass [19].

As a shock front travels through a medium, there is a sudden change in state across the shock front of the particle velocity, density, internal energy and pressure, not a gradual increase. They essentially change from an unshocked front to a shocked front. Figure 4.2 demonstrates this process in a simple diagram across a square shock front.

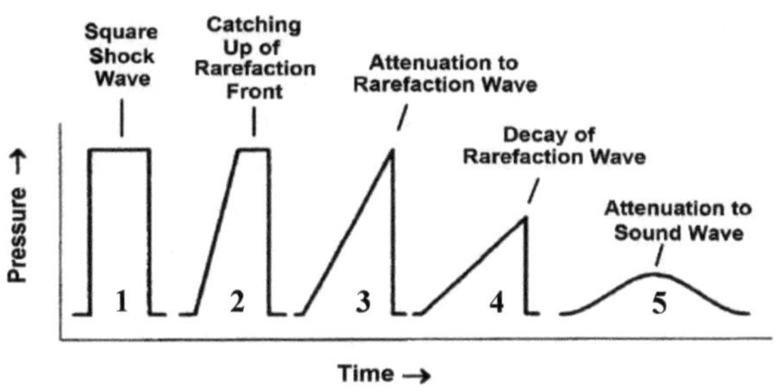

Fig. 4.1 Progression of a shock wave, after [19]

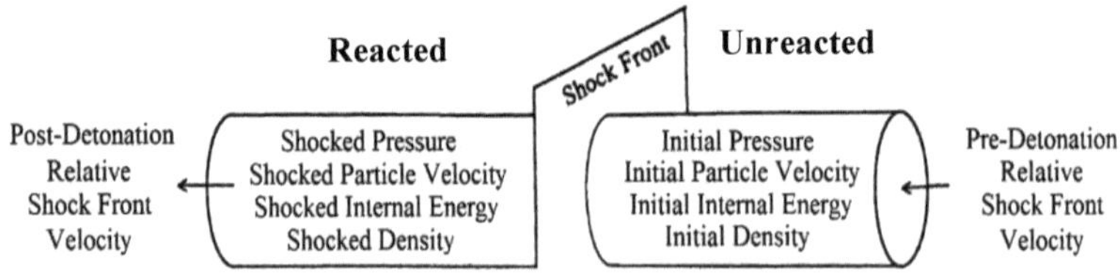

Fig. 4.2 Mass moving through a shock front, after [19]

4.3.3 Colliding Shock Waves in Strata

Measuring and observing two shockwaves collide in strata is difficult. The laws of shock wave propagation in media other than that of air or water are not well known at short distances [18]. Considerable data is available at greater distances through the use of seismographs and ground vibration analysis. It is assumed that the same basic relationships occur, but how the additional parameters of shear and tensile strengths, density and unavoidable fractures in the rock mass, affect the process is more difficult to quantify. Yamamoto et al. propose that when two shock waves collide and no free face is present, the tensile waves will meet, increase in magnitude, and form cracks. For simultaneously detonating charges, this happens at the mid-section of the spacing of these blast holes; for a delayed charge, it occurs away from the centerline [1]. This delay has a very short range of only a few milliseconds dependent on the speed of sound in the rock being blasted. Therefore, delay timing has to be selected appropriately for any interaction to occur.

The interaction of these stresses has been difficult to measure as the pressure that builds up in the borehole depends not only upon explosive composition, but also the physical characteristics of the rock. Strong competent rock will result in higher pressures than weak, compressible rock. Delay time, wave speed in the rock mass, shape of the wave pulse, and acoustic impedance mismatch have become decisive parameters in advanced blast design [2]. As a blast hole detonates, it creates a leading compressive pulse and a trailing tensile pulse, the length of this is dependent on the explosive type and type of rock. The wave speed can vary between 2000 meters per second (m/s) for soft sandstone to 6500 m/s for granite [18]. According to Rossmanith, maximum fragmentation is achieved in those sections where the two tensile trailing sections of the blast wave meet. When two adjacent blast holes of spacing, s, are considered, the elementary event is the interaction between the two stress waves: P_1–P_2, S_1–S_2, P_1–S_2 , and S_1–P_2. In the close vicinity of the blast, the P and S waves will overlap, but will separate with distance from the blast due to their varying speeds. Figure 4.3 shows a Lagrange diagram in one dimension between the stress wave interaction when two holes are detonated, simultaneously. Subscripts F and E denote the front and end of the compressive waves, respectively [2].

If the detonation of the second wave is delayed, the regimes in Fig. 4.3 will move closer to the delayed blast hole, consequently controlling the fragmentation

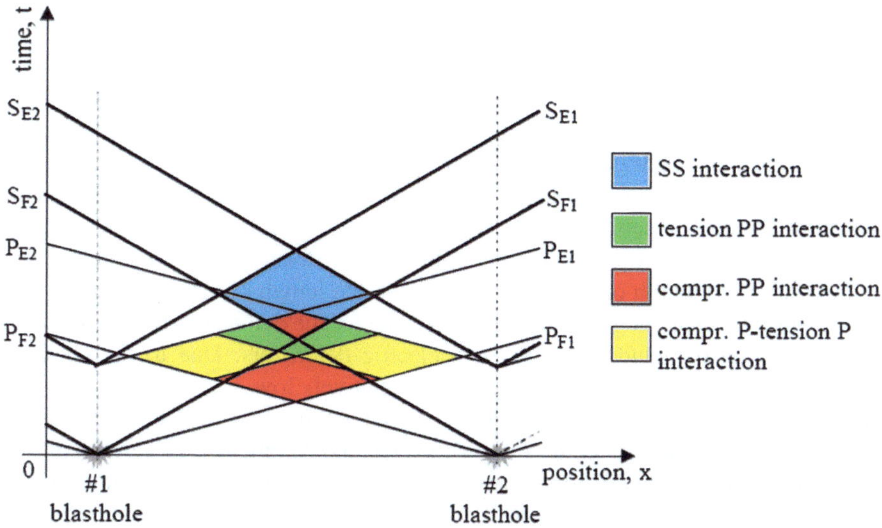

Fig. 4.3 Lagrange diagram showing interaction of stress waves emerging from two simultaneously detonated blast holes, after [2]

pattern. Conversely, if the stress waves occur in the center of the two blast holes, overlapping at the greatest magnitude, the potential for fragmentation is increased. Longer wave pulses can also be obtained using an explosive with a lower VOD and brisance. These explosives are usually associated with larger volumes of produced gas, generating differing opinions on the precise cause of the enhanced fragmentation. Rossmanith's theory, however, has come under scrutiny for being too simplistic [6]. Experimental tests by Katsabanis et al. and Johansson and Ouchterlony do not agree with the interaction defined by Rossmanith [2, 6, 20–22]. Computer simulations in LS-DYNA by Sjoberg et al. also disagree with the Rossmanith theory, concluding greater fragmentation is achieved at greater time delays where no wave interaction occurs [23]. Results from these studies will be discussed further in Sect. 4.4.2.

4.4 Timing Effects on Fragmentation

The effect of timing with regards to blasting is not a new concept. Timing has, for many years, been thought to have an influential control on muck pile placement and heave, and the control of adverse environmental effects such as ground vibration and air blast. It is only since accurate timings have been available through electronic detonators that timing effects have been associated with fragmentation.

4.4.1 Historical Improvement in Accuracy of Detonators

Detonators have come a long way since the introduction of safety fuses in 1831 utilizing a cord of black powder. In the early 1920s, blasters used electric detonators that consisted of electric wires connected to an aluminum shell cap and uses a bridge wire to ignite the charge. In the 1960s and 1970s, a nonelectric detonator was introduced. A shock tube with a light explosive dusting on the inside replaces the electric bridge wire of an electric detonator. A delay element was introduced using a specified length of pyrotechnic element, the length of time required to burn the pyrotechnic element governed the delay time for the detonator. This pyrotechnic element was also included in the early electric detonators. Due to the chemical makeup of the delay element, the accuracy is relatively low. Larsson and Clark [24] published accuracies of 1.5–2.5% of the total delay time and also stated that inaccuracy increased with delay time. Bajpayee and Mainiero [25] agree with this statement and concluded that there is increased scatter with increased delay time for electric and nonelectric detonators.

Electronic detonators include an electronic circuit and bridge wire to create a delay with a small microchip that can currently be programmed in 1 ms increments, rather than the length of pyrotechnic element seen in electric and nonelectric detonators.

Tables 4.1 and 4.2 show results published by Lusk et al. on the accuracy of electronic detonators against nonelectric [26]. Detonators A and B simply refer to two different unnamed manufacturers. These results agree with the observations of Bajpayee et al. and Larsson et al. that the scatter of all detonators increases with delay time, demonstrated by a larger standard deviation in Table 4.2 than 4.1. The results also show that nonelectric detonators have a larger scatter range than electronic detonators with maximum standard deviation of 19.054 ms compared with that of 3.751 ms for electronic detonators.

Timing is crucial for blast design, fragmentation, and maximum scaled distance requirements. There is clear evidence that, when using pyrotechnic detonators, the time you design your blast to and what your final blast shot produces are two very different things [24–26].

Table 4.1 Electronic detonator accuracy [26]

	Electronic detonators A			Electronic detonators B		
Programmed delay (ms)	10	1000	8000	10	1000	8000
Number of detonators tested	53	43	50	51	52	47
Delay average (ms)	9.95	1000.54	8003.38	9.99	999.80	7998.59
Standard deviation	0.09	0.32	3.75	0.03	0.11	0.85
Maximum (ms)	10.20	1001.12	8015.63	10.05	999.95	7999.40
Minimum (ms)	9.82	999.96	7995.19	9.91	999.46	7995.80
Percent error	−0.501	0.054	0.042	−0.130	−0.020	−0.018

Table 4.2 Nonelectric detonator accuracy [26]

	Nonelectronic detonators A			Nonelectronic detonators B		
Programmed delay (ms)	9	1000	1400	25	100	700
Number of detonators tested	68	60	67	59	65	59
Delay average (ms)	11.34	1125.50	1418.77	27.75	102.73	715.71
Standard deviation	4.59	6.56	19.05	0.77	11.25	6.20
Maximum (ms)	15.76	1146.79	1462.38	29.30	123.19	730.58
Minimum (ms)	1.53	1114.70	1367.04	26.16	79.84	697.93
Percent error	26.023	12.550	1.340	11.005	2.730	2.244

4.4.2 *Review of Timing Research with Relation to Fragmentation and Throw*

Electronic detonations have many proven benefits including, but not limited to, operational and energy efficiency, vibration reduction, consistency of results, and safety. A relationship between the effects of fragmentation and timing has been examined by researchers since the introduction of electronic detonators. This section presents results from previous researchers in this area.

Stagg and Rholl showed that fragmentation is improved when timing is greater than 3.3 ms/m of burden [27]. The same study showed that short (1 ms/m) and long delays (26 ms/m) had adverse effects on fragmentation. The very short delays suggest periods where wave propagation will influence fragmentation. Rossmanith [2] states that wave interactions are of vast importance to rock fragmentation. These interactions are that of the wave itself with natural features in the rocks, such as cracks or free faces, implying that short delays are going to be most influential with many mines utilizing this concept with electronic detonators.

Katsabanis et al. [21] reviewed certain literature on timing and fragmentation relationships. Some of his earlier work with Liu used a 6.5 ft bench with a burden of 2.6 ft and hole diameter of 1.5 into establish the effect of delay time [6]. Since they manually analyzed fragmentation from videos captured with a high-speed camera, a method that is not very accurate and cannot capture small fragments, they could only observe large differences [6]. Though they observed that optimum fragmentation (smallest D50 size) occurred at 8 ms/m, discrepancies in the testing means, we cannot put a lot of confidence in these results. McKinstry, on the other hand, showed that a time of 3 ms between holes on the same row, to utilize the collision of stress waves, proved beneficial at Barrick [28].

Katsabanis observed from experiments that fragmentation increases with delay until the delay becomes so great it is equivalent of firing each hole independently. At times greater than 22 ms/m, no increase in fragmentation is present [20]. The results of Katsabanis' work agree with those of Stagg and Rholl and indicate that improved fragmentation is now available using electronic detonators. However, it is more likely that the reason for this is improved accuracy of timing, not the ability to detonate at very short delay times where there is a possibility for stress wave interaction [20, 27].

Johansson and Ouchterlony [22] attempted to eliminate some of the primary concerns with previous experimental data including additional reflection zones around the block and few points covering the range of timings in question. The novel experimental technique used by Johansson and Ouchterlony [22] was the addition of a magnetic mortar yoke around the sample to reduce the reflection zones not present in a typical bench blast. They used two blast rows, blasted independently of each other, and analyzed fragmentation after each. All times used were below 2 ms/m of burden, below the optimum found by other authors but within the shock interaction times [21, 28]. Results display a bimodal boulder and dust character that does not fit well with the Kuz-Ram model due to the normal distribution from the Rosin-Rammler curve [22]. For row 1, scatter was much greater than in row 2, where the material had already been preconditioned. The least back break and smoothest wall after a row was shot was for instantaneous detonation, essentially representing a presplit blast.

Katsabanis et al. used a similar technique to Johansson and Ouchterlony with a stronger yoke around the back and sides of the concrete to be tested [21, 22]. The material to be blasted was positioned within the three walls. Timing between holes of spacing 10.5 cm had a nominal delay of 2000 ms. This is equal to 27 ms/m of burden based on the 7.5 cm burden. Their results show that the worst fragmentation is achieved with instantaneous detonation, both for average particle size and fit to the Swebrec and Rosin-Rammler distributions. The results plateau at the optimum fragmentation between 400 and 1000 μs, so an exact optimum could not be determined. Mean fragmentation size increases again after 1000 μs.

Due to the supersonic speed of a shockwave, the time between two adjacent holes would have to be almost instantaneous for this interaction to occur. Numerous studies have published results stating that a reduction in fragmentation, or increase in average particle size, occurs with timing of less than 2 ms/m of burden [6–8, 20–22]. Timings greater than 2 ms/m of burden are far too slow for any shock wave collision to occur. Results by Johnson [13] on shock wave collisions indicated that a potential reason for the reduced fragmentation observed when holes are initiated simultaneously is that the directional particle movement actually increases the density at the point where the two shock waves meet.

In scaled concrete model experiments, Johnson [13] found that when two primers are simultaneously detonated within a blast column, detonation wave collisions reflect pressure towards the face rather than back towards the top or bottom of a blast column. These collisions occur at the maximum VOD of the explosive towards the center of the blast column. The high velocity and change in shock direction towards the face demonstrated an increase in throw away from the face. This author knows of no other investigations on detonation wave collisions in strata other than the study presented in Sect. 4.5. However, through discussions with blasters, simultaneous initiation of two or more primers in a single hole is a common practice due to the observed increased fragmentation and improved diggability, though no fragmentation data has been recorded or published. (Personal communication with Brett Richter, Buckley Powder, MO, USA, July 7th, 2017). Delay time influences how the blasted rock will move, and blasts can be designed so that the desired throw is achieved. Since explosive energy is already required to fragment rock, it may be of interest to design for optimum throw so that less energy is necessary later in the mining cycle through the use of excavating and hauling equipment for material handling.

4.5 Full-Scale Experiments on Shock and Detonation Wave Collisions at a Granite Quarry

It has already been established within this chapter that timing can influence the fragmentation and throw of a production bench blast. From previous literature, timing between holes that are longer than any shock collision zone demonstrates the best fragmentation. When investigating throw, researchers and blasters have observed that when two detonators within a hole are initiated at the same time, it leads to improved fragmentation and diggability in the field and scaled experiments show an increase in the distance fragments are thrown from the blast column. Any increase in either fragmentation or throw will have an effect on the energy efficiency of a mine or quarry operation. The following section outlines the fragmentation and throw results of a two stage investigation at a granite quarry in Talbotton, GA, USA conducted and published by this author [29].

4.5.1 Blast Design and Experimental Setup

The investigation had two stages; the first investigation looked at shock wave collisions between holes, discussed as inter-hole timing, while the second investigation examined detonation wave collisions, or intra-hole timing.

Stage one altered the inter-hole delay and consisted of four test blasts on a bench of an active granite quarry. The test blasts were full-size production shots conducted between April 16, 2015 and September 15, 2015. Each blast shot approximately 37,000 m^3 of rock. Maximum rock strength for the granite was recorded to be 114 MPa. The bench used for analysis can be described as “massive” with few major bedding planes and joints. The mine’s standard inter-hole delay time was 16 ms and the inter-row delay was 142 ms. Each shot consisted of two rows of 14.605 cm diameter holes with a total of 85 holes per blast. The burden and spacing were 4 and 5 m, respectively. The bench height was approximately 21 m and holes were drilled with a 1 m sub-drill, at a 5° angle. The shots had only one open face. The typical stemming height was 2.75 m and the stemming material used was good quality angular 1.905 cm crushed rock. Holes were loaded with emulsion and initiated by dual electronic detonators. Detonators with boosters were placed near the top and bottom of the powder column and had a 2 ms delay between the bottom and top detonators. The bottom detonator was fired first. The top detonator served as a backup in case of a misfire; otherwise, it was engulfed in the charge column as the VOD was faster than the timing delay. The research only modified the inter-hole delay times, investigating 0, 1, 4, 10, 16, and 25 ms times. The 0 and 1 ms timings allowed for shock wave collisions while the 16 ms time was the quarry's original timing delay.

Stage two kept the inter-hole delay constant and altered the intra-hole delay time. The research collected data from three separate, consecutive test shots at the same quarry on July 5, August 9, and September 20, 2016. The shots were full-scale blasts, blasting on average 40,500 m^3 of rock using 28,500 kg of bulk emulsion. The remaining blast parameters were the same as the stage one tests. Two primers were loaded into each hole, one towards the top and the other towards the bottom. The delay times between top and bottom primer and order of initiation were variable from shot to shot; the first was simultaneous (dual initiation), the second was 1 ms bottom-initiated (single initiation), and the third was 1 ms top-initiated (also single initiation). In the latter two shots, the researcher chose 1 ms delay between primers so that the second primer detonated before being engulfed in the explosive column, unlike that in stage one of the investigation.

4.5.2 Discussion of the Results

The study analyzed seven full-scale test blasts for fragmentation, using digital photographic analysis, and throw. The first four test blasts investigated shock wave interactions between holes at delay times between 0 and 25 ms. This author found that 25 ms delays between holes resulted in the greatest fragmentation. Another 3 test blasts investigated the effects of detonation wave collisions within the blast column on rock throw and fragmentation using this optimum inter-hole delay timing. These blasts consisted of two primers down each blast hole located at the top and bottom, with initiation sequences consisting of simultaneous, bottom-initiated, and top-initiated.

In stage one, the 25 and 10 ms delay times resulted in the best fragmentation. Through photographic fragmentation analysis, the study found that the 25 ms delay had the smallest mean, D50, and D90 sizes. This delay time was too long for shock wave collisions to affect fragmentation. Short hole-to-hole delay times did not improve rock fragmentation in full-scale bench blasting. The best performing delay times were outside of the short delay range and the worst performing delays were the shortest.

Stage two attempted to maintain the improvements to fragmentation from the 25 ms delay while increasing throw from wave collisions. Detonation wave collisions caused by simultaneous detonation of the top and bottom primers in the blast column resulted in increased throw, approximately 19 m over both single initiation timings.

This investigation demonstrates that shock wave collisions between holes do not improve fragmentation results but detonation wave collisions within a blast hole can increase throw. The increase in throw can also lead to additional secondary fragmentation where fragments break further as they hit each other and the floor. However, further investigations are necessary regarding this aspect. Full-scale blast analysis has several uncontrollable factors to consider over lab scale tests. However, the aim of this investigation was to investigate whether results found in lab scale experiments could be mimicked successfully in the field and lead to improved blast performance and energy efficiency of mine and quarry operations.

4.6 Conclusions

Timing is now a crucial part of the blast design process, along with traditional parameters such as powder factor and blast geometries (burden, spacing and hole diameter). Often, timing has to be altered to accommodate vibration control, for example. It is important to understand how any change in timing can affect the overall efficiency of a blast and, consequently, the overall mine operation. Precise timing of electronic detonation provides a mechanism for controlling the collision of shockwaves or detonation waves when the blaster understands strata characteristics. This chapter has given an overview of the theory behind shock and detonation pressures produced during a bench blast. Additionally, the chapter explored research showing how timing can influence fragmentation and throw patterns, demonstrating to readers the knowledge on how time can be used to optimize the energy and operational efficiency of a blast without altering powder factor and, consequently, monetary cost. When interaction of these shock or detonation waves occurs and pressures increase at the collision point, further utilization of the explosive energy occurs, ultimately resulting in improved energy efficiency of the mining operation.

References

1. Yamamoto M (1999) Experimental and theoretical study on smooth blasting with electronic delay detonators. Fragblast 3:3–24
2. Rossmanith HP (2003) The mechanics of electronic blasting. In: Proceedings of the 33rd conference on explosives and blasting technique, ISEE
3. ISEE (2011) Blasters' handbook, 18th edn. International Society of Explosives Engineers, Ohio, USA
4. MacKenzie AS, Cong J (1966) Cost of explosives—do you evaluate it properly? pp 32–41
5. Kanchibotla SS, Valery W (2010) Mine to mill process integration and optimization—benefits and challenges. In: Proceedings of the annual conference on explosives and blasting technique
6. Katsabanis PD, Liu L (1996) Delay requirements for fragmentation optimization, measurement of blast fragmentation. Belkema

7. Rossmanith HP (2002) The use of Lagrangian diagrams in precise initiation blasting, part 1: two interacting boreholes. Fragblast 6(1):104–135
8. Vabrabant F, Espinosa A (2006) Impact of short delays sequence on fragmentation by means of electronic detonators: theoretical concepts and field Validation. In: Proceedings of the 8th international symposium on rock fragmentation by blasting. Fragblast, vol 8, pp 236–331
9. Bauer F (2014) Blast optimization through long term fragmentation analysis. in: Proceedings of the 40th annual conference on explosives and blasting technique. International Society of Explosive Engineers, Denver, February
10. Motion Metrics (2015) Portable analysis portametrics. http://www.motionmetrics.com/portable/. Accessed Oct 2015
11. Split Engineering (2015) Split-desktop software. http://www.spliteng.com/products/splitdesktop-software/. Accessed Oct 2015
12. Sanchidrian JA, Segarra P, Ouchterlony F, Lopez LM (2009) On the accuracy of fragment size measurement by image analysis in combination with some distribution functions. Rock Mech Rock Eng 42(1):95–116
13. Johnson CE (2014) Fragmentation analysis in the dynamic stress wave collision regions in bench blasting. Dissertation, University of Kentucky
14. WipFrag (2015) WipFrag manual. North Bay, Ontario, Canada. Accessed Oct 2015
15. Onederra I, Bailey V, Cavanough G, Torrance A (2012) Understanding main causes of nitrogen oxide fumes in surface blasting. Institute of Materials, Mineral and Mining
16. Broek D (1988) The practical use of fracture mechanics. Kluwer Academic Press, Berlin
17. Clark GB (1987) Principles of rock fragmentation. Wiley, New York
18. Cook MA (1974) The science of industrial explosives. Graphic Services and Supplies, Inc.
19. Cooper PW (1996) Explosives engineering. Wiley-VCH, USA
20. Katsabanis PD, Tawadrous A, Braun C, Kennedy C (2006) Timing effects on fragmentation. In: Proceedings of the 32nd conference on explosives and blasting technique, Dallas, 29 Jan–3 Feb
21. Katsabanis P, Omidi O, Rielo O, Ross P (2014) Examination of timing requirements for optimization of fragmentation using small scale grout samples. Blasting Fragmentation 8(1)
22. Johansson D, Ouchterlony F (2013) Shock wave interactions in rock blasting: the use of short delays to improve fragmentation in model scale. Rock Mech Rock Eng 46:1–18
23. Sjoberg J, Schill M, Hilding D, Yi C, Nyberg U, Johansson D (2012) Computer simulations of blasting with precise initiation. Eurock
24. Larsson B, Clark DA (1982) Cost savings and improved stability through optimized rock blasting. In: First international conference on stability in underground mining, SME-AIME, Vancouver, 16–18 Aug
25. Bajpayee TS, Mainiero RJ (1990) Firing accuracy of electric detonators. In: International Society of Explosives Engineers, General proceedings
26. Lusk B, Silva J, Eltschlager K (2012) Field testing and analysis of blasts utilizing short delays with electronic detonators. OSM Final Report: S09AP15632
27. Otterness RE, Stagg MS, Rholla SA, Smith NS (1991) Correlation of shot design parameters to fragmentation. In: Proceedings of the 7th annual symposium of explosives and blasting research, ISEE, Las Vegas
28 McKinstry R, Bolles T, Rantapaa M (2004) Implementation of electronic detonators at Barrick Goldstrike Mines, Inc. In: Proceedings of the 30th conference on explosives and blasting technique, ISEE, New Orleans
29. Cahill PG, Hettinger MR, Nawrocki J, Johnson CE (2017) Full scale testing of delay timing effects on rock fragmentation in the shock collision regions. Blasting Fragmentation J 11(1)

Chapter 5
Energy Efficiency of Drilling Operations

Celal Karpuz

Abstract This chapter presents a review of the literature on energy efficiency of drilling operations in mineral industries. It introduces the drilling systems, factors affecting drilling work, and general relations between type of drilling systems and rock properties. First, the basic features of rock drilling and the role of drilling work in energy efficiency of mineral industry are generally described. Then the importance of energy concept which is usually expressed in terms of specific energy and related approaches as well as the related studies is reviewed. The author makes recommendations for future research directions to enhance the energy efficiency of drilling operations in mineral industries. The papers included in the review were mainly selected through searches in major abstract databases of web of science.

Keywords Drilling operations · Specific energy · Energy efficiency
Mineral industry · Future research

5.1 Introduction

Drilling has an important role in the mining industry and greatly affects the economics of mining. It is a fundamental stage of surface mining and, together with blasting, accounts for around 15% of total cost [11]. Drilling efficiency largely depends on the rate of penetration, which affects power consumption and bit life. The operating cost can be reduced significantly by selecting both the most suitable bit type and the most efficient operating conditions. Therefore, any effort which aims to increase efficiency in drilling by reducing power consumption and bit wear will improve the economics of mineral projects. In mine engineering, drilling may be used for exploration including resource definition (to obtain geological

C. Karpuz (✉)
Mining Engineering Department, Middle East Technical University, Universiteler Mah., Dumlupinar Bulvari, no. 1, 06800 Ankara, Turkey
e-mail: karpuz@metu.edu.tr

K. Awuah-Offei (ed.), *Energy Efficiency in the Minerals Industry*, Green Energy and Technology, https://doi.org/10.1007/978-3-319-54199-0_5

information including nature and strength of the materials and type of ore body, coalbed methane (CBM) detection, and production), mine site investigation, rock bolt and foundation drilling, blast hole drilling, ventilation, drainage, shaft construction, and dewatering and disposal wells. Among those, blast hole and exploration (sometimes called diamond drilling) drilling are the most commonly applied drilling systems. This chapter discusses, in detail, the basic properties and efficiencies of drilling in these applications.

Drilling systems can generally be divided into two categories: rotary percussive and rotary methods. The most frequently used one for all rock types is rotary percussive type, and top hammer and down-the-hole hammer are the commonly used equipment. Rotary methods are divided into two subgroups depending upon the type of bits used as rotary cutting and rotary crushing. Rotary crushing with tricone bits may be used for medium to hard rocks while rotary cutting systems with drag bits may be used for soft rocks. The coordination of percussive, rotation, and cutting actions with the geometric properties of the bits enables these bits to penetrate the rocks. Tamrock suggested the relations between the type of drilling system, corresponding hole diameter, type of equipment used, and property of rock (Fig. 5.1) [31].

An important component of the drilling system is the bit, which is mostly made of steel, tungsten carbide, diamond, and polycrystalline diamond (PDC). Bits are designed for percussion, rotary crushing, and rotary cutting actions to penetrate the rock by the forces imposed on them. The main parameters to consider, in the

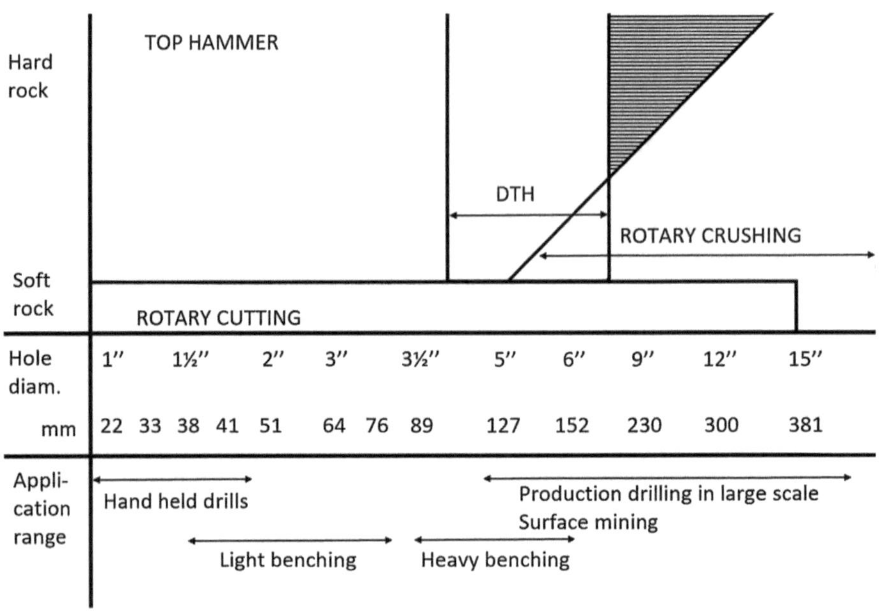

Fig. 5.1 Relations between type of drilling system, corresponding hole diameter, type of equipment used, and property of rock [31]

selection of a suitable bit, are rock properties, the angular speed of the drill string (measured in revolutions per minute (rpm)), circulation medium, diameter, and depth of hole. The penetration rate, at the end of the bit life, can be reduced by as much as 50–75% compared to a new one in tricone bit, due to bit wear.

Penetration rate mainly depends on rpm, thrust, torque, circulation medium, and type of bit. Optimizing these parameters for a given condition increases the efficiency of drilling work. They are automatically monitored during drilling in both laboratory research and field drilling work. This system, called "monitor-while-drilling" (MWD), has been provided by drill rig manufacturers and used widely in the last three decades. Also, remote control systems (RCSs), which use computer and information technology to facilitate and increase quality drilling, are used in surface blast hole rigs as a form of MWD [3].

The effect of thrust on bit wear and cost on bit life and production are presented in Figs. 5.2 and 5.3 [3, 7], respectively.

The goal of this chapter is to present a literature review of energy efficiency of drilling operations, which establishes the current knowledge on the subject and makes recommendations for future research. The review mostly focused on peer-reviewed journal publications identified through search in major abstract database Web of Science using relevant keywords such as "rock drilling" and "efficiency in drilling". In a few cases, the handbooks and web pages of some worldwide drill manufacturing companies and relevant papers in peer-reviewed conference proceedings were included in the list of references. In all, the author reviewed 60 papers and after review, the author included 41 references to address the research objective and those related to Petroleum Engineering were disregarded.

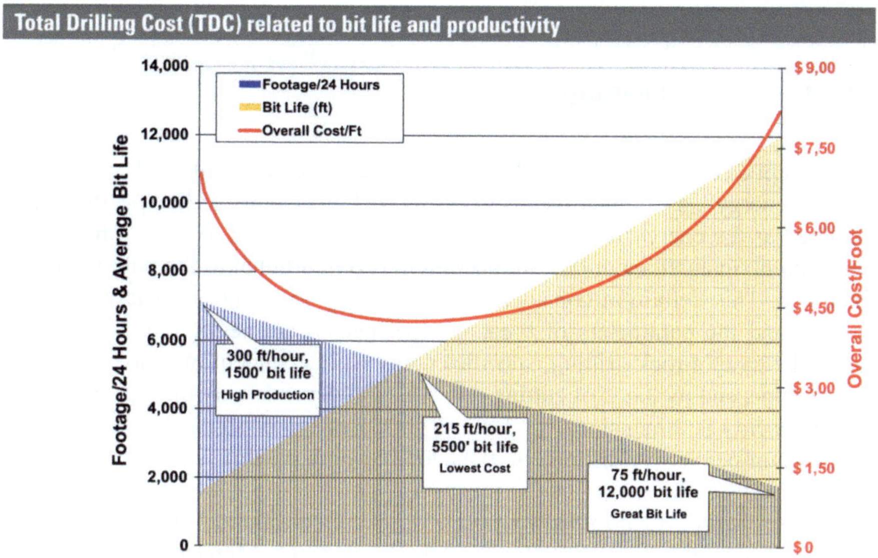

Fig. 5.2 Bit life versus cost and production [3]

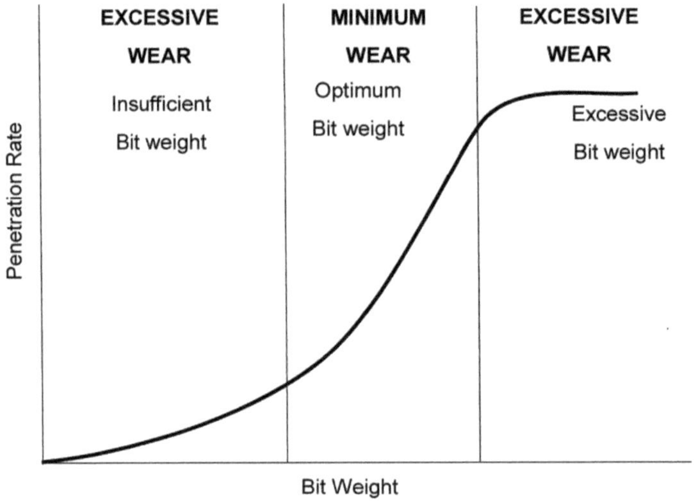

Fig. 5.3 Penetration rate—bit weight and wear relation [7]

The review concentrated mainly on research in the last decade in order to establish the state-of-the-art; however, the sources related to basic properties of the drilling extended till 1950s. The author hopes that this will be a useful reference for mine managers and engineers as a reference on energy efficiency in drilling. Additionally, this overview should be a useful reference for beginning graduate students and researchers who are interested in optimizing energy consumption in drilling operations.

5.2 Energy in Drilling

The concept of energy in rock drilling was first proposed by Teale and expressed as "specific energy", as a quick means to assess rock drillability [33]. Specific energy is defined as the energy required to remove a unit volume of rock. In other words, energy input, in an efficient drilling operation, is proportional to the volume of rock penetrated. This definition considers specific energy to be an intrinsic property of the rock and the parameters of breakage mechanism are negligible. This implies that, in the case of rock drilling, specific energy is independent of shape of the drill bit, drill type, and methods of cutting removal, and depends only on rock properties. Specific energy is dimensionally identical with pressure or stress. Taele stated that in rotary non-percussive drilling, work is done by both the thrust, F (N), and torque, T (N m) [33].

Specific energies determined for rotary drills are considerably higher than those determined for percussive drills for the same rock types. Moreover, specific energies of rocks are sensitive to bit types and specific energies determined in the laboratory and values determined in field experiments differ by up to 83% [29]. This shows that there are other factors, besides rock properties, that affect the energy efficiency of drilling. No one rock property completely defines breakage characteristics and prediction of drill performance can best be obtained by using empirical equations based on relevant rock properties [29].

Using some assumptions, based on the United States Bureau of Mines (USBM) researchers' data, Rabia [29] suggested the following specific energy equation for percussive drill bits:

$$\mathrm{SE} = 4\mathrm{Tr} \times (\mathrm{Power\,output})/\pi D^2 \times \mathrm{PR}$$

and for rotary drill bits:

$$\mathrm{SE} = 2.35\, W.\mathrm{RPM}/D.\mathrm{PR}$$

where W is the weight on the bit (kg), RPM is revolutions per min, D is the diameter of bit (mm), PR is the penetration rate (m/h), and $\mathrm{Tr} = 0.7$ is the transfer ratio between energy transferred to the rock and the energy available for each blow.

Penetration rate of both rotary and percussive drilling is predicted by the ratio of energy input to rock strength. The most important rock strength parameters are uniaxial compressive strength of rock (UCS), tensile strength, and hardness of rock [29].

There are many other empirical models for penetration rate. For example, by considering thrust, RPM, and UCS, Bauer and Calder established an empirical penetration rate equation for rotary blast hole drilling [4]. Also, Fish conducted a model study which concluded that penetration rate is directly proportional to thrust and inversely proportional to UCS [9]. The empirical penetration rate equation presented by Clark was based on RPM, bit diameter, weight on bit, hardness, and triaxial compressive strength [6].

Researchers have observed a parabolic relationship between specific energy, thrust, and bit wear for impregnated bits [24] (Fig. 5.4). Obviously, minimum specific energy corresponds to balanced bit wear; while the extremes on either side of the minimum show either low-friction wear flats or high-friction stalling or seizure with high specific energy values.

The drillability work on rotary blast hole drilling for both tricone and drag bits produced penetration rate equations related to thrust, RPM, and rock properties such as UCS, tensile strength, cohesion, and density [19]. Researchers have concluded that drag bits have higher penetration rates than tricone bits for the same rock type, since the contact area is higher in tricone bits.

Modulated specific energy (MSE) is a novel measure of energy efficiency developed to characterize drilled material in open-pit coal mining using MWD technology of rotary rigs [17]. MSE is defined as the product of specific energy of drilling and a logistic function of the "rotational work fraction (RWF)". This proposal utilizes a hypothesized link between a derived drill performance indicator (rotation-to-thrust power ratio) and geomechanical properties of sedimentary rock

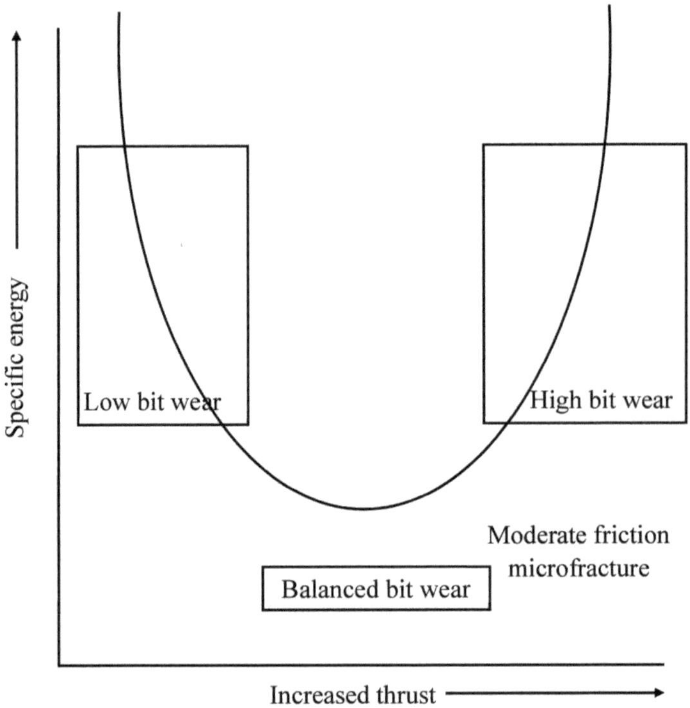

Fig. 5.4 Thrust-specific energy bit wear relations [24]

strata (shear and compressive strengths) to increase the coal discriminative power of MSE relative to Teale's specific energy measure. Its efficiency is demonstrated using mutual information, a simple threshold strategy, and an artificial neural network. By reducing the detection uncertainty, MSE is able to provide consistent feedback while drilling and eliminate trial-and-error.

On the other hand, Li and Itakura introduced "effective specific energy", defined as the sum of specific energy consumed by cutting and feeding, to develop a model which evaluates drilling efficiency for rotary drag bits using MWD [22]. The drilling process is divided into cycles, each of which includes two motions: feeding and cutting, where feeding is treated as an indentation motion. In the model, drilling torque consists of four parts generated, respectively, from cutting, friction, feeding, and idle running. Similar to torque, specific energy also has four parts from cutting, friction, feeding, and idle running. In producing the model, the SE equation of Teale and the study of Li were considered [21, 33]. Li [21] showed that the value of the second term of the SE equation of Teale [33] is about 10–200 times greater than the first one. It means energy resulting from torque is the main part of the drilling specific energy. Therefore, they neglected the energy resulting from thrust and show that the drilling specific energy, SE, is inversely proportional to the cutting depth, z, and effective specific energy, Ees, is defined as

$$\mathrm{Ees} = 2\pi\,\mathrm{Te}/Az$$

The relationship relates effective specific energy to UCS without including the cutting depth "z" and includes only the effective part of the SE. They claimed that effective specific energy is more reasonable than total specific energy when evaluating the drilling efficiency. Figure 5.5 shows that specific energy obtained from field tests is inversely proportional to the penetration rate, whereas effective specific energy is independent of it, if UCS is constant. Similar trend was also obtained from laboratory experiments. They concluded that this verifies the proposed model and shows that effective specific energy is a better index for evaluating the UCS of rocks (Fig. 5.6).

The expressed relation between specific energy and UCS (Uniaxial compressive strength) is given below:

$$\mathrm{Ees} = 2\pi/A\left(k_c\mathrm{UCS} + k_f\mathrm{UCS}\right)$$

where

k_f: constant function of bit geometry and friction angle between bit and rock,

k_c: constant function of bit geometry, friction angle between bit and rock and internal friction angle of rock.

There are many parameters related to energy efficiency of drilling. Some of them are changeable, while some are not. Rock properties are not changeable parameters, since they are inherent properties of the rock and are divided into two main groups. These are rock material properties such as UCS, hardness which is closely related to UCS, tensile strength, cohesion, density, texture, and abrasivity and mass properties such as joints, bedding, and schistosity, which are mostly represented by RQD

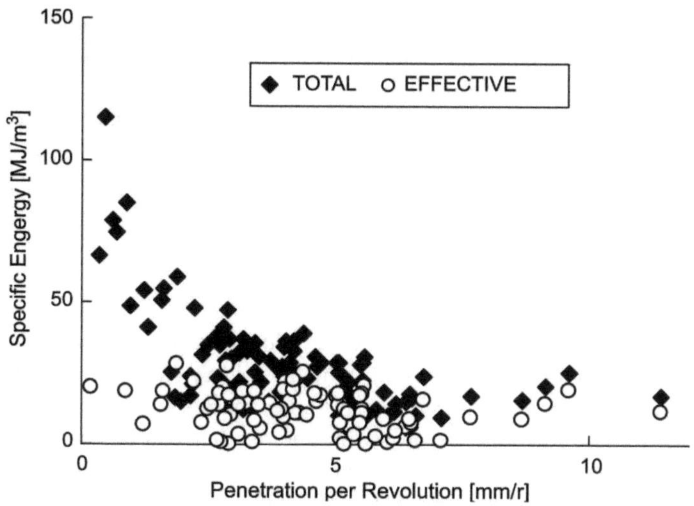

Fig. 5.5 Total and effective specific energy versus penetration rate (field experiments) [22]

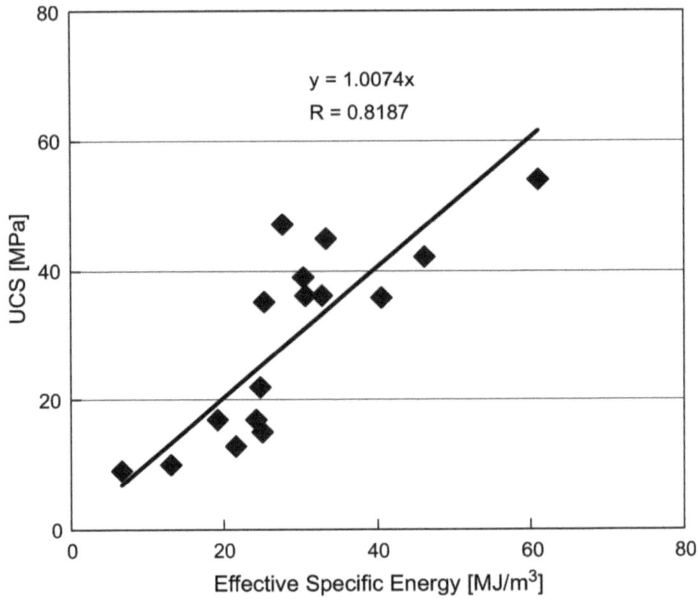

Fig. 5.6 Effective specific energy versus UCS [22]

(Rock Quality Designation) or fracture spacings which disturbs the continuity of rock masses. Abrasivity affects bit wear and depends on mineral composition of rock and quartz content is the dominant abrasive composition. Texture is related to grain structure of rock. Based on some rock material properties, Tamrock suggested "drilling rate index" (DRI), which is a kind of penetration rate and "bit wear index" (BWI) [31]. BWI describes the rock abrasiveness and effect bit life. The details of DRI and BWI can be obtained in the related reference. On the other side, operational parameters bit load, bit rotation (torque), and circulation medium are changeable parameters and chosen by operators. Bit loads and rotation mainly depend on changeably parameters, type and diameter of bit and rig capacity, and unchangeable parameters rock properties and rig capacity.

5.3 Drillability Studies

There are also many studies in the literature that focus on characterizing the ease with which a formation can be drilled, which is a key factor contributing to the energy consumption during drilling. These studies mainly seek to propose measures of drillability (ease of drilling) by proposing metrics that can be used as proxies (in lieu of specific energy) to measure drillability. The research mainly focuses on models (both empirical and analytical) to predict penetration or drilling rate as it is a good indicator of drillability. These studies are relevant to a discussion of the energy efficiency of drilling because they provide further insights into factors that control energy efficiency of drilling.

For example, Ergin et al. attempted to determine optimal operating parameters for various roller cone bits to attain the highest penetration rate and bit life on a large-scale laboratory horizontal drilling rig [8]. They recorded the penetration rate, torque, and power consumption using a data acquisition system. They suggested optimum operational parameters to maximize penetration rate and bit life are bit weight and RPM of 130 kN and 60 RPM, respectively, for the rocks with UCS around 78 MPa. This study shows that such optimum parameters exist that will optimize drilling performance and could potentially increase energy efficiency.

Some researchers have proposed drillability indices to predict penetration rates for rotary blast hole drills of spherical and conical button tungsten carbide bit teeth as a function of rock properties [18]. There is a close relationship between such drillability indices and UCS, tensile strength, N-type Schmidt hammer value, impact strength, P-wave velocity, elastic modulus, and rock density for the rocks with UCS of 40 MPa.

Other field drillability studies stressed the importance of optimizing operational parameters with respect to rock properties to obtain higher penetration rate with minimum bit wear. For example, Akun and Karpuz, in a study using a diamond drill with cored–noncored and impregnated–surface set bits, showed that penetration rate is correlated to rig operational parameters and rock properties [1]. Their work also showed the importance of optimizing circulating fluid so that all chips are removed and all the energy consumed is used for rock penetration. They also found that lower (less than 5000 N) and higher (more than 15,000 N bit loads than required for NQ size hole drilling with wire-line system (NQWL) surface set bits, may cause more bit wear and decrease the efficiency of the bit. This results in higher specific energy and lower penetration rate.

Another issue that researchers have focused on is the effect of bit wear on penetration rate and, therefore, energy efficiency [10]. Ghosh et al. [10] assumed a linear trend for production degradation and used it to calculate initial penetration rate and a production degradation coefficient for each studied bit. The normalized initial penetration rate varied between 0.4 and 1.2 m/min while the bit life length varied from 300 to 1400 m. The large variations in the production degradation profiles are observed for individual bits. The whole population showed an inverse nonlinear relationship between the production degradation coefficient and bit life, indicating that the lower the degradation coefficient, the higher the bit life, and vice versa (Fig. 5.7). This is not because of variations in rock strength since they observed a definite positive correlation between penetration and bit life.

Ghosh et al. also used principal component analysis (PCA) to show that rock mass characteristics represented by penetration rate and specific energy are not correlated to bit life length [10]. Instead, the bit life length is negatively correlated to operational parameters such as torque, rotation speed, and, to a minor extent, feed force. This negative correlation suggests that magnitude of operational parameters can possibly be reduced in order to increase operating life length of the bit, and, in turn, it reduces bit cost and further it decreases total drilling cost to a certain extent.

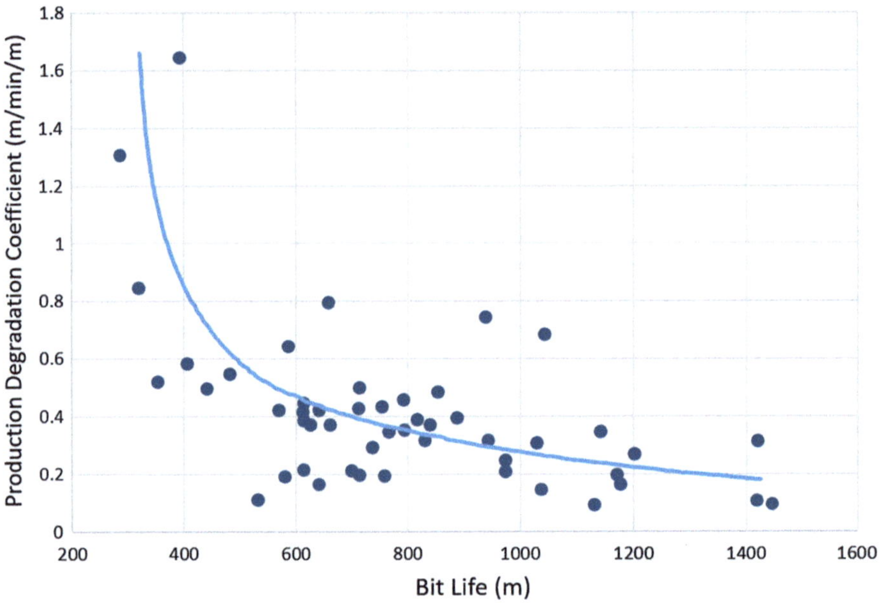

Fig. 5.7 Bit life versus production coefficient [10]

Akun and Karpuz evaluated the effect of bit wear (karat loss/meter), based on diamond salvage, on penetration rates of diamond drilling and then optimized the operational parameters to give minimum bit wear for drilling of Zonguldak sandstone with a UCS around 60 MPa [2]. After an optimization study, they established the relation between penetration rate and specific energy (Fig. 5.8). They found the maximum bit life, for surface set bits in sandstone, to be 112.7 m with 0.058-carat/m loss. However, they observed some extreme results, which they attributed to low Rock Quality Designation (RQD) (17–30%), high number of discontinuities (15–22 #/m), and high (15–20%) quartz content. By optimizing operational parameters, they were able to triple the net penetration rates.

Other researchers have evaluated the performance of polycrystalline diamond compact (PDC) percussion bits using laboratory drilling tests where PDC bits were compared to conventional percussion bits equipped with cobalt (Co)-containing cemented tungsten carbide (WC–Co) tips [25]. It was found that the PDC percussion bit is superior to widely available WC–Co percussion bit in terms of durability and drilling efficiency. Total drilling cost when using the PDC percussion bit is considered to be lower when drilling long holes because the work period, including the round-trip time for replacing the bit can be shortened.

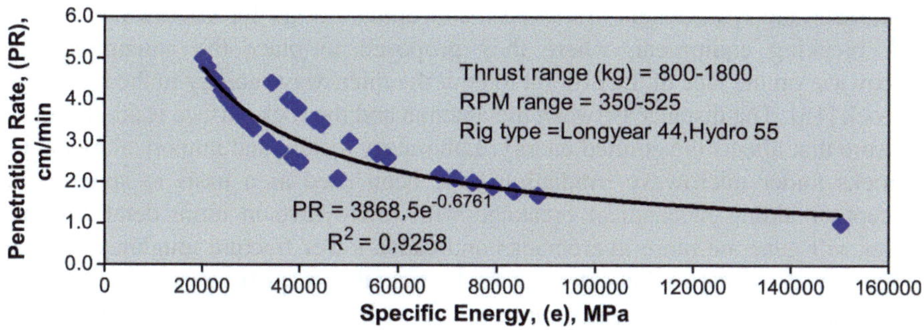

Fig. 5.8 Specific energy penetration rate relation [2]

5.4 Future Techniques

Automation is one area that has the potential to increase the overall efficiency and energy efficiency of drilling operations. Available technologies utilizing mechanical power, sensors, and Global Positioning System (GPS) are being used to automate drilling operations by providing real-time information about location and operational parameters, such as drilled depth, torque, rotational speed, and penetration rate in modern mines [3]. Fully automated drills that drill a given pattern on their own are available on the market. Future advances in automation are bound to make drilling more energy efficient.

Besides automation of drilling equipment, novel techniques to penetrate rock or assist mechanical drilling systems have been studied by various researchers. Rock penetration methods can be broadly classified as traditional and novel [12].

Novel techniques in drilling, such as microwave irradiation, improvements in directional drilling, MWD and RCS, high-pressure water jet, and abrasive waterjet applications are promising techniques. Better rotary percussion systems and the use of PDC bits may also increase the efficiency in drilling. The combination of new techniques such as computational fluid dynamics (CFD), support vector machine (SVM), and neural network in drilling research will facilitate new discoveries and greater efficiency in drilling.

Thermal rock weakening is one of the promising novel techniques as certain minerals within rock have the potential to be heated effectively by a source to induce intergranular or transgranular fractures by thermal expansion [20]. The heterogeneous structure of rocks with varying thermal expansion coefficients based on mineral composition can be taken advantage of to generate stress within the rock [16]. Minerals that are transparent to microwave irradiation behave different from minerals that absorb the energy so that transgranular cracks might be formed to assist mechanical breakage of rocks. Hassani et al. used microwaves to trigger thermal expansion in an innovative way that could support drilling and rock cutting applications [15].

Hassani and Nekoovaght presented a conceptual design for microwave-assisted rock breaking equipment, where they proposed to place the antenna of the microwave on the face of the drill bit to emit the microwave energy to the surface of the rock [13]. The distance between the antenna and the rock surface is an important measure that affects the emitted energy. Laboratory testing and numerical modeling of rocks under microwave irradiation have been used as a basis to support the concept of aiding mechanical breakage with microwave in more detail. Higher power will generate more microcracks and cause lower fracture toughness values, which are also used as indicators of rock strength [26]. Microwaves penetrating rocks mainly increase the temperature of the rock surface due to the low thermal conductivity. Therefore, higher power is suggested to heat the rock surface in a short time [27]. Various rock types have been tested for their strength behavior under microwave irradiation [28] and researchers have found that the presence of water prevents the available microwave energy from being reflected and, therefore, increases the efficiency of microwave irradiation [14].

There are numerous factors that affect the performance of microwave-assisted drilling, such as mineral composition, moisture content, and operational conditions. The exact manner in which these factors affect performance and energy efficiency is yet unknown. This requires further research to ensure optimal design and operation. Also, the implementation of microwave irradiation might increase the cost of drilling as bit wear is one of the main cost components of drilling and a microwave antenna would increase the unit cost of this consumable. The financial status of corporate mining companies depends on various economic drivers, besides commodity prices, so that research budgets for development and implementation of novel methods in drilling could be suspended based on the priority assigned by management. Further research in this field is required to advance the technology toward prototype drill bit for testing on site with different types of rocks. If well designed, this technique could significantly improve the energy efficiency of drilling. This technique would also be advantageous in space mining applications.

Directional drilling has been applied in mining over the last 6 decades and presents another promising drilling technique for the future to increase efficiency in drilling. Recent research has developed techniques to obtain data from a single shot camera survey system and other advanced monitoring systems to provide a rapid and easy underground borehole survey approach using MWD [34]. An additional benefit of this system is that geologic features, such as discontinuities, intercepted during in seam directional drilling, can be monitored by drilling fluid pressures, changes in thrust, vibration, rate of penetration, and inspection of cuttings. However, there still remains four major problems associated with directional drilling: sticky drilling in complex conditions, sensitivity of downhole probe, in-hole stability, and drill depth capacity. Further research is required to resolve these challenges.

Other new developments include the use of radio wireless and wired communication systems to measure parameters, such as inclination angle of the hole, azimuth, and tool face as well as displaying drilling parameters, trajectory and the relationship between designed and actual performance. These facilitate prediction and control of directional drilling trajectory using MWD [35]. The importance of the future development of integrated directional drilling services software which combines MWD with drilling trajectory design, prediction, and control cannot be overemphasized.

Researchers have been long explored using water jets to aid drilling [30]. Recent research has tried to extend water jet application into more competent rocks [23]. In this application, abrasive waterjet erodes the "boss" for a "pilot hole" and, hence, lowers rock tensile and shear strength compared to compressive strength. The exposed wall of the "pilot hole" provides a large area of free surface which can guide the cracks to subsequent rock breaking and promote the cutting edges to overcome the rock tensile strength to complete the tensile and shear failure of the rock around the "pilot hole". In this way, the required drilling force decreases and drilling is more efficient with low bit wear in the hard rock. Lu and coworkers designed and manufactured such a hard rock breaking bit and a set of hard rock drilling equipment system with abrasive water jet assistance and used it to conduct experiments to compare its performance to conventional techniques [26]. Their results showed that with the assistance of abrasive water jet, drilling depth has increased by about 63%, thrust force and torque reduced by about 15% and 20%, respectively, bit wear reduced, and bit life increased, significantly.

Tang et al. showed that it is better to combine drilling with high-pressure water jet for soft rocks, and combining drilling with abrasive water jet is feasible for the hard rock drilling because of higher drilling efficiency and performance [32]. Compared to existing technology, Tang and coworkers found that drilling depth increased around 65%, axial force and torque reduced around 14% and 17%, respectively, and bit wear also reduced. This is a very promising drill technique for the future that is bound to increase the energy efficiency of drilling (reduced triaxial force and torque will lead to lower energy consumption be unit length of drilling). Further research is required to verify the applicability of the system at field scale and to expand the industrial implementation areas.

Another area that will require attention in future is the effectiveness and energy efficiency of drilling for space mining. Some work has been initiated by some researchers in this area. For example, Chen et al. proposed a multistate control strategy for autonomous lunar drilling based on online recognition of drilling media (soil and rock) by SVM and wavelet transform methods [5]. Lunar drilling is categorized into three drilling states. These are the interface detection, initiation of drilling parameters for recognition, and drilling medium recognition. After numerical modeling, drilling experiments are implemented with multilayered drilling media constructed by lunar soil simulant and lunar rock simulant to verify the effectiveness of the multistate control strategy. The results revealed that the multistate control method is capable of detecting drilling state variation and adjusting drilling parameters timely under vibration interferences. Further work is required in this area to develop technology suitable for the mining methods that will be used for space mining. These should be applicable to lunar mining and also to mining of asteroids and Mars.

5.5 Conclusions

This chapter aimed to review the literature on energy efficiency of drilling operations in mineral industries. It discusses drilling systems, factors effecting drilling work, and the general relations between the type of drilling systems and the properties of rock. Additionally, the chapter makes recommendations for future research directions on drilling work to enhance energy efficiency in mining.

The energy concept expressed as "specific energy", which is often used to assess rock drillability, was first presented. Then, the related theoretical and empirical studies carried out at both laboratory and field scales were presented, with regards to energy efficiency in mineral industries.

The promising future techniques that have the potential to increase the overall efficiency and energy efficiency of drilling operations were also presented. These include (i) automation, utilizing mechanical power, sensors, and Global Positioning System (GPS); (ii) novel techniques, such as microwave irradiation; (iii) improvements in directional drilling; (iv) monitor—while drilling and remote control systems; (v) high-pressure water jet and abrasive waterjet applications; (vi) improvements in rotary percussion systems and the use of polycrystalline diamond bits; (vii) the combination of new and facilitated techniques, such as computational fluid dynamics (CFD), support vector machine (SVM), and neural network; and (viii) a multistate control strategy for autonomous lunar drilling based on online recognition of drilling media (soil and rock) may be a technique for the effectiveness and energy efficiency of drilling for space mining.

References

1. Akun E, Karpuz C (1991) Penetration rate behaviour of rock formations during exploration drilling operations in Turkey. Rock mechanics as a multidisciplinary science, In 32nd US rock mechanics symposium, Oklahoma, pp 415–422
2. Akün E, Karpuz C (2005) Drillability studies of surface-set diamond drilling in Zonguldak region sandstones from Turkey. Int J Rock Mech Min Sci 42(3):473–479
3. Atlas Copco (2016) Blasthole drilling in open pit mining, 3rd edn. http:\\www.atlascopco.com . Accessed 20 Jun 2016
4. Bauer A, Calder NP (1967) Drilling in open pit iron mines. Am Min Congr J 53:76–80
5. Chen C, Quan Q, Shi X, Deng Z, Tang D, Jiang S (2016) Multi-state autonomous drilling for lunar exploration. Chin J Aeronaut (in press)
6. Clark GB (1979) Principles of rock drilling. Q J Colorado Sch Min 74:91–93
7. Cummins AB, Given LA (1973) Mining engineering handbook. AIME, New York, pp 1124–1156
8. Ergin H, Kuzu C, Balci C, Tunçdemir H, Bilgin N (2000) Optimum bit selection and operation for the rotary blasthole drilling using horizontal drill rig HDR—a case study at KBI murgul copper mine. Int J Surf Min Reclam Environ 14:295–304
9. Fish BG (1968) The basic variables in rotary drilling. Mine Quarry Eng 27:74–81
10. Ghosh R, Schunnesson H, Kumar U (2016) Evaluation of operating life length of rotary tricone bits using measurement while drilling data. Int J Rock Mech Min Sci 83:41–48

11. Gokhale BV (2011) Rotary drilling and blasting in large surface mines. CRC Press/Balkema, p 389
12. Hartman HL, Mutmansky JM (2002) Unit operation of mining. Introductory mining engineering. Wiley, New Jersey, pp 119–129
13. Hassani F, Nekoovaght P (2011) The development of microwave assisted machineries to break hard rocks. In: Proceedings of the 28th international symposium on automation and robotics in construction (ISARC), Seoul South Korea, pp 678–684
14. Hassani F, Nekoovaght P, Gharib N (2016) The influence of microwave irradiation on rocks for microwave-assisted underground excavation. J Rock Mech Geotech Eng 8(1):1–15
15. Hassani F, Nekoovaght P, Radziszewski P, Waters KE (2011) Microwave assisted mechanical rock breaking. In: 12th ISRM International Congress on Rock Mechanics, Harmonising Rock Engineering and the Environment, Beijing, China, October 2011. CRC Press, London, pp 2075–2080
16. Hassani F, Radziszewski P, Ouellet J, Nokkent M, Nekoovaght P (2008) Microwave assisted drilling and its influence on rock breakage: a review. In: ISRM 5th Asian rock mechanics symposium, Tehran, Iran, Nov 2008, pp 87–103
17. Hatherly P, Leung R, Scheding S, Robinson D (2015) Drill monitoring results reveal geological conditions in blasthole drilling. Int J Rock Mech Min Sci 78:144–154
18. Kahraman S, Balcı C, Yazıcı S, Bilgin N (2000) Prediction of penetration rate of rotary blasthole drills using a new drillability index. Int J Rock Mech Min Sci 37:727–743
19. Karpuz C, Pasamehmetoglu AG, Muftuoglu Y, Dinçer T (1990) Drillability studies on the rotary blasthole drilling of lignite overburden series. Int J Surf Min Reclam Env 4:89–93
20. Lauriello PJ, Fritsch CA (1974) Design and economic constraints of thermal rock weakening techniques. Int J Rock Mech Min Sci 11:31–39
21. Li Z (2011) Research on the prediction of roof geostructure using drilling mechanical data. Dissertation, Muroran Institute of Technology
22. Li Z, Itakura I (2012) An analytical drilling model of drag bits for evaluation of rocks strength. Soils Found 52:216–227
23. Lu Y, Peng J, Li L, He J, Gan X, Yin K (2013) Hard rock drilling technique with abrasive waterjet assistance. Int J Rock Mech Min Sci 60:47–56
24. Miller A, Ball A (1990) Rock drilling with impregnated diamond microbits—an experimental study. Int J Rock Mech Min Sci Geomech Abstr 27(5):363–371
25. Miyazaki K, Ohno T, Karasawa H, Ekop A, Takakura S (2016) Performance evaluation of polycrystalline diamond compact percussion bits through laboratory drilling tests. Int J Rock Mech Min Sci 87:1–7
26. Nejati H, Hassani F (2013) Application of discrete element method for validating, fracture toughness testing of rocks subjected to microwave illumination. In: 23rd world mining congress, Montreal, Canada, Aug 2013
27. Nejati H, Radziszewski P (2014) Microwave heating applications for in-situ resource utilization and space mining. In: 14th earth and space conference (ASCE), Missouri, USA, Oct 2014
28. Nekoovaght P, Gharib N, Hassani F (2014) Microwave assisted rock breakage for space mining. In: 14th earth and space conference (ASCE), Missouri, USA, Oct 2014, pp 414–423
29. Rabia H (1982) Specific energy as a criterion for drill performance prediction. Int J Rock Mech Min Sci Geomech Abstr 19:39–42
30. Summers DA, Lehnhoff TF, Weakly LA (1978) Development of a water jet drilling system and preliminary evaluations of its performance in a stress situation underground. In: Proceedings of the 4th international symposium on jet cutting technology, Kent, Jan 1978
31. Tamrock (1984) Handbook of surface drilling and blasting. Painofaktorit, Tampere, Finland
32. Tang J, Lu Y, Ge Z, Xia B, Sun H, Du P (2014) A new method of combined rock drilling. Int J Min Sci Technol 24:1–6
33. Teale R (1965) The concept of specific energy in rock drilling. Int J Rock Mech Min Sci Geomech Abstr 2:57–73

34. Wang F, Ren T, Hungerford F, Tu S, Naj A (2011) Advanced directional drilling technology for gas drainage and exploration in Australian coal mines. Proc Eng 26:25–36
35. Yao N, Zhang J, Jin X, Huang H (2014) Status and development of directional drilling technology in coal mines. Proc Eng 73:289–298

Chapter 6
Energy Efficiency in Rock Blasting

José A. Sanchidrián, Pablo Segarra and Lina M. López

Abstract Ten production blasts and one single-hole confined blast were monitored in two quarries in order to assess the measurable forms in which the energy delivered by the explosive is transformed in rock blasting. The seismic wave energy, the kinetic energy, and the fracture energy transferred in the blasting process were determined using the seismic field from seismograph records, the initial velocity of the blasted rock face obtained from high-speed video footages, and the fragment size distributions from image analysis of the muck pile material, respectively. The maximum total energy measured accounts for not more than 26% of the available explosive energy, if this is rated as the heat of explosion, though lower figures are usually obtained. The values measured for each of the energy components range from 2 to 6% of the total energy available for the fragmentation energy, 1–3% for the seismic energy, and 3–21% for the kinetic energy. For the confined shot hole, the seismic energy was 9% of the heat of explosion.

Keywords Explosive · Blasting · Rock fragmentation · Vibrations
Energy balance

6.1 Introduction

Explosives are the primary source of energy for rock breaking in the mining, quarrying, and construction industries. Explosive energy does work to transform rock into a mound of fragments and displaces them so that they can be conveniently loaded and hauled for further comminution and processing. Although the energy density of explosives is not particularly high, they are compact energy sources that

---

This chapter is an adapted version of Sanchidrián et al. [45]. With permission of the copyright owner.

J. A. Sanchidrián (✉) · P. Segarra · L. M. López
Universidad Politécnica de Madrid, Madrid, Spain
e-mail: ja.sanchidrian@upm.es

K. Awuah-Offei (ed.), *Energy Efficiency in the Minerals Industry*, Green Energy and Technology, https://doi.org/10.1007/978-3-319-54199-0_6

are able to deliver their energy autonomously at a very fast rate. This results in reaction products at high pressure that can perform mechanical work in deforming and breaking the material in their vicinity. This is what makes explosives useful and in many cases irreplaceable for rock excavation. This fast energy delivery produces a number of transformations other than the desired fragmentation and throw, such as the seismic wave into the rock.

Explosives energy is rated in a variety of ways, obtained either from calculation or from experimental tests. However, the question of what amount of that explosive energy is transferred to the rock and what fraction of it is converted into efficient work in rock blasting remains largely unresolved. Although the measurement of some of the effects of explosive in rock is customary (vibration, fragmentation, and, to a minor extent, rock movement), they are usually conducted for blast control purposes and the results are rarely cast in terms of their energy content. This is because the emphasis lies in the end effects, i.e., degree of fragmentation, throw and vibration levels, and not the energy consumption. Data on and estimates of energy components in rock blasting are thus limited to a few prior researchers. Berta [1], Spathis [2], and Ouchterlony et al. [3] calculated the amounts of energy transformed into kinetic energy of the rock, fracture generation, and seismic wave. Seismic energy has received special attention since earlier times and several authors [4–9] have reported calculations of seismic energy and its comparison with explosive energy.

Spathis [2] suggested the practical use of the energy balance to enable blast designs that direct the available energy into the desired work and, hence, control the explosive energy partition between fracture energy, kinetic energy, and radiated seismic energy, resulting in a more efficient use of it. This chapter assesses the feasibility of this energy approach and aims to establish, through new experimental data and a thorough review of the published work on the matter, the fraction of explosive energy transferred to the rock in its various components, with particular attention to their variability, and the reasonable ranges that could be expected in quarry blasting.

We first describe the basic theory and experimental background for the determination of some of the energy components in rock blasting. We then apply these to ten production blasts and one confined (without rock movement) blast hole. Eight of the production blasts and the confined shot were conducted in a limestone quarry (El Alto, Spain), while two additional production blasts were conducted in an amphibolite quarry (Eibenstein, Austria). We calculated the various energy terms using measurements of the seismic field, the initial velocity of the blasted rock face, and the fragment size distribution obtained from seismographs, high-speed video camera, and fragmentation monitoring systems, respectively. We measured vibrations in the single confined hole.

6.2 The Energy Balance of Blasting

The energy released by the explosive is converted into heat and work to the surroundings according to the first principle of thermodynamics. Some of these forms of energy become apparent during the blast, namely (i) fracture work that ultimately appears as new surface in rock fragments; (ii) work transferred as shock wave into the rock that attenuates ultimately to elastic waves, appearing as seismic waves or ground vibrations; and (iii) work to displace the rock and form the muck pile that appears as kinetic energy imparted to the rock.

This energy partition is to some extent arbitrary and is based on the end effects of the blasting. For instance, part of the fracture work is in the early stages intimately connected to the shock wave flow in the vicinity of the hole and, in the later stages, also to the rock movement which begins as the fractures burst open. Such a partition is, however, convenient inasmuch as the physical magnitudes related with each component can be measured. Other, less apparent, energy transfers include (a) expansion work of the fractures that is absorbed as elastic and plastic deformation of rock at the surface of fractures as they are penetrated by the gases; (b) heat transferred to the rock from the hot detonation products; and (c) heat and work conveyed as enthalpy of the gases venting to the atmosphere through open fractures and stemming.

The energy balance of the blast can thus be expressed as follows [2]:

$$E_E = E_F + E_S + E_K + E_{NM}, \tag{6.1}$$

where E_E is the explosive energy, E_F is the fragmentation energy, E_S is the seismic energy, E_K is the kinetic energy, and E_{NM} represents the energy forms not measured, a term required to close the balance. The terms fragmentation, seismic, and kinetic efficiency are used hereafter for the ratios of the respective energies to the explosive energy.

6.2.1 *Fragmentation Energy*

A specific amount of energy is required to create a new fracture surface [10]. Let this energy, per unit surface area, be G_F. The fragmentation energy can be calculated by

$$E_F = A_F G_F, \tag{6.2}$$

where A_F is the surface area of the fragments generated by the blast. The specific fracture energy, G_F, can be calculated from mechanical comminution tests under controlled energy input, leading to the Rittinger coefficient (a crushing efficiency, the surface area created per unit energy input) or derived from material properties of the rock—the fracture toughness and the elastic modulus. The first method involves millions of fractures in the rock, while the fracture toughness is obtained from tests in which only one fracture is formed. To estimate the fragmentation efficiency by blasting, where a great amount of fines is produced, the inverse of the Rittinger coefficient is used here as the specific fracture energy. The crushing efficiency concept and Eq. 6.2 assume that such efficiency is constant for all fragment sizes.

The surface area of the fragments may be estimated from the muck pile size distribution, assuming spherical or cubic particles of diameter or edge length x [3]:

$$A = 6V \int_0^\infty \frac{f(x)}{x} \mathrm{d}x, \tag{6.3}$$

where V is the volume of the fragmented rock and $f(x)$ is the density function of the fragment size distribution in volume. An incremental version of Eq. 6.3 may be used when the size distribution curve is known in a discontinuous form (as is usually the case) with the material grouped in classes of volume fractions p_k:

$$p_k = 6V \int_{x_k^I}^{x_k^S} f(x)\mathrm{d}x = P\left(x_k^S\right) - P\left(x_k^I\right), \tag{6.4}$$

where x_k^I and x_k^S are the mesh size limits of class k $\left(x_k^S = x_{k+1}^I\right)$ and $P(x)$ is the cumulative size distribution of the fragments. Equation 6.3 can be written as a sum of the integrals for each class:

$$A = 6V \sum_{k=1}^{C} \int_{x_k^I}^{x_k^S} \frac{f(x)}{x} \mathrm{d}x, \tag{6.5}$$

where C is the number of classes. The function $f(x)$ is not known, but integrals for each class are, via Eq. 6.4. A mean value f_k for each class can be calculated from its integral value:

$$f_k = \frac{p_k}{\left(x_k^S - x_k^I\right)}. \tag{6.6}$$

The integrals in Eq. 6.5 can be approximated using the f_k values:

$$A = 6V \sum_{k=1}^{C} \int_{x_k^I}^{x_k^S} \frac{f_k}{x} \mathrm{d}x = 6V \sum_{k=1}^{C} \frac{p_k}{\left(x_k^S - x_k^I\right)} \int_{x_k^I}^{x_k^S} \frac{\mathrm{d}x}{x} = 6V \sum_{k=1}^{C} \frac{p_k}{\left(x_k^S - x_k^I\right)} \ln \frac{x_k^S}{x_k^I} \tag{6.7}$$

or

$$A = 6V \sum_{k=1}^{C} \frac{p_k}{x_k}, \tag{6.8}$$

where x_k is the logarithmic mean of the size limits of class k:

$$x_k = \left(x_k^S - x_k^I\right) / \ln \frac{x_k^S}{x_k^I}. \tag{6.9}$$

The in situ area of the natural discontinuities of the rock mass, though small compared to the surface area created in the blast, is subtracted from the muck pile fragments area to obtain the newly formed surface area [3, 11]. The area of fragments originating from the blast is

$$A_{\mathrm{F}} = A - A_{\mathrm{IS}}, \tag{6.10}$$

where A_{IS} is the in situ specific area.

6.2.2 Seismic Energy

The energy transferred to the rock in the form of seismic wave is estimated as the integral of the energy flow past a control surface at a given distance from the blast. The energy flux (the power or rate of work, per unit area) is the scalar product of the stress at the surface and the particle velocity [12]:

$$\Phi = \vec{t} \cdot \vec{v}, \tag{6.11}$$

where $\vec{t}$ and $\vec{v}$ are the stress and particle velocity vectors, respectively. The stresses are obtained from the stress tensor τ by the Cauchy formula:

$$t_j = \tau_{ij} n_i, \tag{6.12}$$

where n_i are the components of the unit normal vector; the summation convention is used. The energy flux is then

$$\Phi = \tau_{ij} n_i v_j. \tag{6.13}$$

In order to relate the velocities, known from the seismographs records, to stresses, some assumptions must be made. If the seismic wave is considered as longitudinal spherical in an infinite homogeneous medium, the stress tensor in its principal components is, in spherical coordinates [13],

$$\begin{aligned} \tau_{11} &= (\lambda+2\mu)\frac{\partial u_1}{\partial r}+2\lambda\frac{u_1}{r} \\ \tau_{22} = \tau_{33} &= \lambda\frac{\partial u_1}{\partial r}+2(\lambda+\mu)\frac{u_1}{r} \\ \tau_{ij}(i\neq j) &= 0, \end{aligned} \tag{6.14}$$

where u_1 is the radial component of the particle displacement and r is the distance from the source; λ and μ are the Lamé constants. For a spherical surface coincident with the wavefront, its normal unit vector in the principal axes is (1,0,0). Substituting Eq. 6.14 in Eq. 6.13 yields

$$\Phi = \left[(\lambda+2\mu)\frac{\partial u_1}{\partial r}+2\lambda\frac{u_1}{r}\right]v_1, \tag{6.15}$$

where v_1 is the radial component of the particle velocity. The total power, assuming a uniform flux, across the surface of radius r is

$$P = 4\pi r^2\Phi. \tag{6.16}$$

Thus, the energy is

$$E_{S1} = \int_0^\infty 4\pi r^2\Phi\,\mathrm{d}t = 4\pi r^2\int_0^\infty \left[(\lambda+2\mu)\frac{\partial u_1}{\partial r}+2\lambda\frac{u_1}{r}\right]v_1\mathrm{d}t. \tag{6.17}$$

The velocity is measured by vibration monitoring conducted on the ground surface. The radial component may be assumed equal to the longitudinal velocity in seismic measurements. The particle displacements can be calculated by the time integral of the velocity:

$$u_1(t) = \int_0^t v_1(t)\mathrm{d}t. \tag{6.18}$$

The spatial derivative of the displacement can be approximated by the following relation:

$$\frac{\partial u}{\partial r} = -\frac{v}{c}, \tag{6.19}$$

where c is the wave velocity. Equation 6.19 applies when $v \ll c$, which is the case here. The equation for the calculation of the seismic energy is

$$E_{\mathrm{S}1} = 4\pi r^2 \int_0^\infty \left[-(\lambda + 2\mu)\frac{v_1^2}{c_\mathrm{L}} + 2\lambda \frac{u_1 v_1}{r} \right] \mathrm{d}t, \tag{6.20}$$

where the longitudinal wave velocity, c_L, is used. The time integral of the second summand of the integrand is negligible for a harmonic wave; in our seismic records, it is usually less than 0.05% of the integral, so it can be neglected without much error. Then,

$$E_{\mathrm{S}1} = -4\pi r^2 \rho\, c_\mathrm{L} \int_0^\infty v_1^2 \mathrm{d}t, \tag{6.21}$$

where the relation $c_\mathrm{L}^2 = (\lambda + 2\mu)/\rho$ has been used, ρ being the rock density. The negative sign in Eq. 6.21 indicates that the energy is leaving the control sphere. Equation 6.21 is also obtained for a plane longitudinal wave, which can approximate a spherical wave at large radii.

The seismic energy, as calculated by Eq. 6.21, is that of a spherical or plane P-wave in an elastic medium with radial component of the particle velocity, v_1, the longitudinal component measured in the field. Assuming that the transverse and vertical components of the velocity (v_2 and v_3) belong to transverse waves, and using the plane wave approximation, the following expressions are readily obtained:

$$\begin{aligned} E_{\mathrm{S}2} &= -4\pi r^2 \rho\, c_\mathrm{T} \int_0^\infty v_2^2 \mathrm{d}t \\ E_{\mathrm{S}3} &= -4\pi r^2 \rho\, c_\mathrm{T} \int_0^\infty v_3^2 \mathrm{d}t. \end{aligned} \tag{6.22}$$

Adding Eq. 6.22 to the P-wave energy (Eq. 6.21) and using the absolute value gives

$$E_\mathrm{S} = 4\pi r^2 \rho \left[c_\mathrm{L} \int_0^\infty v_1^2 \mathrm{d}t + c_\mathrm{T} \int_0^\infty (v_2^2 + v_3^2) \mathrm{d}t \right]. \tag{6.23}$$

Equation 6.23 is usually further simplified [2, 3, 9], using a unique wave velocity, to

$$E_S = 4\pi r^2 \rho\, c_L \int_0^\infty v^2 \mathrm{d}t, \tag{6.24}$$

where v is the magnitude of the vector sum of velocities, $v^2 = v_1^2 + v_2^2 + v_3^2$. If there is sufficient characterization of the material so that the longitudinal and transverse velocities are known, and three-component records are available, then one should prefer Eq. 6.23. This is used in the present work.

The measured velocity functions do not belong to spherical waves from a point source, but from a number of line (or cylindrical) sources at varying distances from the measuring point. They are a composite of longitudinal, transverse, and surface waves resulting from reflections and refractions at the strata, joints, and free surfaces—the bench face and ground surface. As the composite wave comes from different sources, the assumption that each of the three components belongs to separate longitudinal or transversal waves is obviously unrealistic. Even for the direct waves, the radial component of the velocity measured (pointing toward the center of the blast) is not collinear with the directions to the different blast holes. Finally, the domain boundary is a hemisphere closed by a circle; the use of a sphere surface area in Eq. 6.16 and subsequent rests on the assumption that what was found at the position of the measurements represents the seismic field in a complete sphere. As the velocity was measured on the surface—not a good sample of velocity in an infinite medium—some of the energy is reflected downward, resulting in an outgoing flow smaller than what would be encountered with the gage embedded in the medium at depth. Thus, Eq. 6.23 is only a rough approximation of the seismic energy from a blast.

6.2.3 Kinetic Energy

The kinetic energy is calculated from the initial velocity of the rock face measured at different heights along the highwall, $V_0(y)$. High-speed film and radar measurements showed [2] that the face velocity distributions for many blasts were relatively narrow, and that the rocks behind the face generally move in unison with the face (this behavior is typical of competent brittle rocks, according to Chiappetta and Mammele [14]). Assuming that the velocity of the entire rock mass is constant in a horizontal section of the burden, the kinetic energy, E_K, of the rock displaced by a blast hole is

$$E_K = \frac{1}{2} S B_h \int_0^H \rho(y) \, V_0^2(y) \, dy, \tag{6.25}$$

where a variable rock density has been considered, $\rho(y)$, to account for varying lithology along the height, assuming horizontally layered rock (this is used for El Alto, where the density of the overburden and rock differ). H is the bench height, S is the spacing between holes, and B_h is the mean horizontal burden, obtained from the face profile.

6.3 Measurements and Calculation of Energy Components

6.3.1 *Field Experiments*

We conducted field experiments, which involved ten production blasts and one single-hole confined blast, at the El Alto and Eibenstein quarries.

El Alto quarry belongs to Cementos Portland Valderrivas, a cement and aggregate producer located in the province of Madrid, Spain. The deposit is of Miocene age and lacustrine origin. The geology is simple and essentially uniform. In the upper two to six meters, there is an overburden of weathered clayey marl of sandy nature (with a maximum particle size of 14 mm) and low cohesion, underlying a clayey soil of some tens of centimeters. The limestone formation below has a thickness of 12–19 m; the bedding planes are horizontal or subhorizontal and crossed by some nearly vertical faults. The floor of the limestone is clay that is usually not mined. The quarry is mined in one bench.

Eibenstein quarry is owned by Hengl Bitustein, an aggregate producer in Austria. The formation is a metamorphic body including amphibolite, schist, mica schist, and marbles. The deposit is heavily folded and faulted with a complex structure; schistosity and foliation are present in all rock types. The quarry is mined in several benches.

A comprehensive rock mass description of both quarries has been reported by Hamdi and du Mouza [15]. The main properties of the rocks are given in Table 6.1.

The main characteristics of the blasts are shown in Tables 6.2 and 6.3. The numbering of the blasts used by the quarries has been retained (CB2 is the confined shot, fired in El Alto). The mean and standard deviation (following the ± sign) of the parameters that were measured for every hole are given. Burden B for each hole and bench height were obtained from laser profiles of the bench face.

Blasts in El Alto were carried out in La Concha, one of the two pits of the quarry. Blasts numbers 29/02, 37/02, 45/03, 54/03, and 58/03 were shot in the same area, while blasts 15/02, 43/03, and 50/03 were in different areas of the pit. In

Table 6.1 Rock properties

	Limestone, El Alto	Amphibolite, Eibenstein	References
Density (kg/m^3)	2560	2950	[16]
Tensile strength (MPa)	7.6	7.5	[16]
Elastic modulus (GPa)	64	83	[16]
Poisson's ratio	0.26	0.22	[16]
Rittinger coefficient (cm^2/J)	58	34	[17]
In situ characteristic size[a] (m)	3.01	2.63	[15]
In situ uniformity index[a]	2.86	2.13	[15]
P-wave velocity, in situ rock mass (m/s)	2994	2450	[18, 19]
P-wave velocity, laboratory specimen (m/s)	4314	5726	[15]

[a]Characteristic size (63% passing) and uniformity index of a Rosin–Rammler [20–22] distribution assumed for the in situ block size distribution

Eibenstein, blasts 420-11 and 440-04 were located in levels 420 and 440 of the quarry, respectively.

The blast hole diameter was 142 mm in El Alto and 92 mm in Eibenstein. All production blasts consisted of a single row of blast holes. One charge per hole was blasted in all rounds, except in blast 15/02, where two decks were fired separated by 1 m of stemming with 50 ms top-bottom delay. The delay and detonator types used in each blast are given in Table 6.3. The delay in blast 37/02 was basically 67 ms, although it was variable along the blast since some of the holes were decked; the holes with intermediate stemming are not included in the data we used to estimate the statistics given in Table 6.3.

Gelatine cartridges were used as bottom charge (Goma 2 ECO, abbreviated G2 in Table 6.3, and Danubit 4, abbreviated GD); aluminized ANFO (Alnafo, abbreviated AL), a high-density aluminized ANFO (abbreviated HA), and standard ANFO (Nagolita, abbreviated AN, and Dap 2, abbreviated AD) were used as column charges. Table 6.4 shows the properties of these explosives. Detonation velocities were measured in all blasts in one or two holes. Down-hole initiation was used in El Alto; detonating cord side initiation was used at Eibenstein.

Explosive energy can be rated using either thermodynamic codes or experimental measurements such as the underwater [23, 24] and cylinder [25] tests. Thermodynamic calculations are generally accepted for assessing the energy of explosives. The energy of the explosives used in El Alto has been calculated using the W-Detcom code [26, 27]. The heat of explosion at constant volume, E_{Q}, and the useful work to an expansion pressure of 100 MPa, E_{Wu}, have been used as energy values. We used the BKW-S equation of state [28, 29] for the calculations. For the explosives used in Eibenstein, the heat of explosion given by the manufacturer [30] is the only available figure of explosive energy since their exact compositions are unknown. The useful work has been estimated for the gelatine and ANFO used in Eibenstein assuming the same fractions of energy lost below 100 MPa as for the El Alto's gelatine and ANFO, respectively.

Table 6.2 Characteristics of the blasts, geometrical

Blast no.	Site	*N*	*i* (°)	*H* (m)	h_{ob} (m)	*J* (m)	*B* (m)	B_h (m)	*S* (m)	*S*/*B*	Vol-*t* (m^3)	Vol-*r* (m^3)
CB2	El Alto	1	0	16.5								
15/02	El Alto	22	6	19.6 ± 1.1	3.0	2.7	5.5 ± 0.3	5.5	5.9	1.07	640	542
29/02	El Alto	21	6	19.9 ± 2.6	3.4	1.1	4.9 ± 0.4	4.9	6.0	1.22	588	488
37/02	El Alto	26	6	19.5 ± 2.5	2.4	2.2	5.0 ± 0.5	5.0	5.8	1.16	569	499
43/03	El Alto	10	6	18.0 ± 1.0	1.9 ± 0.5	1.5	4.5 ± 0.3	4.5	6.5 ± 0.1	1.44	529	474
45/03	El Alto	10	6	16.3 ± 0.4	3.9 ± 0.5	1.6	4.6 ± 0.4	4.6	6.5 ± 0.1	1.41	490	373
50/03	El Alto	10	6	18.2 ± 1.0	2.2 ± 0.5	1.7	4.6 ± 0.1	4.6	6.6 ± 0.2	1.43	556	488
54/03	El Alto	11	6	18.5 ± 0.3	2.3 ± 0.5	2.7	5.0 ± 0.4	5.0	6.4 ± 0.1	1.28	595	521
58/03	El Alto	10	6	17.1 ± 0.6	3.4 ± 0.7	1.5	4.4 ± 0.1	4.4	6.4 ± 0.4	1.45	484	388
420-11	Eibenstein	7	21	10.1 ± 0.4	0	1.5	3.6 ± 0.7	3.9	3.3 ± 0.1	0.92	129	129
440-04	Eibenstein	9	19	12.9 ± 0.6	0	0.1	4.1 ± 0.4	4.3	3.3	0.80	185	185

N Number of holes; *i* Hole inclination (nominal in El Alto, measured in Eibenstein); *H* Bench height (hole length for the confined shot); h_{ob} Overburden thickness; *J* Subdrill; *B* Burden; B_h Horizontal burden; *S* Spacing; *S*/*B* Spacing-to-burden ratio; Vol-*t* Total volume blasted per hole; Vol-*r* Volume of rock (excluding overburden) blasted per hole

Table 6.3 Characteristics of the blasts, explosives

Blast no.	Expl.	M_{ET} (kg)		E_{ET}, W_u (MJ)	E_{ET}, Q (MJ)	M_{EH} (kg)		q (kg/m^3)	E_{EH}, W_u (MJ)	E_{EH}, Q (MJ)	t (ms)
		Cartr.	Bulk			Cartr.	Bulk				
CB2	G2/AN	2.5	75	203	302	2.5	75		203	302	–
15/02	G2/AL	550	4664	15,524	25,243	25	212 ± 35	0.37	706	1147	84-N
29/02	G2/AL	546	4326	14,523	23,560	26 ± 1	206 ± 48	0.39	692	1122	67-N
37/02	G2/AL	650	5798	19,181	31,243	25	223 ± 26	0.44	738	1202	67-N
43/03	G2/AL	250	2000	6706	10,883	25	200 ± 13	0.43	671	1088	67-E
45/03	G2/AL	250	1830	6210	10,044	25	183 ± 25	0.42	621	1004	17-E
50/03	G2/AL	250	2130	7085	11,523	25	213 ± 13	0.43	709	1152	30-E
54/03	G2/HA	275	2849	10,421	15,299	25	259 ± 17	0.48	947	1391	67-E
58/03	G2/HA	250	2100	7846	11,470	25	210 ± 10	0.49	785	1147	67-E
420-11	GD/AD	255	125	1302	1633	36.4 ± 3.1	17.8 ± 6.4	0.42	186	233	20-e
440-04	GD/AD	267	275	1731	2263	29.7 ± 9.7	30.6 ± 10	0.33	192	251	20-e

M_{ET} Total mass of explosives; E_{ET} Total explosive energy, useful work (W_u) or heat of explosion (Q); M_{EH} Mass of explosives per hole; q Powder factor; E_{EH} Explosive energy per hole; t In-row delay and type of initiation; N Non-electric, E Electronic, e Electric

Table 6.4 Properties of the explosives

Name	Type	Manufacturer	Density (kg/m^3)	VOD (m/s)	E_{Wu} (kJ/kg)	E_Q (kJ/kg)
Nagolita	ANFO	UEE	800	3941[a]	2591	3893
Dap 2	ANFO	Istrochem	800	3867 ± 85	2529	3800
Alnafo	ANFO + Al	UEE	800	4029 ± 93	2918	4930
High-density Alnafo	ANFO + Al	UEE	950	3424 ± 98	3322	4975
Goma 2 ECO	Gelatin	UEE	1450	6321 ± 118[b]	3480	4090
Danubit 4	Gelatin	Istrochem	1450	5933 ± 197	3871	4550

[a]One-shot value; [b]Values include data for 65 and 85 mm diameter cartridges

6.3.2 *Fragmentation Energy*

The fragment size distribution was determined in both quarries by digital analysis of the images recorded by a camera installed at the hopper of the primary crusher. The procedure used to obtain the size distribution in El Alto is described in [31].

The overburden in El Alto complicates the assessment of the fragmentation, as it contributes to the fines present in the muck piles. We used an adjusted fines correction factor, which varied with the amount of fines originating from the loose overburden (natural fines). The fragment size distributions of the limestone from blasting were obtained by subtracting the fraction of natural fines (estimated as the ratio of the overburden thickness to the bench height) from the raw muck pile fragmentation curves. This was done using laboratory screening data of the overburden material and of the muck pile fine fraction. We assumed the fines tail of the limestone had the same distribution in all the blasts for sizes below 14 mm (maximum size of the overburden), following the natural breakage characteristic principle [32, 33].

In Eibenstein, no manual correction was done. Fragmentation was measured with the image analysis code *Fragscan* [34], calibrated with muck pile sieving data in order to overcome the problem of fines detection [35, 36]. We used the Swebrec function [37] for extrapolating the measurements outside the range of resolution of the image analysis (63 mm down to 10 mm). Further extrapolation would be risky, as a second function would probably be required.

Figure 6.1 shows the size distribution curves. The smallest sizes of the curves are 0.25 mm for El Alto and 10 mm for Eibenstein (downward extrapolations are shown in Fig. 6.1 as dashed lines). El Alto's data corresponds to broken limestone (i.e., natural fines discounted). The kink at 14 mm apparent in some of the curves is the junction of the fines tail to the fragmentation measured with *Split*.

We used Eq. 6.8 to calculate the surface area of the fragments, except for the finer class, $[0, x_{min}]$. For that, a Rosin–Rammler–Weibull [20–22] distribution is assumed:

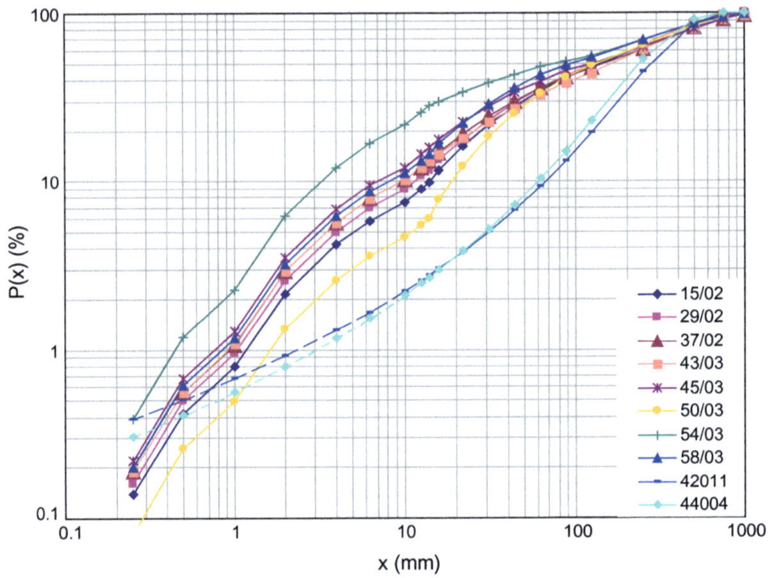

Fig. 6.1 Size distribution curves

$$P(x) = 1 - e^{-(x/x_c)^n}, \quad 0 \leq x \leq x_{min} \tag{6.26}$$

where x_c and n are the parameters of the Rosin–Rammler–Weibull distribution (characteristic size and uniformity index, respectively). Following the technique of [38], the two parameters are determined from the last two known data points. The probability density function is

$$f(x) = \frac{n}{x_c^n} x^{n-1} e^{-(x/x_c)^n}, \quad 0 \leq x \leq x_{min} \tag{6.27}$$

By substituting Eq. 6.27 into Eq. 6.3 and integrating between 0 and x_{min}, the contribution to the specific surface area of the undefined fines tail can be estimated by:

$$\frac{A_{tail}}{V} = 6 \int_0^{x_{min}} \frac{f(x)dx}{x} = 6 \int_0^{x_{min}} \frac{n}{x_c^n} x^{n-2} e^{-(x/x_c)^n} dx = \frac{6}{x_c} \gamma \left[1 - \frac{1}{n}, \left(\frac{x_{min}}{x_c} \right)^n \right] \tag{6.28}$$

γ being the lower incomplete Gamma function. We successfully used this approach for the El Alto data, as the lower part of the size distribution is reasonably of Rosin–Rammler–Weibull type, but it proved unfeasible for the Eibenstein distributions. This is due to the use of the Swebrec function in Eibenstein, so that the Rosin–Rammler–Weibull functions obtained for the lower part of the distribution have a very low uniformity parameter n (less than 1), which makes the integral in Eq. 6.28 divergent. Hence, the Swebrec functions were extended down to 0.25 mm (the dashed portion of the curves in Fig. 6.1) in order to make the calculations comparable with El Alto, although such size range is probably beyond the validity of the Swebrec. The surface area of the lower class [0, 0.25 mm] was in this case calculated as one more summation term of Eq. 6.8, with a mean size $x_k = 0.125$ mm. Comparing the two methods of calculation for El Alto curves, the latter results in an underestimation of the total fragmentation energy of about 3–7% with respect to the integral formulation.

Table 6.5 shows, for each blast, the specific surface area of the fragments, A/V, and the new specific area, A_F/V (by subtraction of the in situ block size area, calculated from the Rosin–Rammler–Weibull parameters in Table 6.1: 2.8 and 3.8 m^2/m^3 for El Alto and Eibenstein, respectively). The fragmentation energies per hole, E_F, are obtained from the new area, the volume of rock blasted per hole in Table 6.2 (column Vol-r) and the specific fracture energy (inverse of the Rittinger coefficient in Table 6.1). The energies and the fragmentation efficiencies, η_F, with respect to useful work and heat of explosion, are also listed in Table 6.5. The efficiencies in Eibenstein are within the range of efficiencies estimated for El Alto. The range of efficiencies is 1.9–6.0% of the heat of explosion and 3.1–8.8% of the useful work. The mean efficiencies in El Alto are 5.5 and 3.5% with respect to useful work and heat of explosion, respectively. In Eibenstein, it is 6.3 and 4.9%.

6.3.3 *Seismic Energy*

The seismic field was measured with vibra-tech multisets plus seismographs. The ranges of velocity and frequency are 0–254 mm/s and 2–300 Hz, respectively, with an accuracy of 3% at 15 Hz. The seismographs were coupled to the ground surface in El Alto by excavating a shallow hole to remove some tens of centimeters of the

Table 6.5 Fragmentation energy and efficiency

Blast no.	A/V (m^2/m^3)	A_F/V (m^2/m^3)	E_F, per hole (MJ)	η_F, W_u (%)	η_F, Q (%)
15/02	382.2	379.4	35.4	5.0	3.1
29/02	432.1	429.4	36.1	5.2	3.2
37/02	488.7	486.0	41.8	5.7	3.5
43/03	483.6	480.9	39.3	5.9	3.6
45/03	561.9	559.2	35.9	5.8	3.6
50/03	259.8	257.1	21.6	3.1	1.9
54/03	934.8	932.0	83.8	8.8	6.0
58/03	529.0	526.3	35.2	4.5	3.1
420-11	285.7	281.9	10.7	5.7	4.6
440-04	246.5	242.7	13.2	6.9	5.3

upper layer of soil and the devices were spiked in the bottom. Sandbags (ANFO bags filled with sand) were fitted in the hole to secure the mount to prevent horizontal and vertical movements. In Eibenstein, the seismographs could not be spiked

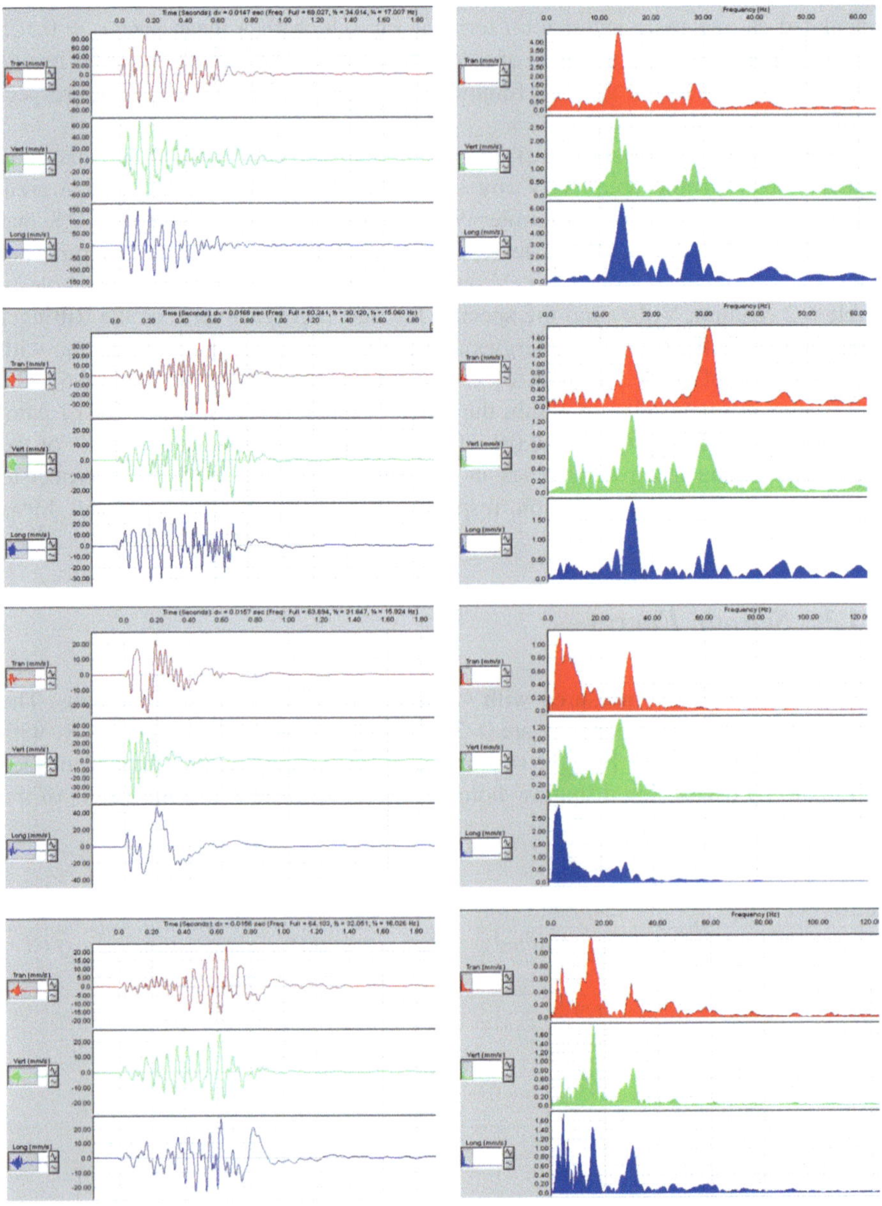

Fig. 6.2 Sample seismograms and frequency spectra. From top to bottom: blast 54/03, units 7640 and 6783 (acceptable records); blast 50/03, unit 7837 and blast 58/03, unit 7101 (rejected records)

to the ground and they were placed in a shallow hole, of a depth slightly larger than the seismograph height, covered with gravel and the surrounding earth compacted. Given that this mounting might not be optimal [39, 40], care was exercised to evaluate the recorded signals by inspection of the time records and the frequency analysis. Where most of the energy was packed at very low frequencies in one or more components of the velocity, the measurement was rejected [41]. This happened in two measurements. Figure 6.2 shows two examples of acceptable records and the two records rejected.

The distances of the seismographs to the blast (average of distances to each of the holes) were kept, as much as possible, similar from blast to blast. The distances were 51.8 m for the confined shot and a mean of 65.5 m for the production blasts in El Alto (standard deviation 4.4 m); in Eibenstein the mean distance and standard deviation were 36.9 and 1.3 m, respectively. The seismographs were placed around the block to be blasted in both top and floor levels as close as possible to the blast, but to a distance large enough so as not to be too close to any of the holes. The sensors in the bottom level were placed slightly farther in order to allow room for the muck pile. The sensors in El Alto were located around the blast at approximately even angles. The confined blast hole, CB2, was drilled in the upper level of the quarry, away from the bench face; all measurements were made at that level. The berms at Eibenstein were narrow, which restricted the position of the seismographs to directions close to the blast holes line. Figure 6.3 shows two examples.

Table 6.6 shows the means of the distances to the blast holes for each sensor unit, r and the peak vector sum particle velocity, PPV_S. Data measured on the top and floor of the bench are shown separately.

We estimated the integrals in Eq. 6.23 numerically from the seismic records' particle velocity data v_1 (longitudinal), v_2 (transverse), and v_3 (vertical) using a step size equal to the sampling interval, 1/1024 s, in all the records. The rock mass longitudinal wave velocities measured on-site (Table 6.1) are used. The transverse wave velocity used is obtained from the following relation:

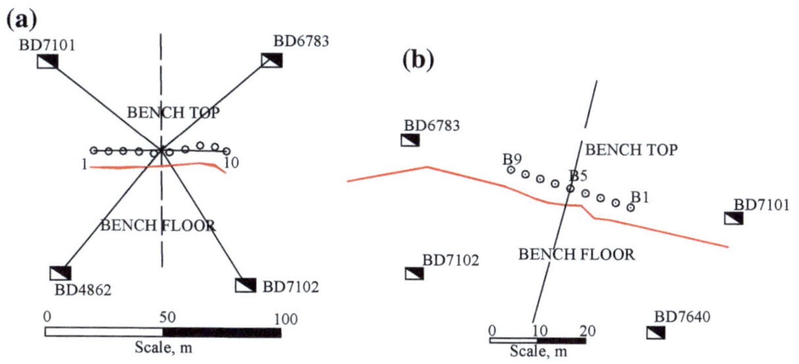

Fig. 6.3 Sample seismograph location. **a** Blast 58/03 (El Alto); **b** Blast 440-04 (Eibenstein)

Table 6.6 Basic vibration data and seismic energy

	Benchtop						Bench floor					
Blast no.	Unit #	r (m)	PPV_S (mm/s)	E_S (MJ)	η_S, W_u (%)	η_S, Q (%)	Unit #	r (m)	PPV_S (mm/s)	E_S (MJ)	η_S, W_u (%)	η_S, Q (%)
CB2	7640	51.8	56.3	30	14.6	9.8						
	7102	51.8	53.8	20	10.0	6.7						
	7101	51.8	42.9	28	14.0	9.4						
15/02	7102	65.6	51.5	416	2.7	1.6	7101	77.9	26.7	92	0.6	0.4
29/02	7101	60.4	94.6	507	3.5	2.2	7102	66.1	89.2	338	2.3	1.4
	6783	61.3	93.4	358	2.5	1.5	7640	66.3	46.4	225	1.6	1.0
37/02	7101	61.5	102	479	2.5	1.5	6783	69.8	43.8	226	1.2	0.7
	7640	63.6	138	450	2.3	1.4	7102	62.5	36.7	108	0.6	0.3
43/03	7101	69.1	36.0	60	0.9	0.6	4862	68.5	53.7	327	4.9	3.0
	7640	60.6	95.9	383	5.7	3.5	6783	68.5	90.1	390	5.8	3.6
45/03	7101	61.0	146	203	3.3	2.0	6783	67.3	62.2	143	2.3	1.4
	4862	59.3	158	314	5.1	3.1	7640	66.1	47.1	98	1.6	1.0
50/03	7640	60.8	150	467	6.6	4.1	7102	68.1	31.1	31	0.4	0.3
	7837	Rejected					7101	69.4	62.1	59	0.8	0.5
54/03	7640	62.6	169	937	9.0	6.1	7102	67.5	29.6	61	0.6	0.4
	6783	58.5	41.1	78	0.8	0.5	7101	69.8	36.8	105	1.0	0.7
58/03	6783	63.3	100	310	4.0	2.7	7102	69.7	42.5	133	1.7	1.2
	7101	Rejected					4862	68.6	54.7	234	3.0	2.0
420-11	7101	34.9	74.9	20	1.5	1.2	7102	38.5	41.0	9	0.7	0.5
	6783	36.7	31.9	6	0.4	0.3	7640	37.4	43.4	5	0.4	0.3
440-04	6783	36.2	63.9	13	0.7	0.6	7640	37.3	66.8	31	1.8	1.4
	7101	35.7	111	42	2.4	1.9	7102	38.8	54.7	14	0.8	0.6

$$c_T = (\mu/\rho)^{1/2}, \tag{6.29}$$

where μ is the shear modulus, determined from the in situ P-wave velocity used, the density and the Poisson's ratio in Table 6.1. For El Alto limestone, μ = 7.44 GPa, c_T = 1705 m/s; for Eibenstein amphibolite, μ = 6.36 GPa, c_T = 1468 m/s.

Table 6.6 shows seismic energy obtained at each seismograph's location. Seismic efficiencies are obtained from the seismic energies and the total explosive energies given in Table 6.3.

The scatter of energies and efficiencies is, in some cases, large among sensors in the same blast. Such dispersion is not unusual in seismic measurements. Given that the range of distances in each site was very narrow, we cannot establish any significant functional relationship between seismic energy efficiency and distance (Fig. 6.4). Thus, all energies recorded at each site can be analyzed together without an attenuation correction for distance. Using lognormal distributions for the seismic efficiencies (the hypothesis of lognormality cannot be rejected at a 95% significance), their means in El Alto are different in the top and floor levels at a 95% significance. They are 2.5% in the top and 1.2% in the floor with respect to the heat of explosion. The mean seismic efficiency of the confined shot is 8.6% with respect to the heat of explosion, more than three times the mean value of production blasts in the top level. Though the distance of measurement in the confined shot was about 20% shorter than in the production blasts, such a high energy value must be explained by the absence of interference between waves from different holes, the

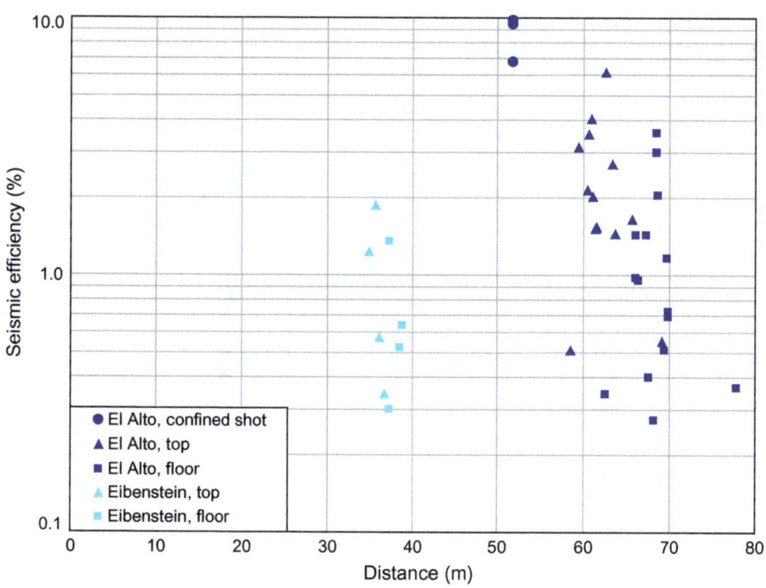

Fig. 6.4 Seismic efficiency with respect to distance

Table 6.7 Summary of seismic efficiencies

	Useful work		Heat of explosion	
	Range	Mean	Range	Mean
El Alto, confined hole	10.0–14.6	12.9	6.7–9.8	8.6
El Alto, production blasts, top level	0.8–9.0	3.9	0.5–6.1	2.5
El Alto, production blasts, floor level	0.4–5.8	1.9	0.3–3.6	1.2
Eibenstein, both levels	0.4–2.4	1.1	0.3–1.9	0.9
All production blasts	0.4–9.0	2.5	0.3–6.1	1.6

confinement of the charge and especially, according to Blair and Armstrong [42], the undamaged nature of the rock mass around the hole.

The efficiencies in the top and floor levels in Eibenstein (mean values of the lognormal distributions 1.1 and 0.7%, respectively) are not different at a 95% significance. Taking all values as one distribution, the mean efficiency in Eibenstein is 0.9% with respect to the heat of explosion. Table 6.7 gives a summary of the results.

6.3.4 *Kinetic Energy*

We used a Motion Meter 1000 high-speed digital camera, of Redlake Imaging, with a recording velocity of 250 frames per second to record the blast. We used wooden targets, hanging on the bench face, to determine the rock displacement. We used four fixed points with known coordinates as a reference system. Figure 6.5 shows a sketch of the setup. Details of the system used in each quarry are given by [43].

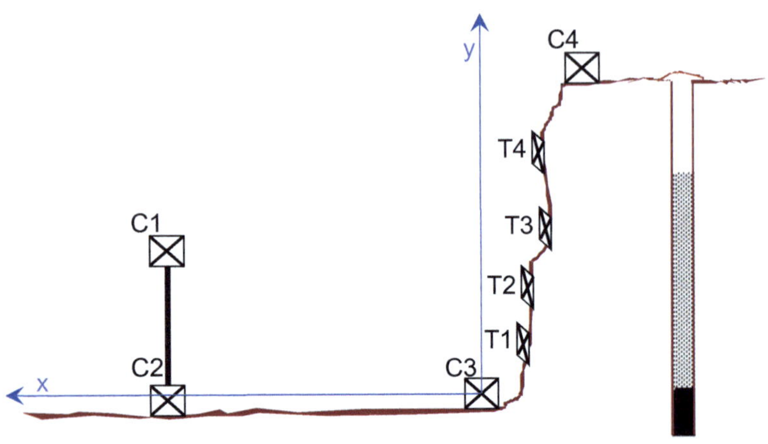

Fig. 6.5 Experimental setup for measuring the face movement. *C* Control points; *T* Targets

The targets were placed approximately in front of the explosive column. In blast 15/02 with two decks, the targets were located in front of the upper deck area. Two targets were used in Eibenstein and in blasts 15/02, 29/02, and 37/02 of El Alto. In the other five blasts monitored in El Alto, the number of targets was increased to four (though not all of them could be tracked in every blast). Table 6.8 gives the positions of the targets (hole in front of which they were located and height from grade). The targets are numbered from the floor to the top of the bench face. As the features of the rock movement in a vertical section depend basically on the local geometry and charging values and not on the average values of the whole blast, we used the precise drilling and charging data of the blast hole behind the targets to calculate the kinetic energy term (Table 6.8).

We used Motion Tracker™ 2D software [44] to obtain the raw path of the targets. The initial velocity of the targets (Table 6.8) is obtained from a trajectory model in which the two components of the initial velocity are modified until the calculated trajectory best fits the measured flight points [43].

In our experiment, we measured the velocity at a maximum of four different heights and in many cases for only two. We used an average velocity of the measured values for each blast since a sound, statistically significant, fitting of a function $V_0(y)$ to be used in Eq. 6.25 is not possible for most of the blasts. Equation 6.15 becomes simply, for two layers of thickness h_1, h_2 and densities ρ_1, ρ_2:

$$E_{\mathrm{K}} = \frac{1}{2}\bar{V}_0^2 S B_{\mathrm{h}}(\rho_1 h_1 + \rho_2 h_2) = \frac{1}{2}\bar{V}_0^2 S B_{\mathrm{h}} H\left[\rho_1 \frac{h_1}{H} + \rho_2\left(1 - \frac{h_1}{H}\right)\right]. \quad (6.30)$$

We used the burdens, spacings, and bench heights in Table 6.8. Table 6.2 shows the overburden thicknesses, h_1. For the El Alto quarry, the overburden and limestone densities are estimated at 1600 and 2560 kg/m^3, respectively.

Table 6.9 shows the kinetic energies and efficiencies. Since kinetic energy depends on the square of velocity, moderate variations in velocity lead to important variations in the energy. The efficiencies in El Alto range from 3.3 to 10.3% with respect to heat of explosion and 5.3–16.5% with respect to useful work. In Eibenstein, the results from the two blasts are quite different; the velocities measured in blast 420-11 are the highest of all blasts, due to the small local burden in the rock that was tracked (2.5 m horizontal burden; the average in that blast was 3.3 m), which resulted in an overcharged hole.

Table 6.8 Characteristics of the blast holes behind the targets and high-speed camera monitoring results

Blast no.	15/02	29/02	37/02	43/03	45/03	50/03	54/03	58/03	420-11	440-04
Hole no.	9	11	15	2	8	6	5	2	3	6
Bench height (m)	20	17.2	17.2	16.7	15.7	17.9	18.5	16.5	10.4	12.8
Subdrill (m)	4.8	−2.2	7.2	1.3	2.2	1.4	3.3	2.5	1.1	0.6
Mean (horizontal) burden (m)	5.7	4.4	4.9	4.5	4.0	4.6	5.4	4.4	2.5	4.5
Spacing (m)	5.9	6	5.8	6.4	6.4	6.7	6.4	6.4	3.3	3.3
Target height (m)										
T4				11.1			12.8			
T3				8.9	9.1	11	11.3	8		
T2	16.1	10.3	11.6	6.4	5.5	6.3	9.6	5.5	5.9	9.4
T1	11.1	5.5	5.9	4.7		1.9	5.1	1.3	2.7	4.8
Initial velocity (m/s)										
T4				10.4			6.4			
T3				12.8	7.3		9.2	14.8		
T2	9.8	10.3	6.5	9.6	9.2	9.8	8.7	16.1	18.9	9.3
T1	8.3	15.9	12.4	5.2		10	8.8	13	23.8	2.8
Explosives (kg)										
Cartridged	25	25	25	25	25	25	25	25	35	35
Bulk	258	138	250	181	178	200	275	213	21	20
Energy, W_u (MJ/hole)	840	490	817	615	606	671	1001	795	189	186
Energy, Q (MJ/hole)	1374	783	1335	995	980	1088	1470	1162	239	235

Table 6.9 Average velocities, kinetic energies, and efficiencies

Blast no.	Av. veloc. (m/s)	E_K, per hole (MJ)	η_K, W_u (%)	η_K, Q (%)
15/02	9.1	66.9	8.0	4.9
29/02	13.1	92.9	16.5	10.3
37/02	9.5	53.2	6.5	4.0
43/03	9.5	53.5	8.7	5.4
45/03	8.3	31.9	5.3	3.3
50/03	9.9	66.4	9.9	6.1
54/03	8.3	53.7	5.4	3.7
58/03	14.6	118.2	14.9	10.2
420-11	21.4	49.5	26.2	20.7
440-04	6.1	8.8	4.7	3.7

6.3.5 *Uncertainty Analysis*

We estimated the uncertainties of the calculated energies based on the uncertainties in the experimental data using error propagation through the applied formulae. The uncertainties of the measured variables are described by the relative standard error (standard deviation of the mean, expressed as a fraction of it). These have been calculated for each blast and the average values in all blasts used as an estimate of the uncertainty. The relative standard errors, influential factors, and main assumptions for the energy calculations are reported in Table 6.10. Details of the error analysis are contained in [45].

Table 6.10 Errors, influential parameters, and assumptions in the energy calculations

Energy	Relative std. error[a]	Main influential parameters	Main assumptions
Fragmentation	0.09 0.11	Fracture-specific energy Smaller fragment size considered	Cubic or spherical fragments Constant specific fracture energy for all sizes
Seismic	0.48 0.49	Wave velocities Particle velocities	Spherical wave Only body waves Radial velocity belongs to a P-wave; transverse and vertical velocities to s waves
Kinetic	0.34 0.35	Rock ejection velocity	Uniform initial velocity of the rock mass
Explosive	0.07	Composition Calculation method Pressure cut-off for useful work	All energy delivered (heat of explosion) or some energy discarded (useful work to a certain pressure)

[a]The two values given for the relative standard error refer to energy (upper) and efficiency (lower)

6.4 Discussion

Table 6.11 summarizes the results of fragmentation, kinetic, and seismic efficiencies estimates. Values of seismic energy, for each blast, are the average of the averages in the top and floor levels. Considering that the heat of explosion is the energy available in the blast, only 8–26% of it has been measured through rock fragmentation, seismic wave, and rock movement. The useful work already discards a portion of the total energy, so that the efficiency with respect to it results in somewhat higher values (13–33%). The blasting efficiency (sum of the fragmentation and kinetic) ranges from 7 to 25% of the heat of explosion.

The energy components can be described by lognormal distributions (the hypothesis of lognormality cannot be rejected at 95% confidence level). Using lognormal distributions, relative standard errors of the efficiencies have been calculated as the errors of the logarithmic efficiencies [46]:

$$\delta \ln \eta = \frac{\delta \eta}{\eta} = \sqrt{\frac{s^2}{N} + \frac{s^4}{2(N-1)}}, \tag{6.31}$$

where N is here the number of production blasts and s^2 is the variance of the logarithmic efficiencies. These errors (Table 6.11) are generally lower than the errors given in Sect. 6.3.5 (Table 6.10), as these were obtained from the propagation of standard errors of the data in individual blasts while the errors in Table 6.11, from Eq. 6.31, are errors of the means.

The standard errors from Eq. 6.31 have been used to calculate the 95% confidence intervals by a modified Cox method [47]. They are also shown in Table 6.11.

The value of seismic energy measured for the confined hole was 12.9% (useful work) or 8.6% (heat), which is much higher than the mean value in the production bench blasts and largely out of their confidence interval. Figure 6.6 shows a chart with the energy efficiencies with respect to heat of explosion; mean values and those corresponding to the blasts of minimum and maximum energies are plotted. The seismic energy of the confined shot is also shown for comparison.

The energy not measured (the term E_{NM} in Eq. 6.1) accounts on average for almost 87% of the heat of explosion or 81% of the useful work.

Energy data in blasting comprising the three components have been reported in [1–3], while others [4–7, 9] have specifically reported seismic energy. These are compiled in Table 6.11. A detailed discussion of how our results compare with these works can be found in [45]. It draws the following highlights:

1. Fragmentation efficiency. The fines tail of the size distribution curve has a significant influence on the resulting fragmentation energy efficiency, as fines have a large specific surface and consequently a relatively high amount of the fracture energy is required for their generation. The integral formulation for the area of the smaller size class in the present work includes all this area. As a result, our fragmentation efficiency values are (with the exception of Berta's,

Table 6.11 Summary of energy efficiencies

	Useful work				Heat of explosion			
	Fragment.	Seismic	Kinetic	Total	Fragment.	Seismic	Kinetic	Total
Bench blasts								
15/02	5.0	1.6	8.0	14.6	3.1	1.0	4.9	9.0
29/02	5.2	2.5	16.5	24.1	3.2	1.5	10.3	15.0
37/02	5.7	1.6	6.5	13.8	3.5	1.0	4.0	8.5
43/03	5.9	4.3	8.7	18.9	3.6	2.7	5.4	11.6
45/03	5.8	3.1	5.3	14.1	3.6	1.9	3.3	8.7
50/03	3.1	3.6	9.9	16.6	1.9	2.2	6.1	10.2
54/03	8.8	2.8	5.4	17.0	6.0	1.9	3.7	11.6
58/03	4.5	3.1	14.9	22.5	3.1	2.2	10.2	15.4
420-11	5.7	0.8	26.2	32.7	4.6	0.6	20.7	25.9
440-04	6.9	1.4	4.7	13.0	5.3	1.1	3.7	10.1
Range	3.1–8.8	0.8–4.3	4.7–26.2	13.0–32.7	1.9–6.0	0.6–2.7	3.3–20.7	8.5–25.9
Mean	5.7	2.6	10.7	18.8	3.8	1.6	7.2	12.6
Rel. std. error	0.09	0.18	0.19	0.10	0.11	0.16	0.21	0.11
Conf. interval 95%	4.6–6.9	1.7–3.9	6.9–16.5	15.1–23.3	3.0–4.8	1.2–2.3	4.5–11.4	9.8–16.2
Confined shot		12.9				8.6		
Other works								
(1)					9.7	23.4	3.3	36.4
(2) Average	0.57	7.0	36.6	44.2	0.33	4.1	21.3	25.7
Range	0.2–1	2.5–14	13–63	20–69	0.1–0.6	1.5–8	7.5–39	12–40
(3) Average					0.16	7.5	10.4	18.0

(continued)

Table 6.11 (continued)

	Useful work				Heat of explosion			
	Fragment.	Seismic	Kinetic	Total	Fragment.	Seismic	Kinetic	Total
Range					0.10–0.21	3–12	7.2–12.0	15.4–20.4
(4)						5.3		
(5)						5–9		
(6)						2.7		
(7)						1.8–3.7		
(9)						0.2–5		
Average								
Range						0.1–25		

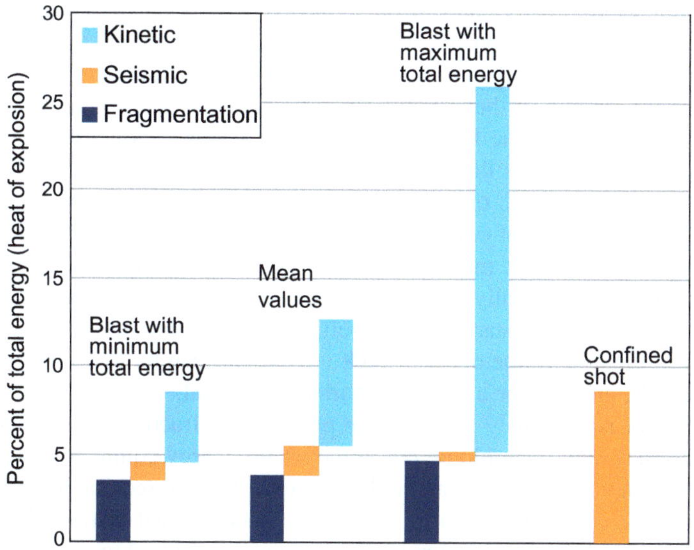

Fig. 6.6 Energy efficiencies

who considers a low specific surface area, 80 m^2/m^3, but uses an abnormally high fracture-specific surface energy, 1470 J/m^2) comparatively high and should be considered more realistic. A range from a few units to about 6% of the available energy seems to be spent in rock fragmentation in usual quarry blasting. This figure could perhaps be higher if the internal microcracking and the fracture surface roughness were considered.

2. Seismic efficiency. The fraction of energy conveyed as seismic wave in production blasts may range from as low as less than 1% to nearly 15% of the explosive energy. Though the confined shot's seismic efficiency measured in the present work, 8.6%, is clearly higher (by a factor of more than 5) than the corresponding values for our production blasts, even higher values have been reported occasionally in production blasts [3]. Conversely, comparable confined shot holes [4, 5, 7] and underground cut holes [9] may give seismic efficiencies well below the 8.6% value.
3. Kinetic efficiency. This usually ranges from about 3% to slightly in excess of 20% of the heat of explosion, though higher values (up to near 40%) have been reported [2].

The average energy lost below 100 MPa is 36% of the heat of explosion (the calculation of such value requires the knowledge of the exact composition of the explosives involved, only available for El Alto blasts). This energy is conveyed as enthalpy (heat and pressure) of the gases venting to the atmosphere without performing any further work on the rock. Though the 100 MPa cut-off value is to some extent arbitrary [48–51], and so is the percentage of energy lost, we can still conclude that approximately half of the explosive energy appears not to cause any measurable physical effect on the rock. Rock blasting is not anything beyond the first principle of Thermodynamics, so what are the mechanisms missing in this balance? There are two of them, very important energy-wise: (i) elastic and plastic work against the rock in the fracture opening that forms the duct network through which the detonation gases flow (the elastic part of it is given back to the medium when the stress is released, or is partially converted to kinetic energy as the fragments detach), and (ii) heat convection from the hot gases to the fracture surfaces, further diffused to the rock mass. Though the detonation process itself is very fast and, consequently, essentially adiabatic, the residence time of the gases within the rock until the movement starts is relatively long. Response times (time from the detonation in the hole to the onset of rock movement) can be up to 60 ms, or even longer [52, 53], enough for a significant transfer of heat.

Heat transfer is obviously a loss of the process, but fracture opening is required to break the rock, and it should be put in the useful energy count. It is possible that explosives are not as inefficient as the low values of measured fragmentation work (accounted as fracture energy only) and kinetic energy spent in rock displacement may, at first instance, indicate. Numerical modeling is required to assess the fracture opening work/heat transfer partition.

6.5 Conclusions

The basic measurable energy components of the blasting process have been determined from production blasts data. Emphasis has been put on describing in detail the calculations and simplifying assumptions required to derive the energy values from the raw data measured. The following ranges of energetic efficiency (given as the 95% confidence intervals of the means of lognormal distributions) have been obtained for bench production blasts: Fracture energy, 3–5%, 5–7%; seismic energy, 1–2%, 2–4%; kinetic energy, 5–11%, 7–16%; total energy measured: 10–16%, 15–23% (for each energy component, the first range applies to heat of explosion and the second to useful work to 100 MPa; values are approximated to the nearest 1%).

We estimated a seismic efficiency for a confined shot hole of 9 and 13% with respect to heat of explosion and useful work, respectively. This value is higher by a factor of more than five if compared to ordinary production blasts with rock movement. Fracture in the rock surely exists in the confined shot—obviously not measurable—which would put the total energy efficiency figure at a level similar to that in production blasts. In other words, the sum of seismic plus kinetic efficiencies in production blasts is close to the seismic efficiency in the confined shot.

The results of seismic efficiency calculated from velocity records measured on the surface are comparable to other results published from measurements with in-hole embedded gages, if the effect of the use of in situ P and S wave velocities (as opposed to the use of a unique, laboratory P-velocity) is discounted.

The variability of the various energy efficiencies is in some cases quite high, though they are not beyond what is typically encountered in usual blast monitoring. Different blasting features, explosives, and rock types, their unavoidable variations within a blast and the inaccuracies of the measurements, propagate through the calculations resulting in standard uncertainties for the energies and efficiencies in a given blast up to 50% of the calculated value. Besides, explosive energies from different sources are sometimes difficult to bring to a common base. In spite of these limitations, it is possible to set the fractions of explosive energy spent in fragmentation, seismic wave, and rock displacement as follows:

1. The fracture energy calculation depends largely on the minimum size of fragments considered. However, it is unlikely, in typical quarry blasting, to obtain a fracture energy much higher than 6% of the total explosive energy.
2. Seismic energy determinations are done generally in an oversimplified manner. A more precise calculation is probably out of reach for field blasting conditions. Accepting a high variability consubstantial with all field blasting measurements, the range of seismic efficiency values reported seems especially wide: values from as low as 0.1% up to, perhaps, 15% of the explosive energy can be expected.
3. The kinetic energy fraction has been found to range from about 3 to 21% of the explosive energy, the higher value corresponding to an overcharged hole. These bounds seem to cover most of the values of kinetic efficiency available in the literature though some sources put this figure as high as 40%.

The total energy measured ranges from 8% to about 26% of the explosive energy. The energy spent in useful work (fragmentation and throw) accounts for 7–25% of the explosive energy. The exact estimate in a particular instance may be in question, but the upper limit is not likely to be much higher in normal blasting practice. It is conceivable, and generally accepted, that explosion gases do not exhaust their energy content in their interaction with the rock, but vent to atmosphere at a relatively high pressure. The amount of energy lost depends on what this pressure is and on the isentropic expansion path, which varies from one explosive to another. It has been calculated to be in excess of 30% of the explosive energy for the explosives used in the present work. Adding up this energy to the measured fractions, about 40–60% of the energy remains unaccounted for. This energy must have been indeed transferred from the gases to the rock, in the form of rock deformation work and heat transfer.

The energetic analysis provides a good understanding of the explosive–rock interaction and blasting process, encompassing a broad range of scientifically challenging subjects. However, the central issues in rock blasting in engineering practice (i.e., steer fragmentation toward coarser or finer distributions, reduce vibrations, etc.) comprise a limited fraction of the total explosive energy, with a variance of the same order of their magnitude. An analysis of such variance in relation to blast design parameters and explosive and rock properties is required before energetic arguments may be implemented into engineering design methods for rock blasting. No such analysis has been attempted with the data presented in this work, or with the other (relatively scarce) data existing in the literature. For that analysis to bear statistical significance, the amount of data needed is much larger than what is currently available.

Acknowledgements The experimental work was partially funded by the European Union under contract no. G1RD-CT-2000-00438, "Less Fines Production in Aggregate and Industrial Minerals Industry".

References

1. Berta G (1990) L'esplosivo strumento di lavoro. Italesplosivi, Milan, pp 31–64
2. Spathis AT (1999) On the energy efficiency of blasting. In: Proceedings of the 6th international symposium on rock fragmentation by blasting, Johannesburg, 8–12 Aug 1999. The South African Institute of Mining and Metallurgy, Johannesburg, pp 81–90
3. Ouchterlony F, Nyberg U, Olsson M (2003) The energy balance of production blasts at Nordkalk's Klinthagen quarry. In: Holmberg R (ed) Proceedings of the 2nd world conference on explosives and blasting, Prague, 10–12 Sept 2003. Balkema, Rotterdam, pp 193–203
4. Howell BF, Budenstein D (1955) Energy distribution in explosion-generated seismic pulses. Geophysics 20(1):33–52
5. Fogelson DE, Atchinson TC, Duvall WI (1959) Propagation of peak strain and strain energy for explosion-generated strain pulses in rock. In: Proceedings of the 3rd US symposium on rock mechanics, Golden, CO, 20–22 Apr 1959. Colorado School of Mines, Golden, pp 271–284
6. Berg JW, Cook KL (1961) Energies, magnitudes and amplitudes of seismic waves from quarry blasts at Promontory and Lakeside, Utah. Seismol Soc Bull 51(3):389–400
7. Nicholls HR (1962) Coupling explosive energy to rock. Geophysics 27(3):305–316
8. Atchinson TC (1968) Fragmentation principles. In: Pfleider EP (ed) Surface mining. The American Institute of Mining, Metallurgical and Petroleum Engineers, New York, pp 355–372
9. Hinzen KG (1998) Comparison of seismic and explosive energy in five smooth blasting test rounds. Int J Rock Mech Min Sci 35(7):957–967
10. Grady DE (1985) Fragmentation under impulsive stress loading. In: Fourney WL, Boade RR, Costin LS (eds) Fragmentation by blasting. Society for Experimental Mechanics, Bethel, CT, pp 63–72
11. Hamdi E, du Mouza J, Fleurisson JA (2001) Evaluation of the part of blasting energy used for rock mass fragmentation. Int J Blasting Fragmentation 5(3):180–193
12. Achenbach JD (1975) Wave propagation in elastic solids. Elsevier, Amsterdam, p 166
13. Rinehart JS (1975) Stress transients in solids. Hyperdynamics, Santa Fe, New Mexico, p 41
14. Chiappetta RF, Mammele ME (1987) Analytical high-speed photography to evaluate air decks, stemming retention and gas confinement in presplitting, reclamation and gross motion applications. In: Fourney WL, Dick RD (eds) Proceedings of the 2nd international symposium on rock fragmentation by blasting, Keystone, Colorado, 23–26 Aug 1987. Society for Experimental Mechanics, Bethel, CT, pp 257–301
15. Hamdi E, du Mouza J (2005) A methodology for rock characterization and classification to improve blast results. Int J Rock Mech Min Sci 42:177–194
16. Goetz D, Rouabhi A, Tijani M (2002) Mechanical behaviour of rocks. Technical Report 35, EU Project GRD-2000-25224. École National Supérieure des Mines, Paris
17. Böhm A, Mayerhofer R (2002) Mechanical fragmentation tests. Technical Report 10, EU Project GRD-2000-25224. University of Leoben, Austria

18. Sanchidrián JA, Segarra P, López LM (2003b) Blasting performance in El Alto quarry (Cementos Portland). Technical report 43. EU Project GRD-2000-25224. Universidad Politécnica de Madrid, Spain
19. Sanchidrián JA, Segarra P, López LM (2003a) Blasting performance in Eibenstein (Hengl Bitustein). Technical report 42. EU Project GRD-2000-25224. Universidad Politécnica de Madrid, Spain
20. Rosin P, Rammler E (1933) The laws governing the fineness of powdered coal. J Inst Fuel 7:29–36
21. Weibull W (1939) A statistical theory of the strength of materials. Ingeniörvetenskaps-akademiens Handlingar 151:1–45
22. Weibull W (1951) A statistical distribution function of wide applicability. J Appl Mech-T ASME 18:293–297
23. Bjarnholt G, Holmberg R (1976) Explosive expansion work in underwater detonations. In: Proceedings of the 6th international symposium on detonation, Coronado, California, 24–27 Aug 1976. Office of Naval Research, Department of the Navy, Arlington, Virginia, pp 540–550
24. Mohanty B (1999) Explosives performance—the underwater test revisited. In: Proceedings Explo'99, Kalgoorlie, 7–11 Nov 1999. The Australasian Institute of Mining and Metallurgy, Carton Victoria, Australia, pp 131–137
25. Nyberg U, Arvanitidis I, Olsson M, Ouchterlony F (2003) Large size cylinder expansion tests on ANFO and gassed bulk emulsion explosives. In: Holmberg R (ed) Proceedings of the 2nd world conference on explosives and blasting technique, Prague, 10–12 Sept 2003. Balkema, Rotterdam, pp 181–191
26. Sanchidrián JA, López LM (2003) Calculation of the explosives useful work-comparison with cylinder test data. In: Holmberg R (ed) Proceedings of the 2nd world conference on explosives and blasting technique, Prague, 10–12 Sept 2003. Balkema, Rotterdam, pp 357–361
27. Sanchidrián JA, López LM (2006) Calculation of the energy of explosives with a partial reaction model. Propellants Explos Pyrotech 31(1):25–32
28. Hobbs ML, Baer MR (1992) Nonideal thermoequilibrium calculations using a large product species data base. Report SAND92-0482. Sandia National Laboratories, Albuquerque, NM
29. Hobbs ML, Baer MR (1993) Calibrating the BKW-EOS with a large product species data base and measured C-J properties. In: Proceedings of the Tenth International Detonation Symposium, Boston, MA, 12–16 July 2003. Office of Naval Research, Arlington, Virginia, pp 409–418
30. Istrochem (2002) Dap 2 and Danubit 4 data sheets. Istrochem, Bratislava
31. Sanchidrián JA, Segarra P, López LM (2006) A practical procedure for the measurement of fragmentation by blasting by image analysis. Rock Mech Rock Eng 39(4):359–382
32. Moser P, Cheimanoff N, Ortiz R, Hochholdinger R (2000) Breakage characteristics in rock blasting. In: Holmberg R (ed) Proceedings of the 1st world conference on explosives and blasting technique, Munich, 6–8 Sept 2000. Balkema, Rotterdam, pp 165–170
33. Moser P (2003) Less fines production in aggregate and industrial minerals industry. In: Holmberg R (ed) Proceedings of the 2nd world conference on explosives and blasting technique, Prague, 10–12 Sept 2003. Balkema, Rotterdam, pp 335–343
34. Schleifer J, Tessier B (2000) Fragmentation assessment using the Fragscan system: quality of a blast. In: Holmberg R (ed) Proceedings of the 1st world conference on explosives and blasting technique, Munich, 6–8 Sept 2000. Balkema, Rotterdam, pp 111–115
35. Ouchterlony F (2003) Influence of blasting on the size distribution and properties of muckpile fragments, a state-of-the-art review. MinFo project P2000-10. Swebrec-Luleå University of Technology, Luleå, Sweden, pp 89–95
36. Latham JP, Kemeny J, Maerz N, Noy M, Schleifer J, Tose S (2003) A blind comparison between results of four image analysis systems using a photo-library of piles of sieved fragments. Int J Blasting Fragmentation 7(2):105–132
37. Ouchterlony F (2005) The Swebrec function. Linking fragmentation by blasting and crushing. Trans Inst Min Metall A 114:29–44

38. Kemeny J, Girdner K, Bobo T, Norton B (1999) Improvements for fragmentation measurement by digital imaging: accurate estimation of fines. In: Proceedings of the 6th international symposium on rock fragmentation by blasting, Johannesburg, 8–12 Aug 1999. The South African Institute of Mining and Metallurgy, Johannesburg, pp 103–109
39. Blair DP (1995) Soil-embedded detector mounts for seismic monitoring. Geophysics 60 (1):120–133
40. ISEE (1998) Blasters' handbook. International Society of Explosives Engineers, Cleveland, pp 613, 732–734
41. Wheeler RM (2005) The importance of proper seismometer coupling. In: Holmberg R (ed) Proceedings of the 3rd world conference on explosives and blasting, Brighton, 13–16 Sept 2005. European Federation of Explosives Engineers, Rochester, UK, pp 237–243
42. Blair DP, Armstrong LW (2001) The influence of burden on blast vibration. Int J Blasting Fragmentation 5(1–2):108–129
43. Segarra P, Sanchidrián JA, López LM, Pascual JA, Ortiz R, Gómez A, Smöch B (2003) Analysis of bench face movement in quarry blasting. In: Holmberg R (ed) Proceedings of the 2nd world conference on explosives and blasting technique, Prague, 10–12 Sept 2003. Balkema, Rotterdam, pp 485–495
44. BAI (2001) Motion Tracker™ 2D program user's manual (Version 5.x). Blasting Analysis International, Allentown, Pennsylvania, 66 p
45. Sanchidrián JA, Segarra P, López LM (2007) Energy components in rock blasting. Int J Rock Mech Min Sci 44(1):130–147
46. Zhou X-H, Gao S (1997) Confidence intervals for the log-normal mean. Stat Med 16:783–790
47. Olsson U (2005) Confidence intervals for the mean of a log-normal distribution. J Stat Educ 13(1)
48. Mohanty B (1981) Energy, strength and performance, and their implications in rating commercial explosives. In: Proceedings of the 7th annual conference on explosives and blasting technique, Phoenix, Arizona, 19–23 Jan 1981. International Society of Explosives Engineers, Cleveland, Ohio, pp 293–306
49. Persson PA, Holmberg R, Lee J (1994) Rock blasting and explosives engineering. CRC Press, Boca Raton, Florida, p 87
50. Cunningham CVB, Sarracino RS (1990) The standardization of explosives ratings by ideal detonation codes. In: Proceedings of the 3rd international symposium on rock fragmentation by blasting, Brisbane, 26–31 Aug 1990. The Australasian Institute of Mining and Metallurgy, Carlton Victoria, Australia, pp 345–351
51. Katsabanis PD, Workman L (1998) The effect of available energy on blast design. In: Proceedings of the 14th annual symposium on explosives and blasting research, New Orleans, Louisiana, 8–11 Feb 1998. International Society of Explosives Engineers, Cleveland, Ohio, pp 87–96
52. Onederra I, Esen S (2003) Selection of inter-hole and inter-row timing for surface blasting—an approach based on burden relief analysis. In: Holmberg R (ed) Proceedings of the 2nd world conference on explosives and blasting technique, Prague, 10–12 Sept 2003. Balkema, Rotterdam, pp 269–275
53. Sanchidrián JA, Segarra P, López LM (2005) On the relation of rock face response time and initial velocity with blasting parameters. In: Holmberg R (ed) Proceedings of the 3rd world conference on explosives and blasting, Brighton, 13–16 Sept 2005. European Federation of Explosives Engineers, Rochester, pp 375–389

Part II
Material Handling

Chapter 7
Energy-Efficient Loading and Hauling Operations

Ali Soofastaei, Elnaz Karimpour, Peter Knights and Mehmet Kizil

Abstract Approximately, 40% of the total energy used in surface mines is related to diesel consumption. Truck haulage is responsible for a majority of this. This chapter introduces the principal equipment used to load and haul materials in mines, namely trucks, electric rope shovels, hydraulic excavators and crushing and conveying systems. The chapter discusses factors that contribute to the energy-efficient operation of such equipment. Based on gross weight hauled per unit weight of payload, belt conveyors appear to be the most energy-efficient means of transporting material in surface mines. However, a number of factors, including large upfront capital expenditure and limited ability to relocate and scale up belt capacities, currently restrict their widespread applicability.

Keywords Energy · Efficiency · Mining · Loading · Hauling

7.1 Introduction

In the 2012–2013 financial year, some 603 PJ of energy was consumed in the mining and quarrying industry in Australia. The three biggest consumers of energy in the Australian mining industry are as follows (in decreasing order) [1]:

A. Soofastaei (✉) · E. Karimpour · P. Knights · M. Kizil
The University of Queensland, Brisbane, Australia
e-mail: ali@soofastaei.net

E. Karimpour
e-mail: elnazkarimpoor85@gmail.com

P. Knights
e-mail: p.knights@uq.edu.au

M. Kizil
e-mail: m.kizil@uq.edu.au

K. Awuah-Offei (ed.), *Energy Efficiency in the Minerals Industry*, Green Energy and Technology, https://doi.org/10.1007/978-3-319-54199-0_7

- Crushing and grinding (40.5%),
- Materials handling (20.6%), and
- Mine ventilation (10.1%)—underground mines only.

This chapter provides an overview of the current knowledge on energy used for materials handling during the extraction phase of mining operations and identifies opportunities to reduce energy consumption associated with these processes. The mining method (shovel/truck, conveyor, etc.) and associated equipment dictate the quantity of energy consumed in any mining operation. In the following, the discussion is organised based on the type equipment.

7.2 Haul Trucks

Approximately, 40% of the total energy used in surface mines relates to diesel consumption [2]. Truck haulage is responsible for a majority of this diesel consumption [3]. Haul trucks are used in combination with other equipment such as excavators, diggers and loaders, depending on the production capacity and site layout.

Trucks in surface mines are used to haul ore and overburden from the pit to a stockpile, dump site or the next stage of a mining process. Trucks are expensive to purchase, operate and maintain and use a major proportion of diesel in surface mines.

Many parameters, such as production rate, age and maintenance of the vehicle, operator practices, payload, speed, cycle time, mine layout, mine plan, idle time, tyre wear, rolling resistance, dumpsite design, engine operating, parameters and transmission shift patterns, affect the productivity of trucks in surface mining. This knowledge can be merged into mine plan costing and design procedures to improve effective process control. The major truck types used in surface mining are shown in Fig. 7.1.

7.2.1 Types of Trucks

There are three main types of trucks: rear, bottom and articulated dump trucks.

In rear dump trucks, the tray is mounted on the truck frame. Dumping is carried out by a hydraulic hoist system raising the tray. These are very flexible units capable of handling all types of material. They have good grade ability and are easily manoeuvred [5]. They are the most common haulage truck globally (see Fig. 7.2).

The standard rear dump haul truck has two axles with two wheels on the front axle and four wheels on the rear axle. The rear wheels are usually the only ones

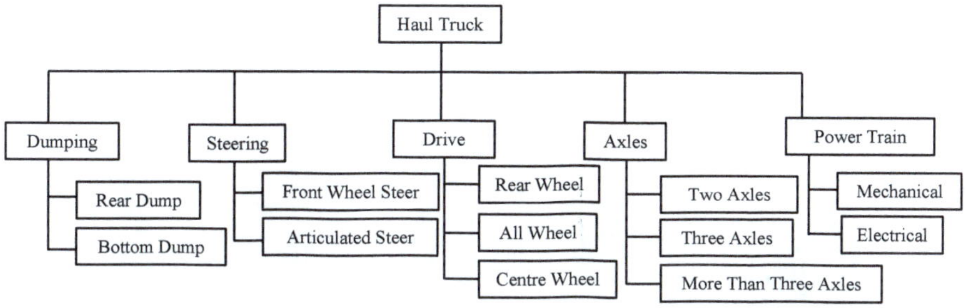

Fig. 7.1 Haul truck classifications [4]

Fig. 7.2 Rear dump truck

driven. Three-axle trucks are less common in mines but are used for on/off highway hauls.

Bottom dump trucks provide faster dump times and higher payload for the same engine horsepower, but at the cost of grade ability and manoeuvrability. This type of truck has three axles, two tyres in front, four drive tyres at the rear of the tractor, and four tyres on the rear of the trailer (see Fig. 7.3) [5].

In general, they are used in strip coal mines where the ramp gradients are kept at five percent or less.

Articulated dump trucks tend to be smaller and of lighter construction. Maximum size is in the order of 50 t (see Fig. 7.4).

The main application of this type of trucks is in wet and poor road conditions. Their lighter construction results in a shorter life [5].

Some advantages and disadvantages of the types of trucks mentioned above have been tabulated in Table 7.1.

Fig. 7.3 Bottom dump truck

Fig. 7.4 Articulated dump truck

7.2.2 Effective Parameters on Truck Productivity

Table 7.2 shows some parameters that influence haul truck productivity in mines.

7.2.3 Haul Truck Fuel Consumption

Haul truck fuel consumption is a function of various parameters, the most significant of which have been identified and categorised into six main groups (see Fig. 7.5).

Table 7.1 Advantages and disadvantages of popular mine haul trucks [4]

Truck type	Advantages	Disadvantages
Rear dump truck	• Mobile and flexible in moving to other working areas • Handle a range of material properties and sizes • Medium transport distance • Can effectively operate gradients up to 12%	• Require good road surface for efficient operation and tyre protection • Higher operating labour component (Compared to conveyors)
Bottom dump truck	• Higher speed on flat hauls • Mobile and flexible in moving around working area	• Better on flatter gradients <5% • Require good roads • Requires a drive over dump hopper to discharge • Suited to lighter and finer materials due to light trailer and dump doors (e.g. Coal, Bauxite)
Articulated dump truck	• Can handle difficult floor conditions—rough, boggy • Handle a range of material properties and sizes • Can handle steeper gradients	• May require higher maintenance • Try can roll sideways safety • Higher capital cost/capacity

Table 7.2 Effective parameters on haul truck productivity [4]

Parameter	Detail
Truck model and type	Each type and model of the truck has special characteristics, and these can affect haul truck productivity
Material	Material which is hauled
Bucket density	The density of the material being loaded
Swell factor	The swell factor is the volume increase after the material has been disturbed
Bucket load	Estimated bucket load that the loading unit can carry in BCM
Calculated passes to fill	Estimate of how many bucket loads (passes) is required to fill the truck to its nominal capacity
Calculated truck payload	Estimated average payload that the truck will carry after considering all the above factors
Load factor	Percentage of truck fill compared to its nominal or rated payload
Time per pass	Time taken for a loading unit to complete one pass
Load time	Time taken to load the truck
Spot time	The time during which the loading unit has the bucket in place to dump, but is waiting for the truck to move into position. Spot time will depend on the truck drivers' ability and the system of loading. Double-side loading should almost eliminate spot time

(continued)

Table 7.2 (continued)

Parameter	Detail
Dump time	Time taken for the truck to manoeuvre and dump its load either at a crusher or dump
Fixed time	The sum of load, spot and dump time. It is called 'fixed' because it is essentially invariable for a truck and loading unit combination
Travel time	Time is taken to haul and return the load
Cycle time	Round trip time for the truck, it is the sum of fixed, travel and wait times
Efficiency	A measure of how much productive time is achieved in 1 h of operating time. The sort of activities that the efficiency factor includes is Clean-up by the loading unit or dozer, Crusher and dump slowdowns, Fuelling, Inspections, Loading unit movement, Operator experience, Under trucking, Unusual delays due to weather
Queue factor	Accounts for time lost due to queuing. It is another measure of wait time
Productivity	Tonnes of production hauled in an operating hour (t/h) Productivity = (Efficiency/Cycle time) × Truck payload × Queuing factor
Physical availability	Measure of time available to work divided by calendar time
Utilisation	Operating time divided by available time
Production	Hourly productivity × operating hours

Of these, the most significant factors affecting haul truck fuel consumption are as follows (see also Fig. 7.6):

- The gross vehicle weight (GVW), which is sum of the weight of an empty truck and the payload;
- The haul truck velocity (V);
- The total resistance (TR), which is equal to the sum of rolling resistance (RR) and the grade resistance (GR) when the truck is moving against the grade of haul road; and
- The rimpull force (RF), which is the force available between the tyre and the ground to propel the truck.

Figure 7.7 illustrates the variation of maximum truck velocity (V_{max}) and fuel consumption (FC) with GVW for six values of total resistance (TR). The results show that for all values of total resistance, truck velocity decreases and fuel consumption increases as the GVW increases. It should be noted that the rate of fuel consumption is calculated based on the best performance of the truck as recommended by the manufacturer (calculate at maximum achievable truck velocity and corresponding rimpull).

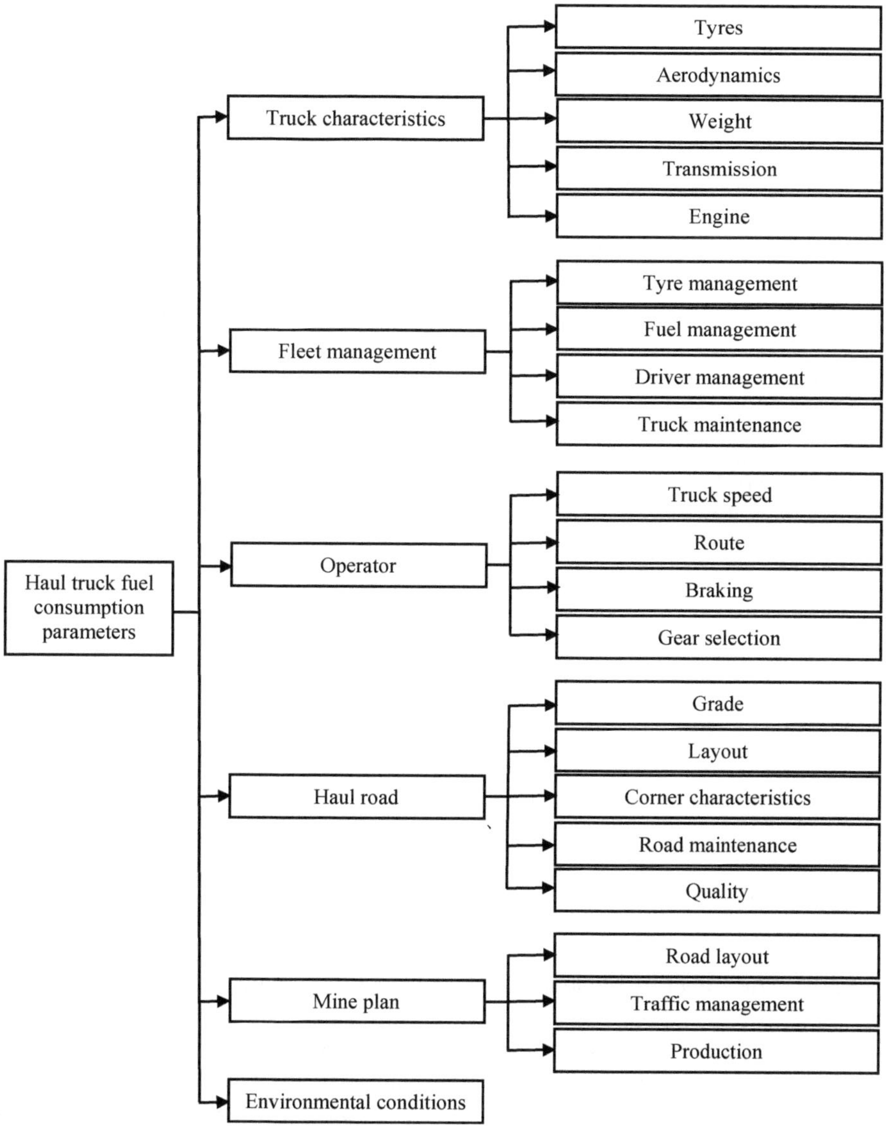

Fig. 7.5 Parameters affecting haul truck fuel consumption [6]

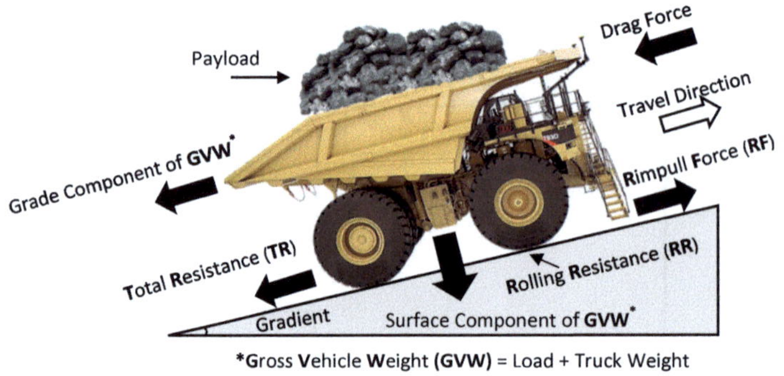

Fig. 7.6 Effective parameters on haul truck productivity and fuel consumption [6]

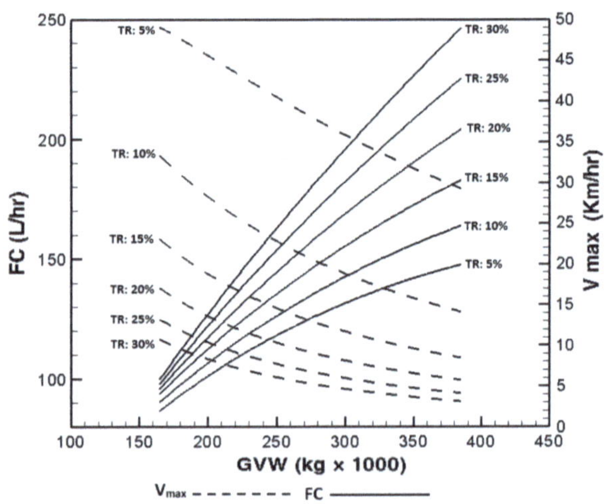

Fig. 7.7 Variation of V_{max} and FC with GVW for different TR (Caterpillar 793D) [4]

7.2.3.1 The Effect of Payload on Haul Truck Fuel Consumption

The loading process in truck and shovel operations is often modelled as a stochastic process due to the high variability. An analysis of the haul truck payload data obtained from some mine sites around the world shows that the payload distribution can be estimated by a normal distribution [7]. The variance associated with haul truck payloads is typically large and depends on some parameters such as particle size distribution, swell factor, material density, truck–shovel matching, the number of shovel passes and the bucket fill factor. Many attempts have been made to reduce the payload variance by using technologies such as on-board truck payload measurement systems, shovel payload management systems and fleet monitoring systems. Also, to load a truck in an effective manner, the shovel operator should load the truck within optimal payload limits using the minimum number of passes. The optimal payload can be defined in different ways, but it is always designed so that

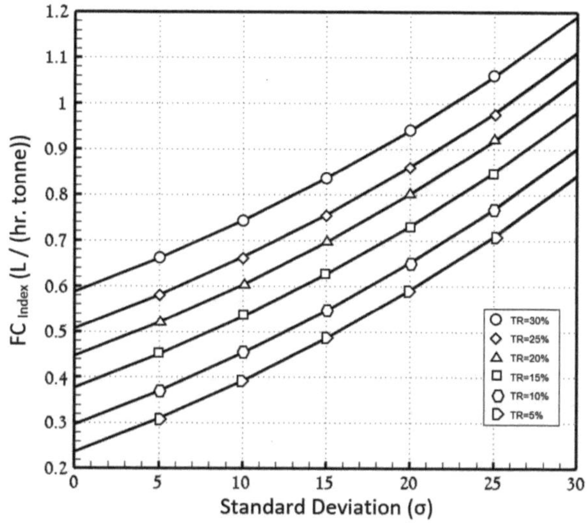

Fig. 7.8 The variation of FC_{Index} with standard deviation (σ) (CAT 793D) [9]

the haul truck will carry the greatest amount of material with lowest payload variance. The range of payload variance can be defined based on the capacity and power of truck. The payload variance in a surface mine fleet can significantly influence productivity due to truck congestion, or "bunching" phenomena,[1] in large surface mines [8].

The increasing of payload variance decreases the accuracy of a scheduled maintenance programme. This is because the rate of equipment wear is not predictable when the mine fleet faces a large payload variance. Minimising the variation of particle size distribution, swell factor, material density and fill factor can decrease the payload variance but it should be noted that it is not always possible to control all these parameters.

The effect of payload variance on haul truck fuel consumption in different haul road conditions is illustrated in Fig. 7.8.

In this chapter, we use the fuel consumption index (FC_{Index}) as a measure of haul truck fuel efficiency. This index represents the quantity of fuel burnt by a haul truck to move one tonne of mined material (ore or overburden) in an hour (L/(h tonnes)).

7.2.3.2 The Effect of Rolling Resistance on Haul Truck Fuel Consumption

The rolling resistance (RR) is a major component of total resistance (TR), and it is one of the main controllable effective parameters for haul truck fuel consumption.

[1]This is where trucks loaded at rated payloads are forced to travel slowly up ramp because they are stuck behind heavily loaded trucks which travel at low speeds.

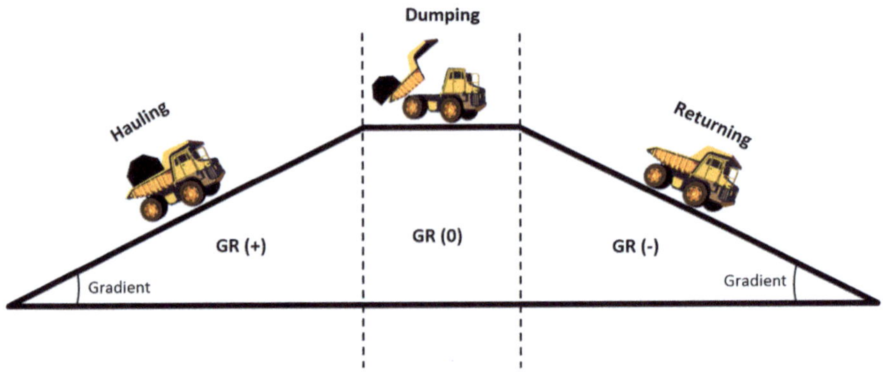

Fig. 7.9 Grade resistance [4]

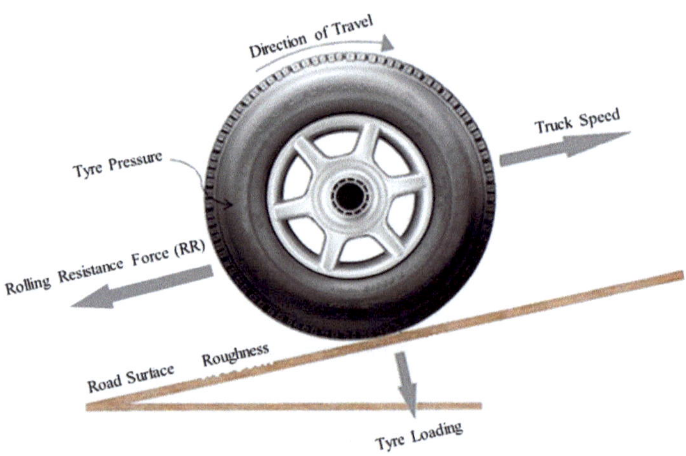

Fig. 7.10 Rolling resistance and the most influential parameters [10]

TR is equal to the sum of RR and grade resistance (GR) when the truck is moving against the grade of the haul road.

RR depends on the tyre and hauling road surface characteristics and is used to calculate the rolling friction force, which is the force that resists motion when the truck tyre rolls on the haul road. GR is the slope of the haul road, measured as a percentage and is calculated as the ratio between the rise of the road and the horizontal length (see Fig. 7.9).

GR is positive when the truck is travelling up the ramp, and it is negative when it travels down the ramp.

RR is defined as a measure of the force required to overcome the retarding effect between the tyre and road. This resistance is predominantly measured as a

Table 7.3 Influential parameters on rolling resistance

Rolling resistance	Group	Category[a]				Parameter
		D	C	O	M	
	Road	✓			✓	Roughness
		✓	✓		✓	Defects
		✓	✓		✓	Material density
				✓		Moisture content
					✓	Road maintenance
	Tyre	✓		✓	✓	Tyre penetration
		✓				Tyre diameter
				✓		Tyre pressure
			✓	✓	✓	Tyre condition
				✓		Tyre loading
				✓		Tyre temperature
	System			✓		Truck speed
				✓		Driver behaviour
	Weather			✓		Humidity
				✓		Precipitation
				✓		Ambient temperature

[a]D: Design C: Construction O: Operational M: Maintenance

percentage of the GVW, but can also be expressed as energy divided by a distance or force.

Tyre RR can also be characterised by a rolling resistance coefficient (RRC), a unit-less number. RR manifests itself predominantly in the form of hysteresis losses described as the energy lost, usually in the form of heat, when a section of vulcanised rubber is regularly deformed, such as during the operation of a haul truck.

The parameters affecting RR can be categorised into four groups: road, tyre, system and weather properties. Figure 7.10 illustrates the most influential parameters on RR.

The effective parameters on RR are also categorised into the design (D), construction (C), operational (O) or maintenance (M) parameters. Table 7.3 illustrates the parameters affecting RR, and their categories.

The surface material of the haul road is a major contributor to RR. Table 7.4 shows the RR associated with different surface types.

Estimating fuel consumption rate requires some assumptions. Figure 7.11 illustrates the relationship between the haulage operation parameters and truck fuel consumption.

The relationships between three main effective parameters on RR and FC_{Index} have been illustrated in Fig. 7.12.

Table 7.4 Surface type and associated rolling resistance [11]

Type of surface	Rolling resistance (%)
In situ clay till	4–6.7
Compacted gravel	2–2.7
Compacted clay gravel	3.9
Subsoil stockpile	4.4–8.3
Compacted clay till	4.1
Subsoil on mine spoil	7.3

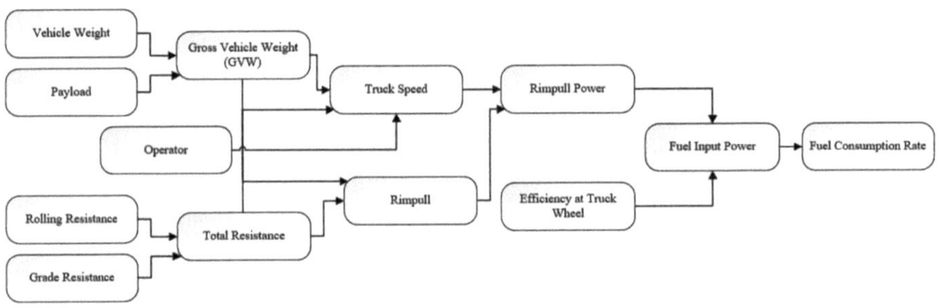

Fig. 7.11 Variable relationships required for truck fuel consumption estimation [4]

7.3 Loading Units

Loading equipments are applied to dig and load material in surface and underground mines. They are often regarded as critical equipment because there is typically no extra loader capacity contrary to trucks that tend to have excess capacity. Therefore, their availability and productivity can constrain production. Efficient loading process can lead to improved production, energy efficiency and decreased costs.

7.3.1 Operations of Major Loading Units

The major loading units are rope shovels, hydraulic excavators and front-end loaders (Fig. 7.13). Other loading equipment includes draglines, surface miners, dozers, scrapers and bucket wheel excavators.

Hydraulic excavators can be configured as either front (or shovel) or backhoe configurations. Face shovels allow for either front or bottom dumping. A backhoe's bucket is typically smaller in volume compared to that of a face shovel on a similar sized machine. Backhoe shovels are capable of loading trucks located either on the same bench level or at a lower bench (elevation) to the shovel.

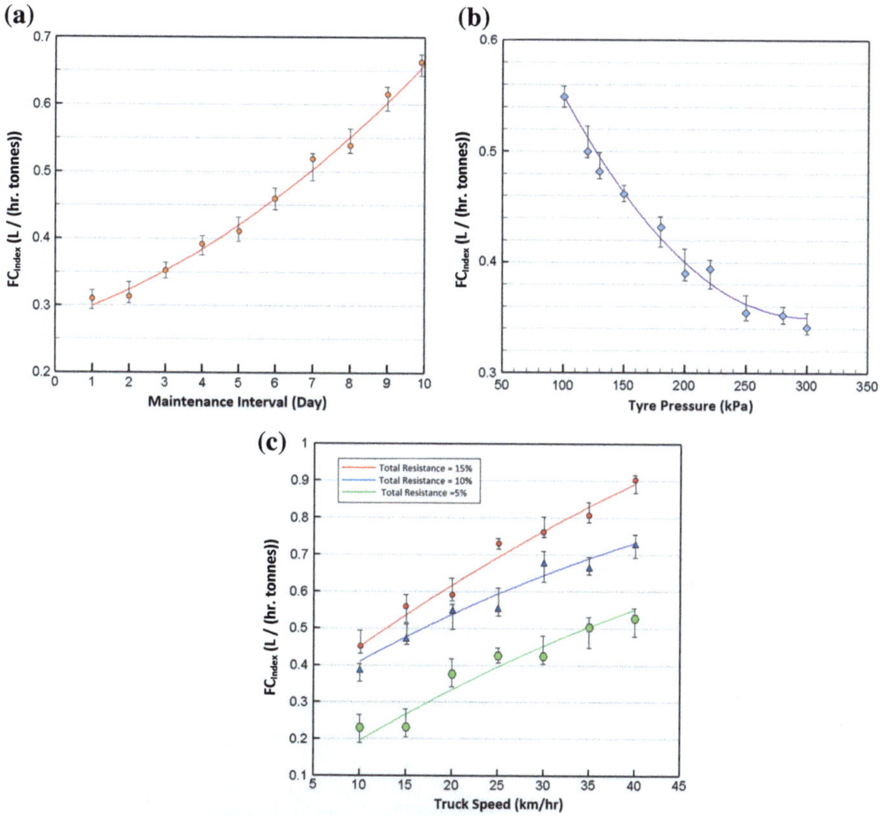

Fig. 7.12 **a** The relationship between maintenance interval and FC_{Index} (Caterpillar 793D) [4]. **b** The relationship between tyre pressure and FC_{Index} (Caterpillar 793D) [4]. **c** The relationship between truck speed and FC_{Index} (Caterpillar 793D) [4]

The hydraulic excavator uses diesel engines or electric motors to drive hydraulic pumps, motors and cylinders that in turn actuate the motions required to dig and load material and propel the machine (see Fig. 7.14).

The electric shovel uses electric motors, gear reducers, drums and wire rope to actuate the motions required for digging, loading and propelling (see Fig. 7.15).

The three primary parts of the hydraulic and electric shovel are the lower, upper and the attachment. A large electric mining shovel is capable of maximum propel speeds of nearly 1.6 km per h (1.0 mph) and a practical grade climbing capability of 20%. The average work cycle of electric mining shovel can take approximately 25–45 s depending on the machine, load, swing angle, bank conditions and operator proficiency [12].

Although hydraulic face shovels provide a high degree of flexibility and can generally produce high digging forces low in the bank, electric mining shovels are inherently more capable of consistently generating higher production rates through

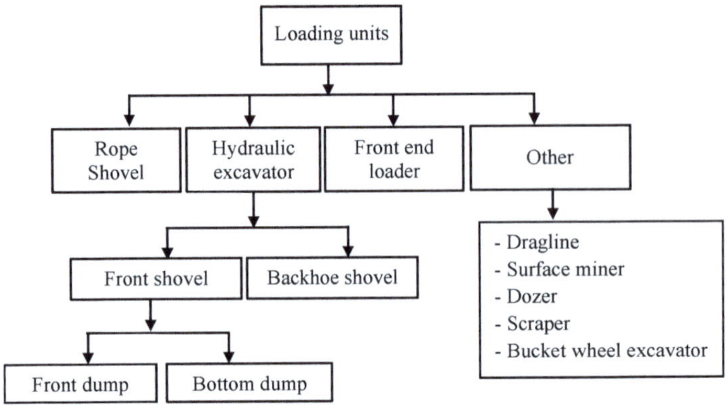

Fig. 7.13 Loading units' classifications

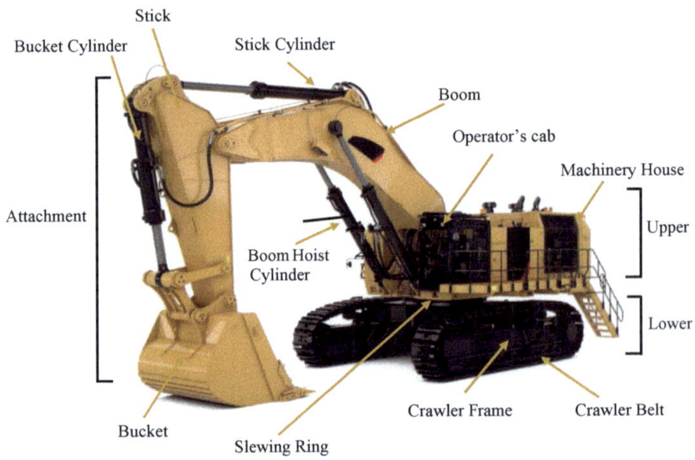

Fig. 7.14 Typical hydraulic mining shovel assemblies

a combination of consistent dig forces throughout the digging phase, high fill factors, low cycle time and reduced operator fatigue [12].

The four primary motions executed by the shovel are propelling, swinging, hoisting and crowding/retracting. In the cable shovel, the crowd and hoist motors attain the crowd/retract and hoist motions, respectively. The hoisting equipment on the cable shovel involves a rope drum which is reeled in or spooled out by the electric motor-driven hoist transmissions.

Figure 7.16 depicts a typical shovel dipper trajectory. The shovel cleans free material from the starting point (A) in the direction of the bench toe (B). Then the position of the shovel dipper teeth changes from (B) to the start point of the coasting phase (C). The task of moving the dipper into the final coasting phase (C–D) is to make the bank clear [13].

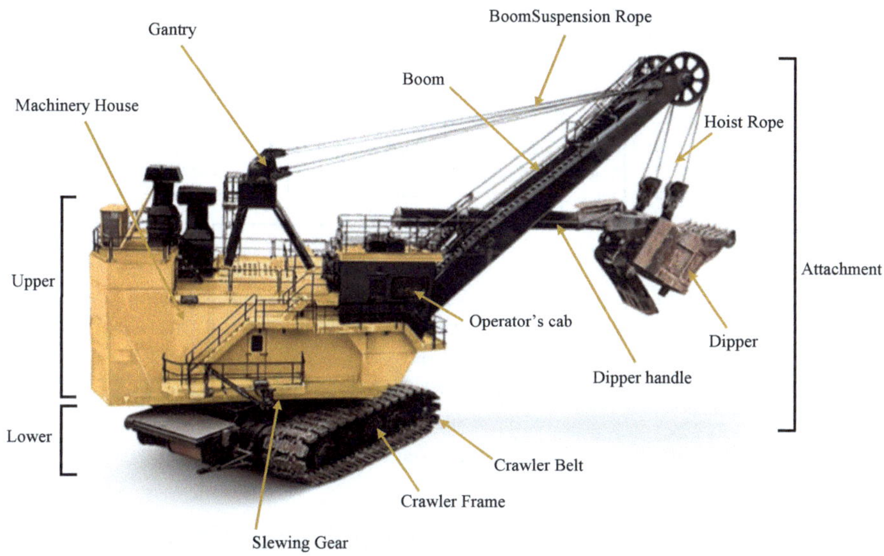

Fig. 7.15 Typical electric mining shovel assemblies

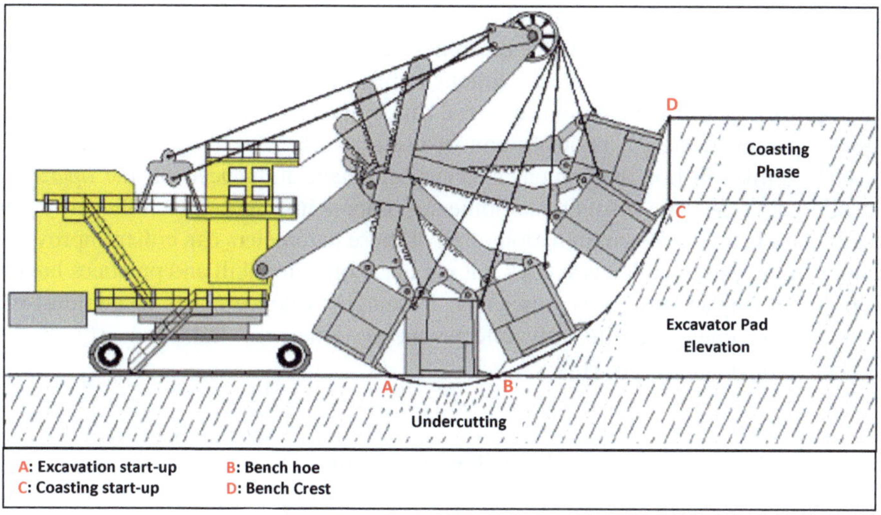

Fig. 7.16 The three separate digging phases accomplished by a cable shovel [14]

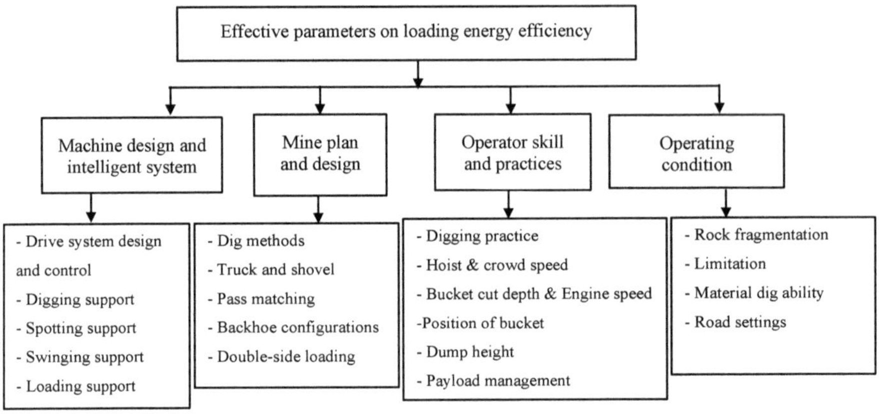

Fig. 7.17 Effective parameters on loading energy efficiency

7.3.2 Effective Parameters on Loading Energy Efficiency

Energy efficiency of loading operations can be improved by enhancing the equipment, ensuring better-operating conditions, selecting proper mine planning and design, and training the operator (Fig. 7.17).

The equipment defines how the machines transform energy efficiently into useful work in specified circumstances, although difficult operating conditions such as sub-optimal rock fragmentation, material digability and road settings lead to considerably higher energy input per unit of useful work. The mine plan and design (which defines in how equipment is organised) have influenced on the operating conditions, relations between equipment components, and how efficiently the equipment is used. The combination of all of these parameters can either improve or diminish the energy efficiency. Last but not least, operator skill and practices have a substantial impact on energy efficiency. Research evidence exists to demonstrate that this parameter significantly affects energy efficiency independently of the influence of other factors [14].

7.3.2.1 Machine Design and Intelligent System

To enhance the energy efficiency of loading operations, intelligent automated digging, spotting, swinging and loading support should be considered.

Typically, for the shovel, three aspects can improve cycle time: greater pull torque, greater peak power and increased speed to swing and to move the dipper. The shovel operator should have greater power and the capability to go faster from side to side in the bank. Therefore, enhancing excavating equipment can lead to machines that more efficiently transform energy into useful work.

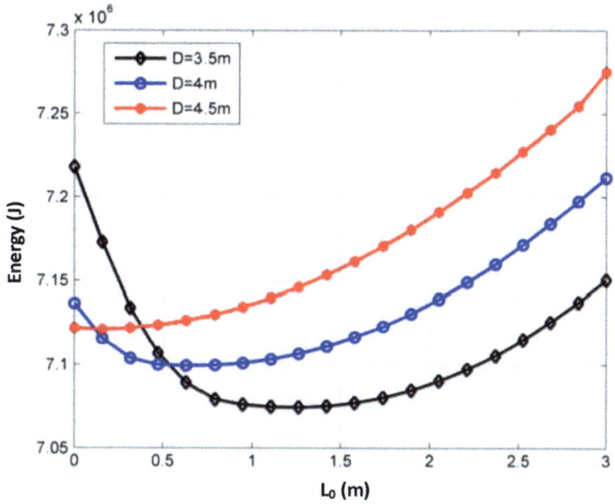

Fig. 7.18 Energy consumption with different starting points [17]

The digging torque has a significant influence on the hoist and crowd movements during the excavation period. Critical energy savings can be reached using regenerative AC drives in mining shovels. Particularly in the swing and hoist movement, the regenerated energy increases to 80 and 22%, respectively. The overall energy savings attained by implementing regenerative AC drives in comparison with non-regenerative drives is approximately 26% [15].

Moreover, it is necessary to raise the torque and horsepower for the propel task to avoid getting stuck. Since the same motors carry out hoist and propel tasks, a structure which reduces hoist–propel switchover time would improve the energy efficiency. The propel task and digging operation of the shovel influence cycle times [16].

Trajectory of the bucket (Digging Trajectory)

When it comes to the digging trajectory, both defining the optimum starting point and choosing the appropriate type of trajectory curves can increase the energy efficiency.

The three curves shown in Fig. 7.18 illustrate the difference in energy consumption with various starting points and consequently three digging distances: $D = 3.5$, 4 and 4.5 m. L_0 is the distance between the new starting point and the end of the soil pile. For every digging distance, energy consumption changes and there is an ideal starting point which minimises energy consumption.

Swing trajectory

It is essential to optimise the position of the shovel in relation to the truck. Large swing angles extend shovel cycle time and waste swing energy, while small swing angle cause swing cycles to become hoist dependent [18]. With the better coordination of the shovel's hoist and crowd tasks, the cycle time could be decreased through the loading period. Passing through the bank and filling the bucket rapidly would be desired, but not so quick that the hoist motors get blocked and halted. In other words, the hoist and crowd tasks should be controlled, simultaneously.

Control the hoist and crowd functions together
During the digging process, the shovel operator retracts the crowd before the hoist motors stall and then crowds as hard as possible against the bank to fill the dipper. If the shovel's hoist and crowd functions are better coordinated, the cycle time could be reduced during the loading phase. The aim would be passing through the bank and filling the bucket quickly, but not so fast that the hoist motors stall. Mines need a system that combines these two motions and can be controlled together. We suggest that further research should be conducted to develop algorithms that could control hoist and crowd together during digging. When the hoist speed falls off, it retracts until hoist speed picks up again [16].

Collision-free trajectory
A semi-automated load assistance system (Auto Load) and a collision avoidance system (Truck Shield) are two sample technologies based on three self-reliant approaches which estimate the position of the haul truck corresponding to the shovel, initially assisted by global positioning system (GPS), ultra-wideband (UWB) ranging receivers and 3D scanning LIDAR [18, 19].

Shovel load assist program (SLAP) technology has numerous benefits such as shovel safety, accessibility, efficiency and lower maintenance as well as faster shovel cycle times, lower machine duty, enhanced material delivery in trucks, fewer influences between truck and shovel and lower operator workload. SLAP is developing equipment that will assist operators of electric mining shovels to load trucks with higher productivity and safety.

An appropriate organisation for supporting the operator of an electric mining excavator to avoid collisions with identified obstacles within the workspace of the excavator is essential. This can be achieved by applying a receding horizon avoidance filter. In this technology, the command provided by the operator is adjusted for collision avoidance. The receding horizon avoidance filter computes the filtered command using a receding horizon control framework. A collision-free trajectory, which is the lowest variation from the operator's proposed trajectory, is considered, and the first command from the trajectory converts the filtered command [20].

7.3.2.2 Mine Plan and Design

The mine plan and design have an impact on the operating conditions, relations between equipment components, and well-organised practice of equipment. This combination either improves or deteriorates energy efficiency. For example, research results show that higher production is achievable by double, rather than single, benching. Similarly, double-side loading is proved to be slightly more complicated than single-side loading but more productive [16]. Although double-sided loading requires additional care to keep safety criteria, it is possible to excavate more material per shift and to increase truck efficiency. For instance, spotting times in trucks loaded with double-sided loading are frequently around 35 s compared to 65 s with single-sided loading. A 30-s decrease in truck cycle time has a non-negligible impact on production rate [21].

Truck and shovel pass matching has a critical influence on loading efficiency. Shovels and buckets should be sized so as to fill a truck tray within 3–4 passes. Each pass beyond this waste cycle time and energy. Cycle times for backhoe configurations can be marginally faster in comparison to front–shovel configurations [16].

7.3.2.3 Operator Skill and Practice

The operator influences vital factors and consequently defines the production rate and energy consumption, for example, bucket fill factor and cycle time. Significant energy inefficiency in loading operations is as a result of operator practices. The best operators use 40% less energy per tonne of production in comparison with the other operators. Probably, extra savings could be achieved since there is nothing that assures that the best operator operates at the optimum energy efficiency [12]. Operator practices have an excessive influence on shovel performance; consequently, operator practices should be optimal. The result of the best operator practice is a greater proportion of cycles in the lower digging energy classifications while keeping an appropriate loading ratio. Digging energy is a function of both muck pile digging states and primarily digging practice. It has been observed that the operator with the lower hoist speed and higher crowd speed accomplishes dipper cuts in the bank and consumes greater energy during digging.

Figure 7.19 illustrates the optimization problem with the objective function being energy per unit loading rate. The objective function reduces when the hoist

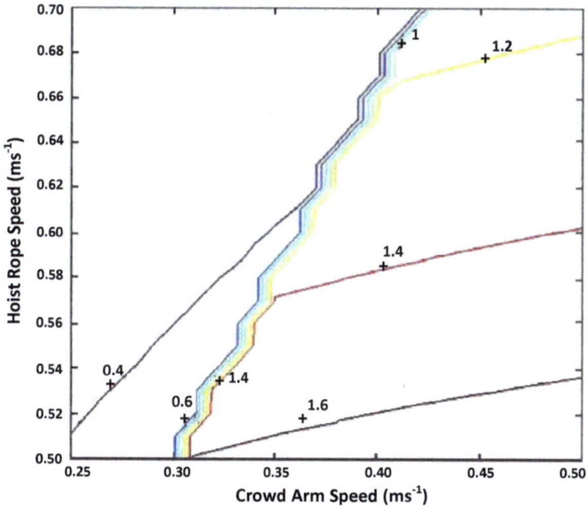

Fig. 7.19 Energy per unit loading rate (kJ s per kg) [23]

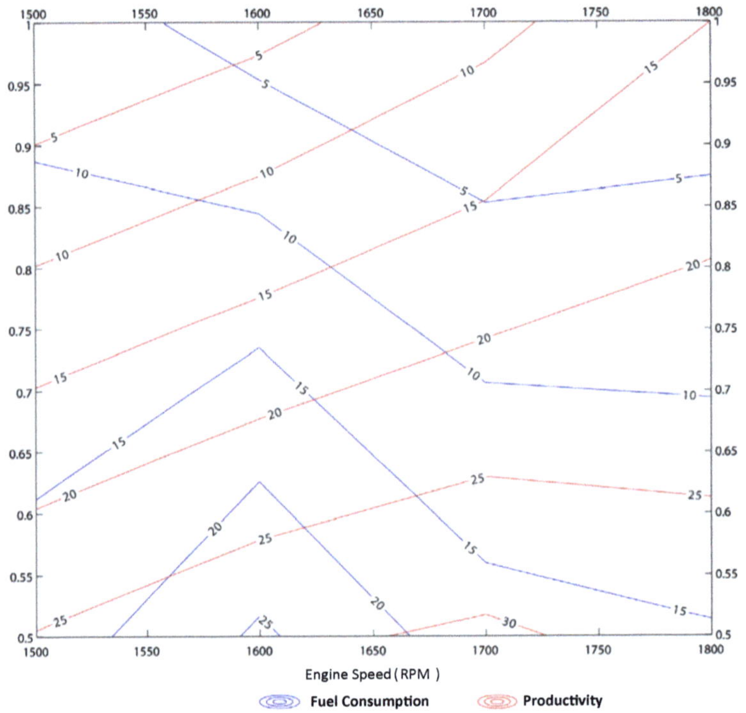

Fig. 7.20 The productivity and fuel consumption [25]

speed increases and crowd speed decreases. The primary inference obtained from Fig. 7.19 is that the best operator practice is accomplished when the dipper is moved at low crowd speeds and high hoist speeds [22].

For draglines, research has suggested that the most effective factors that cause variation in the energy efficiency of operators include dump height and engagement/disengagement position of the bucket. Cycle time, payload and swing in time are the less efficient parameters which lead to variation in operator energy efficiency [23].

It has been observed that engine speed and bucket cut depth (BCD) arrangements have an impact on fuel efficiency and productivity of a hydraulic excavator.

By applying the map showing the influence of RPMs and BCDs on productivity and fuel consumption, manufacturers can provide operators support with an automated system built into the excavator. Thus, the system is enabled to adjust the engine speed and the bucket dig depth during excavating, automatically.

Figure 7.20 demonstrates the result of engine speed and bucket cut depths (BCD) on cycle time, fuel consumption and output. To make the map simpler, cycle time and output were joint into one unit: productivity (m^3/h). The fuel consumption design is laid over productivity to clarify the unexplored correlation between these variables. The *x*-axis and *y*-axis depict the engine speeds, BCDs (in percentage), respectively, and *z*-axis shows two variables, fuel consumption (litres) and productivity (m^3/h). The dependent variables are presented as a percentage gain in 5% intervals. The behavioural configuration of the dependent variables is described with two various coloured lines on the map. Fuel consumption (litres) is characterised in blue and productivity (m^3/h) in red. This map can be implemented as a supportive database in automation process of an excavator. The engine speed and the bucket dig depth during excavating can be arranged. To accomplish the best likely productivity and fuel consumption rate, engine speed and BCD should be adjusted to 1660 RPM and 50%, respectively. However, the operator will have to consciously select between low fuel consumption and high productivity excavation strategies [24].

Payload controlling, or filling the dipper to the proper weight, also affects the digging cycle time. If the shovel operator could know the weight of the material in the bucket throughout the digging procedure, then he/she could stop loading after the desired point, and the break-out and load phase of the cycle could begin. Some digging plans that try to minimise cycle time have a tendency to fill the bucket to only 80% of capacity which results in more passes that would be essential if the bucket was always filled to 100% [16].

7.3.2.4 Operating Condition

Excavator performance is significantly affected by material digability, which depends on blasting. Material digability can be measured by excavator dig time, which is defined as the period from when the bucket engages the muck pile to when it begins to swing the boom across to the truck. Material digability can be influenced by the equipment operator, material characteristics (rill properties, looseness, fragmentation, etc.) and excavator type. It has been reported that a 7–58% improvement in excavator dig time is possible from improvements to the muck pile characteristic and operator skill [26–28].

7.4 Conveyor Transport

Conveyors are a very efficient, low-cost means of transporting sized materials in high volumes. The most common type of conveyor used in mining is the trough conveyor, which consists of a head pulley, drive motor and torque coupling, a take-up mechanism, a belt constructed as a rubber ply structure or using embedded steel cords, a set of troughs with carry and return idlers, and a tail pulley plus transfer chutes to channel material onto and off the belt. Belts can be up to 2 m in width, a few kilometres long and run up to 7 m/s in speed. This enables transport capacities of up to 20,000 tonnes per h, typical of the large German lignite mines.

Fig. 7.21 Semi-mobile IPCC system in an Australian mine

There are many different types of conveyor systems used in mining applications. Conveyors typically handle materials sized below 300 mm, so it is necessary to employ a crusher at the feed end of the conveyor. At the discharge end of the conveyor, there is a need to employ either a stacker or spreader. In between, the feed and discharge stations, a variety of bench, ramp and overland conveyors are employed (Fig. 7.21).

Mining systems are referred to as either in-pit or ex-pit systems, dependent on the crusher location. If the crusher station is located within the pit, there are some alternatives: fixed in-pit crushing and conveying systems (FIPCC), semi-mobile in-pit crushing and conveying systems (SMIPCC) or fully mobile in-pit crushing and conveying (FMIPCC) systems. Fixed crusher stations are designed to be a permanent fixture in the life of mine. Semi-mobile crusher plants are designed in modular architecture so that they can be periodically relocated (at intervals of say, every 3–5 years) as a mine deepens. Truck–shovel systems work in collaboration to feed both fixed and semi-mobile IPCC systems. In fully mobile IPCC systems, the shovel directly feeds a mobile system comprised of a hopper, apron feeder and low-profile crusher, commonly a hybrid roll crusher or sizer. The mobile crusher then feeds a bench conveyor via a stinger conveyor connected to a hopper car, mounted on rails or tracks above a relocatable bench conveyor.

The material is conveyed ex-pit via a ramp conveyor. Dependent on material characteristics, including size distribution, bulk density and moisture content, a ramp conveyor can manage inclinations of between 15° and 18°. This is substantially more than a truck haul road, which averages only 10% gradient. Other than a dedicated conveyor ramp, other pit exit strategies include construction of a dedicated conveyor decline, shared conveyor access via widening an existing truck haulage route, the use of a slot conveyor in the cusp of two of the pit walls or the use of a high angle conveyor. The latter type of conveyor consists of novel pipe or sandwich design and are capable of transporting materials up the slope of around 35°. To date, however, they have been restricted to lower capacity systems not exceeding 3000 tonnes per h.

Once outside of the pit, the material will usually be transferred to an overland (fixed) conveyor. If it is a valuable mineral or energy material, it will proceed to a stacker discharge station that feeds the mill or handling and preparation plant. If it is waste material, it will proceed to the waste dump, where it will be transferred to a bench conveyor linked to a tripper car feeding a spreader. The tripper car provides a variable off-loading point along the length of the bench conveyor, thus enabling the spreader to travel along a waste pad systematically filling using both up-cast and down-cast spreadings.

7.4.1 Power Efficiency

A mining truck with a 327-tonne payload capacity has an empty vehicle mass of around 246 tonnes, dependent on the tray wear packages installed. As part of its duty cycle, the truck needs to expend energy to vertically lift its empty vehicle mass plus payload out of a mine and then to return the empty vehicle weight back into the mine in preparation for the next loading cycle. This means that the total mass moved in one round trip is approximately $2 \times 246 + 327 = 819$ tonnes. Thus the ratio of moved material to total weight moved is 1:2.7, or only 38% efficient.

A troughing conveyor that has a capacity of 10,800 tonnes/h and a belt speed of 5 m per s must deliver 3 tonnes of material per second. These 3 tonnes are distributed over a belt length of 5 m. The mass of 5 m of the belt, plus a 5 m section of the return belt, is around 600 kg. Thus, the ratio of moved material to belt mass for the conveyor is just 1:1.2 or 81%. To this, we can add the fact that the conveyor belt is driven by electric motors, which are around 95% energy efficient.

Belt conveyor power is measured as a function of belt velocity multiplied by equal force. The latter is made up of the sum of main resistances, secondary resistances, slope resistance, special main resistances and special secondary resistances. Of these, the first three are the most important. Main resistances refer to the indentation and rolling resistance of the belt; the flexure resistance of the belt and the rotational resistance of the idlers and bearings. Secondary resistances refer to inertial and frictional resistances due to accelerating material at the loading point and resistance due to friction at the side walls of chutes, pulley bearing resistance and resistance due to the wrap of belts around the pulleys. Slope resistance refers to the potential energy required to lift the load up an inclined slope.

7.4.2 Conveyor Belt Applications in Mining

Apparently, then, conveyor belts offer a more energy-efficient means of transporting bulk materials. Belt conveyor operation also enhances mine safety, as around one-third of all fatalities in Australian surface mines are related to vehicle collisions. Truck operations require a significant logistics chain to supply fuel, tyres and spare parts to a mine. Fewer supply trucks are necessary to support conveyor belt operations, which can be beneficial for mines operating in areas with sensitive community concerns.

With all of the advantages offered by conveying systems, why do we not see them in use more frequently in mining operations? The answer to this question has to do with

- Material type and sizing;
- Upfront capital expenditure;
- Flexibility (mine design limitations, ability to relocate and scale up or down); and
- System reliability.

First, for material to be transported by conveyor belt, it must be sized to below 300 mm. While this is financially viable for ore which needs to be crushed anyway before processing, it is regarded as an unnecessary cost for waste materials.

Second, IPCC systems require significant upfront capital investment. Total system cost can amount to around US$80 million for systems with capacities of up to 10,000 tonnes per h. A crusher station costs around US$15 million, and a typical 10,000 tonnes per h system might employ two such crusher stations at the cost of some US$30 million. A spreader will cost around US$16 million. The remainder of the system costs is then divided between bench, ramp, overland and dump conveyors plus the ancillary hopper and tripper cars. This upfront capital investment has a significant influence on project Net Present Value (NPV), despite IPCC systems offering significantly reduced operational expenditure (in the order of between 15 and 25% less total mining cost, dependent on SMIPCC or FMIPCC system type). Truck systems have the advantage of being able to be scaled up as mining pits grow.

Thus, a mine can begin operations with a relatively small truck–shovel fleet and purchase additional units to add capacity as the mine progresses. Over the life of mine, due to shovel and truck fleet replacements, total capital expenditure is roughly similar to that of the IPCC system. However, for truck–shovel systems expenditure is distributed over the mine life which can lead to an NPV advantage. Financial analysis shows that the NPV of IPCC systems wins out in the case of long life deposits.

Third, IPCC systems require significant changes to be made to a mine plan in comparison to a plan employing truck–shovel systems. For a start, a viable conveyor exit strategy is required. If a ramp conveyor is used, then the wall on which this conveyor is located must be relatively immobile over the life of mine; this is not always the case. Frequent movements of the bench (and dump) conveyors consume large quantities of productive time, and so the optimal mine design for FMIPCC systems will employ long linear benches. This does not suit all deposit types. The capacity of installed IPCC systems is fixed and cannot easily be scaled up or down to suit prevailing economic conditions (unlike truck–shovel operations, for example, where a time of low commodity prices equipment can simply be parked up to save money).

Finally, IPCC systems are complex series connected systems. Downtime any of the system components will cause a system outage and stop material movement. This includes both availability losses due to maintenance and utilisation losses due to the bench or dump conveyor relocations. It is estimated that FMIPCC-effective utilisation is around 64%, equivalent to 5600 productive hours per year [29]. Of this, about one-third of the system losses are due to maintenance or repairs. The remaining losses are utilisation losses.

7.5 Conclusions/Summary

As a result of these limitations, truck–shovel systems continue to be a predominant system of choice for most surface mines. However, around the world, most notably at the Technical University of Freiberg, Germany and at The University of Queensland, Australia, research efforts are underway to address the limitations identified above. In a world becoming more fossil fuel constrained, solving such issues is imperative to continue the level of supply of minerals and energy commodities that the world currently enjoys.

References

1. BREE, Australian energy update (2014) Australian Government, Bureau of Resources and Energy Economics, Canberra, Australia, pp 9–11
2. DOE, Mining industry energy bandwidth study (2012) Department of Energy, USA Government, Washington DC, USA, pp 26–33
3. EEO, Energy-Mass Balance: Mining (2010) Australian Government, Department of Resources Energy and Tourism, Canberra, Australia, pp 21–28
4. Soofastaei A (2016) Development of an advanced data analytics model to improve the energy efficiency of haul trucks in surface mines. The University of Queensland, School of Mechanical and Mining Engineering, Australia
5. Caterpillar (2013) Caterpillar performance handbook, 10 ed., vol 2. US Caterpillar Company, New York City
6. Soofastaei A, Aminossadati SM, Arefi MM, Kizil MS (2016) Development of a multi-layer perceptron artificial neural network model to determine haul trucks energy consumption. Int J Mining Sci Technol 26(2):285–293
7. Soofastaei A, Aminossadati SM, Kizil MS, Knights P (2014) Payload variance plays a critical role in the fuel consumption of mining haul trucks. Aust Resour Invest 8(4):64–64

8. Soofastaei A, Aminossadati SM, Kizil MS, Knights P (2015) Simulation of payload variance effects on truck bunching to minimise energy consumption and greenhouse gas emissions. In: 2015 Coal operators' conference, The University of Wollongong, Wollongong, NSW, Australia, pp 338–347
9. Soofastaei A, Aminossadati SM, Kizil MS, Knights P (2016) A discrete-event model to simulate the effect of payload variance on truck bunching, cycle time and hauled mine materials. Int J Mining Technol 2(3):167–181
10. Soofastaei A, Aminossadati SM, Kizil MS, Knights P (2016) The influence of rolling resistance on haul truck fuel consumption in surface mines. Tribol Int J 2(1):215–228
11. Kecojevic V, Komljenovic D (2011) Haul truck fuel consumption and CO_2 emission under various engine load conditions. SME annual meeting and exhibit, CMA 113th National Western mining conference, SME, USA, pp 186–195
12. MinePro P (2003) Peak performance practices: excavator selection 2(1):185–198
13. Stavropoulou M, Xiroudakis G, Exadaktylos G (2013) Analytical model for estimation of digging forces and specific energy of cable shovel. 2(1):23–51
14. Awuah-Offei Kwame (2016) Energy efficiency in mining: a review with emphasis on the role of operators in loading and hauling operations. J Clean Prod 117(1):89–97
15. Valenzuela GM, Valenzuela, MA (2014) Integrated mechanical-electrical modeling of an AC electric mining shovel and evaluation of power requirements during a truck loading cycle. IEEE industry application society annual meeting, Australia IEEE
16. Fiscor Steve (2007) Productivity considerations for shovels and excavators. Eng Min J 208 (7):38–49
17. Wei B, Gao F (2012) Digging trajectory optimization for a new excavating mechanism of electric mining shovel. In: International design engineering technical conferences and computers and information in engineering conference, American Society of Mechanical Engineers
18. Reid A, Smith Z, McAree P, Cloete S, Horberry T (2013) Shovel load assist. 2(1):145–158
19. Cloete Steven, Horberry Tim (2014) Collision avoidance and semi-automation in electric rope shovel operation. Ergon Australia 4(2):423–439
20. Kearney MP, Smith ZV, McAree PR (2015) Receding horizon collision avoidance for large excavators. In: 14th IFToMM World Congress, Taipei International Convention Center, Taiwan
21. Benchmarking the productivity of Australia's black coal industry, Tasman Asia Pacific Australia, pp 85–101 (1998)
22. Awuah-Offei Kwame, Frimpong Samuel (2011) Efficient cable shovel excavation in surface mines. Geotech Geol Eng 29(1):19–26
23. Oskouei Maryam Abdi, Awuah-Offei Kwame (2016) A method for data-driven evaluation of operator impact on energy efficiency of digging machines. Energ Effi 9(1):129–140
24. Ng Felix, A Harding Jennifer, Glass Jacqueline (2016) An eco-approach to optimise efficiency and productivity of a hydraulic excavator. J Clean Prod 112(2):3966–3976
25. Brunton I, Thornton D, Hodson R, Sprott D (2003) Impact of blast fragmentation on hydraulic excavator dig time. In: Proceedings of the 5th large open pit conference, AusImm (The Minerals Institute), Australia
26. Hawkes PJ, Spathis AT, Sengstock GW (1995) Monitoring equipment productivity improvements in coal mines. In: EXPLO 95, AusIMM, Brisbane, Australia
27. Grant JR, Little TN, Bettess D (1995) Blast driven mine optimisation. In: Explosives in mining conference, Australia
28. McGill M, Freadrich J (1994) The effect of fragmentation on loader productivity. International Society of Explosives Engineers, Cleveland, United States of America, pp 112–121
29. Dean M, Knights P, Kizil MS, Nehring M (2015) Selection and planning of fully mobile In-pit crusher and conveyor systems for deep open pit metalliferous applications. In: Future Mining, AusIMM, The University of New South Wales, Australia

Chapter 8
Energy Efficiency in Cable Shovel Operations

Kwame Awuah-Offei

Abstract This chapter seeks to establish the current knowledge on energy efficiency of cable shovel operations. Additionally, the chapter uses a review of the literature to make recommendations for industrial best practices and for future research to address identified gaps in the literature. The chapter first presents the fundamentals of cable shovel operations and the factors that affect the energy efficiency of shovel operations. Subsequently, the chapter presents an overview of the latest research on cable shovel energy efficiency, which is used as the basis for the recommendations. The chapter recommends that industry practitioners should use the right drive systems for their cable shovels, use data analytics to understand shovel energy efficiency, and carefully evaluate the costs and benefits of energy efficiency initiatives. The chapter also recommends that future research on shovel energy efficiency should: (i) establish theoretical benchmarks for cable shovel operations; (ii) account for human factors in the design of operator guidance systems to assist operators during shovel operations; and (iii) evaluate how effective operator training programs are in improving shovel energy efficiency.

Keywords Cable shovel · Energy efficiency · Data analytics · Human factors

8.1 Introduction

Mine managers and engineers use cable shovels (also referred to as electric rope shovels) in large-scale surface mining operations to excavate and load material into trucks (or at-face-crushers). Engineers regard cable shovels as durable, possessing large breakout forces, and associated with low production costs, high productivity and low ownership costs resulting from long economic operating lives. Due to these advantages, cable shovels are very popular in larger scale surface mining in diverse

---

K. Awuah-Offei (✉)
Mining & Nuclear Engineering, Missouri University of Science & Technology,
326 McNutt Hall, Rolla, MO 65409, USA
e-mail: kwamea@mst.edu

K. Awuah-Offei (ed.), *Energy Efficiency in the Minerals Industry*, Green Energy and Technology, https://doi.org/10.1007/978-3-319-54199-0_8

applications including oil sands, coal, iron ore, and metal mining. However, because of the high initial capital costs and lack of selectivity, mining professionals do not deploy them in smaller mines, those with short mine lives or those that require high mining selectivity.

Cable shovels consume significant amounts of energy in a mine. Most mines operate multiple units and each shovel requires thousands of kilojoules of electricity to execute its motions. Consequently, energy consumption of mines that use cable shovels for material handling, is significantly affected by the cable shovel electricity consumption [1]. Hence, the energy efficiency of such mines can be significantly affected by the energy efficiency of cable shovel operations. As with other equipment units, energy efficiency is inexplicably linked to the productivity of the machine and the associated production costs. Loading costs, as a percentage of the total operating costs of surface mine operating costs, range from 3 to 35%, with a mean of 15% [2]. Hence, an added benefit of increasing the energy efficiency of cable shovels is the reduction in unit production costs that results from lower energy costs per unit of production.

The threefold objective of this chapter is to: (1) establish, using a literature review, the current knowledge on energy efficiency of cable shovel operations; (2) make recommendations for industry best practices based on research in the literature; and (3) make recommendations for future research to fill the gaps in our knowledge regarding energy efficiency of cable shovel operations. The literature review is based on searches in archival databases with emphasis on peer-reviewed journal publications and work published since 2005.

8.2 Energy Efficiency of Cable Shovel Operations

Energy efficiency is the ratio of useful work done (energy output) to energy input. Since cable shovels are powered by electricity from the grid, the energy input is often directly estimated from motor current and voltage signals captured by onboard equipment monitoring systems [3, 4]. In some instances, however, proxies are used to describe the energy input [5]. Since the useful work done during cable shovel loading (i.e., work done in excavating and transporting the load into a truck or hopper) is difficult to estimate in the field [6], the amount of material loaded (payload) is often used as a proxy for useful work done.

The most common measure of energy efficiency that researchers use for cable shovel operations is energy per unit payload (or specific energy), which is actually the inverse of an energy efficiency metric [7]. This practice has its basis in initial research on shovel performance which was motivated by a desire to classify geologic materials or evaluate fragmentation results [8–12]. Hence, researchers intended specific energy to be a measure of how difficult it is to excavate the material or muck pile rather than a measure of the efficiency of the shovel in performing its function. Nevertheless, it is now a common practice for researchers and engineers to use specific energy or payload per unit energy consumed (which is a more theoretically accurate definition of energy efficiency) as a measure of the energy efficiency of a shovel or loader. However, this practice assumes that the rate at which a shovel loads material is of no consequence when describing efficient loading.

On the contrary, mine managers and engineers are very concerned with the loading rate of shovels since it drives the production rate of the entire material handling system. Thus, this author believes the energy per unit loading rate is a better metric for energy efficiency of cable shovel operations [13]. For example, Babaei Khorzoughi and Hall [5] showed that for similar loading rates, a shovel can consume widely varying energy during digging. Hence, to discuss the theoretical basis of energy efficiency of cable shovel operations, let us define energy efficiency of loading as the ratio of loading rate to energy input.

8.2.1 Factors Affecting Cable Shovel Energy Efficiency

The literature includes many efforts to model shovel operations or energy efficiency by accounting for some or all of the factors or variables that affect shovel efficiency [13–15]. A cable shovel operator moves the dipper through the muck by moving the crowd arm and hoist ropes. Once the dipper is out of the muck pile, the operator also uses the swing action to swing the shovel about its axis while still using the crowd arm and hoist ropes to position the dipper over the truck or hopper. Once he or she dumps the load into the truck or hopper, the operator swings the shovel back into position (the crowd arm and hoist ropes are also used to reposition the dipper) for the next cycle.

Perhaps, because of the importance of the digging phase to the overall efficiency of shovel operations (the digging phase is also the most energy-intensive [15]), the literature, almost exclusively, contains models of the kinematics and dynamics of this phase [15]. Equation 8.1 shows that the digging energy of a shovel depends on the crowd (C) and hoist (H) forces, crowd ($\dot{c}$) and hoist ($\dot{h}$) speeds, and the digging time (t_{d}) [13, 16, 17]. The crowd and hoist forces are used to move the crowd arm-dipper assembly through the digging trajectory (usually defined as the trajectory of the dipper tip) and overcome the resistance, offered by the muck, to digging [13].

$$E = \underbrace{\int_0^{t_{\mathrm{d}}} (H)(\dot{h})\mathrm{d}t}_{\text{Hoist energy}} + \underbrace{\int_0^{t_{\mathrm{d}}} (C)(\dot{c})\mathrm{d}t}_{\text{Crowd energy}} \tag{8.1}$$

Based on our working definition, shovel energy efficiency depends on the loading rate and the energy consumed during loading. Therefore, any factor that affects these two parameters will have an impact on the energy efficiency. Research has identified shovel characteristics; operating conditions, mine plan and design, and operator skill and practice as the main factors that affect shovel energy efficiency [13, 16].

The shovel's design and specifications are important in determining whether it efficiently converts energy input into useful work or not. For energy efficient operation, a shovel's drives should be designed such that they efficiently transmit and convert energy into useful work. Perhaps, the most significant innovation in shovel drive systems recently is the introduction of alternate current (AC) drive systems as an alternative to the direct current (DC) drive systems. The AC drives facilitate higher loading rates (due to higher speeds, torque, and power) and reliability [18]. The higher loading rate increases the energy efficiency of shovel operations. However, whether the drives are AC or DC, regenerative drives have a significant impact on the energy consumption, and thus energy efficiency, of a shovel with the potential to regenerate about a quarter of the energy [15].

Both natural and design-imposed operating conditions affect the operational efficiency and, thus, energy efficiency of cable shovels. These include the resistance to digging from the muck pile (whether in situ material or after ground fragmentation), bench profiles, and truck matching (both quantity and sizes) [13, 19–21]. For example, poor ground fragmentation that leads to difficult digging conditions will significantly increase the required crowd and hoist forces (Eq. 8.1), which in turn increase the energy required to dig [22]. This will decrease the energy efficiency of the shovel operation. Similarly, poor truck matching that leads to the shovel waiting for a truck to arrive will increase shovel energy consumed by the shovel while doing no useful work. This will also decrease energy efficiency.

The effect of the operator (skill and practices) is another important factor affecting the energy efficiency of shovel operations. The operator, within the limits of the machine, determines the cycle time and payload, both of which affect the energy efficiency. The cycle time depends on the trajectory, taken by the operator, and rate of travel, which depends on crowd and hoist speeds. Both the trajectory and the rate of travel affect the energy consumed during digging. The trajectory affects the work done to move the crowd arm-dipper assembly and payload through the digging cycle [13, 17]. Trajectories with greater depth of cut have been shown to require more energy than those with lower depths of cut [21]. The rate of travel is directly proportional to the power draw of the crowd, hoist, and swing motors (Eq. 8.1, for hoist and crowd energy). However, higher rate of travel reduces the cycle time which increases the rate of loading and can decrease the overall energy consumed per cycle [16]. Payload, on the other hand, depends on how effective the operator is at filling the bucket (fill factor). The fill factor depends on the depth of cut (trajectory) and material characteristics [23].

These relationships can be illustrated using simulation results from the cable shovel simulator built and presented in other publications by this author [13]. Figure 8.1 is based on simulations of a P&H 4100TS shovel using this simulator. As seen from Fig. 8.1a, increasing hoist speed decreases the required energy. Although this may seem contrary to Eq. 8.1, the reason for this is the decrease in digging time that results from increasing the hoist speed (Fig. 8.1c). This relationship between hoist speed and faster digging has been observed by other

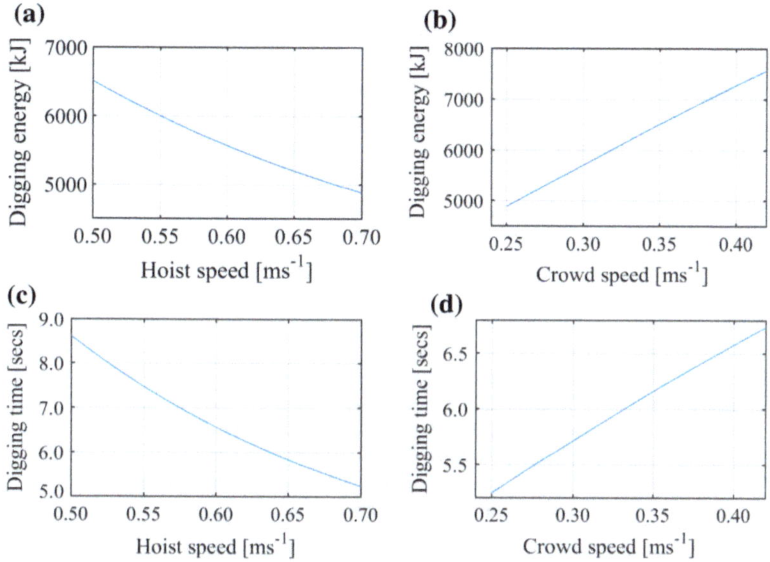

Fig. 8.1 Effect of hoist and crowd speeds on shovel energy consumption: **a** digging energy versus hoist speed at crowd speed of 0.25 m s^{-1}; **b** digging energy versus crowd speed at hoist speed of 0.7 m s^{-1}; **c** digging time versus hoist speed at crowd speed of 0.25 m s^{-1}; **d** Digging time versus crowd speed at hoist speed of 0.7 m s^{-1}

researchers as well [4, 16]. On the contrary, increasing crowd speed increases the digging time (deeper depth of cut trajectories result) and, thus, leads to higher energy consumption (Fig. 8.1b, d). In fact, this author's previous work showed that these increases in energy consumption are substantially worse when the crowd arm is extended so much that it reaches the physical limit [13]. It is important to note that in all these simulations, the resulting payload was the same. Consequently, the relationship between crowd and hoist speed and energy efficiency will be the same as that observed from these results.

8.2.2 *Ongoing Research Initiatives to Improve Cable Shovel Efficiency*

A review of the literature since 2012 reveals four main areas that are relevant to cable shovel energy efficiency: (i) improvements in drive systems for better energy efficiency; (ii) modeling of cable shovel dynamics to better predict shovel trajectories and to control the crowd arm-dipper assembly; (iii) characterization of dipper filling to improve fill factors and payloads; and (iv) better understanding of the role of operators and how to assist operators to do better (Table 8.1).

Table 8.1 Highlights of ongoing cable shovel energy efficiency-related research initiatives

Area	Sample initiatives	References
Drive systems	Investigating the effect of trailing cable length on substation voltage quality Developing fault tolerance strategies for DC micro-grids including cable shovels	[24, 25]
Kinematics and dynamics modeling	Developing more accurate kinematics and dynamics models of the cable shovel front end assembly	[26, 27]
Characterizing of dipper filling	Understanding the effect of digging dynamics on bucket filling Developing superior models for near real-time payload estimation	[23, 28, 29]
Operator and operator assistance	Further exploration of the effect of operators on shovel efficiency and productivity Developing intelligent systems to guide operators during operation Using human factors to design better interfaces for operator guidance systems	[5, 30–33]

There is ongoing research that tries to optimize and improve how energy is delivered to the shovel mechanisms to do excavation and loading of material into trucks and hoppers. This work is crucial to ensure that the drive mechanisms are efficient and energy losses are minimal. Some of the recent work continues to explore even better ways to improve these mechanisms. For example, Abdel-Baqi et al. [24] explored the effect of the long trailing cables, which are highly capacitive, on the voltage quality in mines with cable shovels. Such voltage quality issues can lead to failures that can be expensive from a capital and operational standpoint. Their work shows how engineers can make recommendations for optimal cable length to avoid voltage amplification that can lead to unnecessary downtimes.

Other researchers continue to refine and improve the kinematics and dynamics models of the dipper-crowd arm assembly [26, 27]. Cable shovel kinematics and dynamics models have been improving in the last two decades. Researchers have increasingly used more sophisticated methods to model the motions and forces while also addressing limitations of earlier models. For instance, Awuah-Offei and Frimpong [13] improved upon the model in Hendricks et al. [34] by accounting for the width of the boom point sheave. Other researchers have used more comprehensive methods such as the Newton–Euler approach to provide more detailed models of the dipper-crowd arm assembly [14]. Recent work continues to refine such models for more accurate description of the shovel digging motions [27]. Others have implemented kinematics and dynamics models as a means to evaluate alternative designs to effect more complicated trajectories and motions by the dipper-crowd arm assembly [26].

Other research has sought to understand how the shovel dipper interacts with material (in situ or fragmented) to fill the bucket. The goal of such initiatives is to understand better dipper fill factors so as to maximize payloads [23]. Related research attempts to estimate, in near real time, the shovel payload based on shovel motor current and voltage signals [28].

However, an area of research that has perhaps received the most attention is the development of intelligent algorithms to help operators operate the shovel more efficiently. At the basic level, modern shovels display, in the operator's cabin, various signals that provide intelligence that the operator can act upon. There are various systems (either provided by original equipment manufacturers (OEMs) or after-market installations) are commercially available that provide metrics such as payload, various operator "errors," and machine health information. This area continues to receive research attention as researchers try to better understand the effect of operators on production and efficiency [5, 20, 31]. In addition, other researchers have been exploring the human factors in the design of the interfaces to better make these operator guidance systems useful for improving efficiency [30]. These efforts facilitate the bridging step between unaided operation by a human operator and autonomous operation and constitute a necessary step in the evolution of cable shovel technology for energy and operational efficiency.

8.3 Recommendations

Based on the fundamental understanding of cable shovel energy efficiency and a review of the current literature, the author makes the following recommendations for industrial practice and future research.

8.3.1 Industrial Best Practices

One could make any number of recommendations for industrial best practices that a mine manager or engineer can use to improve the energy efficiency of shovel operations. However, it is the opinion of this author that the best thing mine managers and engineers can do is to use the right technology to deliver energy, use rigorous data analytics to understand the efficiency of shovel operations, and carefully evaluate the costs and benefits of energy efficiency initiatives.

To maximize the energy efficiency of shovel operations, mines should ensure energy is efficiently delivered to the hoist, crowd, and swing mechanisms. It is important to properly assess the mine's electrical system and loads to ensure adding (if new) or the continued use of the shovel(s) will not adversely affect the quality of the voltage delivered to the shovel motors. The electrical substations should be designed to handle the shovel loads in a manner that deliver quality voltage and enough power to the shovels, on demand. Beyond that, the shovel's drive system should be efficient in delivering the breakout forces required to dig the material and hoist and swing the payload to dump into the trucks or hoppers. For existing (or used) shovels, there may be some opportunities to upgrade key systems to improve the energy efficiency of the shovel. In this regard, good preventive maintenance helps to ensure the shovel works efficiently all the time. It is important to ensure that the whole drive mechanism is efficient, if the goal is to make shovel operations energy efficient.

In addition to ensuring the machine's drive mechanism is efficient in delivering energy to the digging tool, mines should take advantage of current equipment monitoring systems that generate lots of data and big data analytics tools to understand the effect of operators and operating conditions on the energy efficiency of shovel operations. With such data and analysis, mines can objectively understand the drivers of energy efficiency and target these to improve energy efficiency. Many commercial monitoring tools provide stock analytical reports and graphs to facilitate this analysis. Some also provide the ability to generate custom reports and graphs that are helpful for achieving the goals of various energy efficiency initiatives. There is work in the literature that shows new analytics tools that can guide mine engineers in developing such custom tools [5, 6, 20].

Mines could also use such analytical tools to guide operator training efforts aimed at improving the efficiency of operators. This provides objective feedback to operators, which can be used to spur improvements as the data can clearly correlate certain behaviors to improved efficiency. In this regard, simulator training can be very useful because the training tool itself (the simulator) can be configured to generate the same data as the data used in the analytics so that the trainer can track the correlation between training outcomes and actual performance [35].

As with other aspects of the mining operation, it is important that the costs and benefits of energy efficiency initiatives to improve shovel energy efficiency are carefully evaluated before mines implement the action plan. For example, though we know that operator guidance systems can improve energy efficiency of shovel operations, the costs and benefits differ for different operations. It is thus important for mines to evaluate these costs and benefits to ascertain how such technology would actually impact operator energy efficiency, the benefits of such improvements, and the costs to the operation. It is only then that the mine's management can make informed decisions on whether to implement such an initiative or not.

8.3.2 Recommendations for Future Research

A review of the literature identifies some gaps in the literature that require more work. First, there is a need for research that will establish theoretical benchmarks for cable shovel operations. In the past, this author has proposed kinematics and dynamics models to predict the energy consumption of shovels during digging [13]. However, this work modeled only the energy consumed during digging and did not include the swinging energy. Also, this work did not include models of the drive systems nor did it account for any losses. However, it is possible to account for the swing and other aspects of the shovel loading cycle beyond the digging phase. Future research work should address this gap in the literature so mines have a theoretical estimate, given specific operating conditions, of the target energy consumption for shovel operations. This will be very useful in guiding shovel energy efficiency initiatives.

Second, further research is necessary to account for human factors in the design of operator guidance systems that assist operators during shovel operations. Often, these in-cabin displays provide too many pieces of information that can lead to information overload. There is already some research in this area [30]. However, more questions remain outstanding. These include: (i) What are the most significant cues that affect operator behavior? (ii) How should these cues be presented to operators to be most persuasive in affecting operator behavior? (iii) How much improvement in energy and production efficiency can be gained by operator guidance systems? These questions can be addressed by further research.

Finally, further research is also required to evaluate the effectiveness of operator training programs in improving energy efficiency of shovel operations. There is only limited work in the literature that attempts to evaluate the effectiveness of operator training in improving energy efficiency and how long any gains in improvement last [35]. Though Dorey and Knights [35] addressed some of these questions for draglines, the number (four operators participated in the training and three operators were used as a control group) of participants is too few to draw any broad inferences about the particular training program and such studies are required for shovel operators too. However, the study shows what needs to be done to evaluate the many operator training programs in order to establish whether and how operator training affects energy efficiency.

8.4 Summary

The objectives of this chapter were to: (1) establish the current knowledge on energy efficiency of cable shovel operations; (2) make recommendations for industry best practices; and (3) make recommendations for future research. The chapter presented the fundamentals of cable shovel operations and the factors that affect the energy efficiency of shovel operations. It also presented an overview of the latest research on cable shovel energy efficiency. Based on those discussions, the work recommends that mine managers and engineers should use the right technology to deliver energy to the cable shovel digging tool, use rigorous data analytics to understand the efficiency of shovel operations, and carefully evaluate the costs and benefits of shovel energy efficiency initiatives. The work also recommends that future research on shovel energy efficiency should establish theoretical benchmarks for cable shovel operations, account for human factors in the design of operator guidance systems to assist operators during shovel operations, and evaluate the effectiveness of operator training programs in improving shovel energy efficiency.

Acknowledgements The author is grateful to Dr. Nuray Demirel who reviewed the original manuscript and made valuable suggestions that improved this manuscript tremendously.

References

1. Houley L, Alahakoon S (2012) Impact assessment of AC and DC electric rope shovels on coal mine power distribution system. In: University power engineering conference (AUPEC), 2012 22nd Australas, pp 1–6
2. Hustrulid W, Kuchta M, Martin R (2013) Open pit mine planning and design, 3rd edn. CRC Press, New York
3. Patnayak S, Tannant DD (2005) Performance monitoring of electric cable shovels. Int J Surf Min Reclam Environ 19:276–294. doi:10.1080/13895260500327912
4. Patnayak S, Tannant DD, Parsons I (2008) Operator and dipper tooth influence on electric shovel performance during oil sands mining. Int J Min Reclam Environ 22:120–145. doi:10.1080/17480930701482961
5. Babaei Khorzoughi M, Hall R (2016) A study of digging productivity of an electric rope shovel for different operators. Minerals 6:48
6. Oskouei MA, Awuah-Offei K (2014) Statistical methods for evaluating the effect of operators on energy efficiency of mining machines. Min Technol 123:175–182. doi:10.1179/1743286314Y.0000000067
7. Hendricks C, Scoble MJ, Peck J (1989) Performance monitoring of electric mining shovels. Inst Min Metall Trans Sect A Min Ind 98:A151–A159
8. Acaroglu O, Ozdemir L, Asbury B (2008) A fuzzy logic model to predict specific energy requirement for TBM performance prediction. Tunn Undergr Sp Technol 23:600–608. doi:10.1016/j.tust.2007.11.003
9. Iai M, Gertsch L (2013) Excavation of lunar regolith with large grains by rippers for improved excavation efficiency. J Aerosp Eng 26:97–104. doi:10.1061/(ASCE)AS.1943-5525.0000221
10. Muro T, Tsuchiya K, Kohno K (2002) Experimental considerations for steady state edge excavation under a constant cutting depth for a mortar specimen using a disk cutter bit. J Terramech 39:143–159. doi:10.1016/S0022-4898(02)00021-6
11. Scoble MJ, Muftuoglu YV (1984) Derivation of a diggability index for surface mine equipment selection. Min Sci Technol 1:305–322. doi:10.1016/S0167-9031(84)90349-9
12. Hadjigeorgiou J, Poulin R (1998) Assessment of ease of excavation of surface mines. J Terramech 35:137–153. doi:10.1016/S0022-4898(98)00018-4
13. Awuah-Offei K, Frimpong S (2007) Cable shovel digging optimization for energy efficiency. Mech Mach Theory 42:995–1006. doi:10.1016/j.mechmachtheory.2006.07.008
14. Frimpong S, Hu Y, Awuah-Offei K (2005) Mechanics of cable shovel-formation interactions in surface mining excavations. J Terramech 42:15–33. doi:10.1016/j.jterra.2004.06.002
15. Guzmán MV, Valenzuela MA, Member S (2015) Integrated mechanical–electrical modeling of an ac electric mining shovel and evaluation of power requirements during a truck loading. Cycle 51:2590–2599
16. Awuah-Offei K (2016) Energy efficiency in mining: a review with emphasis on the role of operators in loading and hauling operations. J Clean Prod 117:89–97. doi:10.1016/j.jclepro.2016.01.035
17. Awuah-Offei K, Frimpong S (2011) Efficient cable shovel excavation in surface mines. Geotech Geol Eng 29:19–26. doi:https://doi.org/10.1007/s10706-010-9366-9
18. Brown GM, Elbacher BJ, Koellner WG (2000) Increased productivity with AC drives for mining excavators and haul trucks. In: Conference record of the 2000 IEEE industry applications conference. Thirty-fifth IAS annual meeting and world conference on industrial applications of electrical energy, vol 1, pp P28–P37. doi:10.1109/IAS.2000.880962

19. Awuah-Offei K, Osei B, Askari-Nasab H (2011) Modeling truck/shovel energy efficiency under uncertainty. Trans Soc Min Metall Explor 330:573–584
20. Abdi Oskouei M, Awuah-Offei K (2015) A method for data-driven evaluation of operator impact on energy efficiency of digging machines. Energy Effic 1–12. doi:10.1007/s12053-015-9353-3
21. Karpuz C, Ceylanoğlu A, Paşamehmetoğlu AG (1992) An investigation on the influence of depth of cut and blasting on shovel digging performance. Int J Surf Min Reclam Environ 6:161–167. doi:10.1080/09208119208944331
22. Singh SP, Narendrula R (2006) Factors affecting the productivity of loaders in surface mines. Int J Surf Min Reclam Environ 20:20–32. doi:10.1080/13895260500261574
23. Rasimarzabadi R, Joseph TG (2016) Particle flow mechanism into cable shovel dippers. J Terramech 64:10–22. doi:10.1016/j.jterra.2015.12.003
24. Abdel-Baqi OJ, Onsager MG, Miller PJ (2016) The effect of available short-circuit capacity and trail cable length on substation voltage amplification in surface excavation industry. IEEE Trans Ind Appl 52:3518–3526. doi:10.1109/TIA.2016.2535162
25. Jahromi MG, Mirzaeva G, Mitchell SD, Gay D (2015) Advanced fault tolerance strategy for DC microgrids in mining excavators. IEEE Int Symp Ind Electron 1502–1507. doi:10.1109/ISIE.2015.7281696
26. Wei B, Gao F (2012) Digging trajectory optimization for a new excavating mechanism. In: Proceedings of ASME 2012 international design engineering technical conferences and computers and information in engineering conference IDETC/CIE 2012. Chicago, IL, USA, pp 1033–1039
27. Raza MA, Frimpong S (2017) Mechanics of electric rope shovel performance and reliability in formation excavation. In: Canbolat H (ed) Lagrangian mechanics, pp 107–133
28. Valenzuela M, Valenzuela A (2016) Payload estimation in AC electric mining shovels using drive signals. In: IEEE transactions on industry application, pp 4470–4479
29. Rasuli A, Tafazoli S, Dunford WG (2014) Dynamic modeling, parameter identification, and payload estimation of mining cable shovels. In: 2014 IEEE Industry Applications Society annual meeting, IAS, Vancouver, BC, Canada, part no. 6978451
30. Onal E, Craddock C, Endsley MR, Chapman A (2013) From theory to practice: how designing for situation awareness can transform confusing, overloaded shovel operator interfaces, reduce costs, and increase safety. In: ISARC 2013–30th International symposium on automation and robotics for the construction, mining, held conjunction with 23rd world mining congress, 30066
31. Vukotic I, Kecojevic V (2014) Evaluation of rope shovel operators in surface coal mining using a multi-attribute decision-making model. Int J Min Sci Technol 24:259–268. doi:10.1016/j.ijmst.2014.01.019
32. Blackwell GH (2013) Remote and semi-automated operation of an electric cable shovel. In: ISARC 2013–30th international symposium on automation and robotics in construction, pp 1526–1541
33. Cloete S, Horberry T (2013) Collision avoidance and semi-automation in electric rope shovel operation. In: 49th Annual human factors ergonomics society Australia conference 2013, HFESA 2013. Perth, WA, Australia, Article number 026
34. Hendricks C, Daneshmend L, Wu S, Scoble M (1993) Design of a simulator for productivity analysis of electric mining shovels. In: Proceedings 2nd international symposium on mine mechanics and automation, pp 329–336
35. Dorey F, Knights PF (2015) Quantifying the benefits of simulator training for dragline operators. Min Technol 124:97–106. doi:10.1179/1743286315Y.0000000007

Chapter 9
Benchmarking Energy Consumption of Truck Haulage

Lalit Kumar Sahoo, Santanu Bandyopadhyay and Rangan Banerjee

Abstract Haul trucks are used for material handling in most surface mines and consume about 32% of the total energy usage in mines that use them. This chapter deals with benchmarking approaches applicable to haul truck operation in mines. The specific fuel consumption (SFC) is used as the energy performance index for benchmarking energy consumption of haul trucks. Benchmarking using a statistical approach estimates the minimum SFC based on the comparison of past aggregate time series data and disaggregate data on fuel consumption and the production rate of haul trucks. A model-based approach calculates the minimum SFC using a mathematical model derived from vehicle dynamics, mass balance, and engine and mine characteristics. This chapter presents an analysis of two case studies of haul trucks operations at different surface mines (coal and limestone) to illustrate the benchmarking methods. The studies revealed that benchmarking of energy consumption in haul trucks using the model-based approach is appropriate for setting the fuel consumption target in an opencast mine and assess the fuel saving potential. The model-based approach results in minimum SFC of 89 g/t and fuel saving potential of 17% for multiple haul trucks operating in a limestone mine. The model-based approach shows a direction for setting rational targets for fuel consumption in haul trucks and result in more energy efficient mines.

Keywords Benchmarking · Energy · Truck haulage · Optimization
Surface mines

L. K. Sahoo (✉)
CSIR-Central Institute of Mining and Fuel Research, 17/C,
Telenkhedi Area, Civil Lines, Nagpur (HQ: Dhanbad) 440001, India
e-mail: klalitsahoo@gmail.com

S. Bandyopadhyay · R. Banerjee
Indian Institute of Technology Bombay, Powai, Mumbai 400076, India
e-mail: santanub@iitb.ac.in

R. Banerjee
e-mail: rangan.banerjee@gmail.com

K. Awuah-Offei (ed.), *Energy Efficiency in the Minerals Industry*, Green Energy and Technology, https://doi.org/10.1007/978-3-319-54199-0_9

9.1 Introduction

Benchmarking is a method to evaluate the best possible energy performance for an industry or process. The minimum energy requirement can be evaluated using statistical approaches, by comparing the energy performance of similar types of industries, or by a model-based approach. Benchmarking techniques can be applied to mining processes like truck haulage and excavation (loading) to assess their best energy performance. This chapter presents a review of techniques for benchmarking energy consumption in truck haulage. Before discussing the application of benchmarking techniques relevant to mines, let us understand the energy requirement of mining.

Mining processes include exploration, excavation, transportation and processing of the ore. In surface mining haul trucks are used to transport material from shovels to waste/overburden (OB) dumps or the process plant or ore/coal stockpiles. Benchmarking truck haulage is important as its energy consumption contributes to about 32% of the total energy consumption in a surface mine [1]. The energy consumption in a mine varies due to the variation of the mine topography, mine equipment characteristics and operational practices. Energy consumed by truck haulage in Indian opencast mines varies between 44 and 85 MJ/t from the total energy consumption of 152–207 MJ/t [2]. The energy consumption in opencast mines of China varies between 90 and 225 MJ/t [3] and Canada varies from 97 to 256 MJ/t [4]. An efficient truck haulage system must optimize the distance, gradient, cycle time and speed to maximize the production rate for benchmarking fuel consumption. The distance and gradient of the haul road depend on the strategic mine plan and design, ground conditions and mine topography. Cycle time includes load, travel, unloading and wait times. Travel time depends on the speed limits set on the haul roads and operating speed of truck. Hence, estimation of optimal operating speed of a truck is important to minimize the fuel consumption. The cycle time of trucks affects theoretical material output rate and specific fuel consumption. Vemba [5] illustrated how to optimize haulage cycle times to maximize production rate.

Benchmarking is a process by which some performance index or efficiency is evaluated or measured and then compared with similar establishments, processes and systems. Performance indices and efficiency metrics for benchmarking may include specific energy consumption (kWh/t or MJ/t), cost per ton ($/t), and specific energy per km (MJ/t/km); and efficiency metrics such as technical and economic efficiency metric. Benchmarking is done by comparing performance indices or technical and economic efficiency of system to the best possible performance. Benchmarking is a continuous process of setting performance targets, achieving the targets, and setting new targets. Sardeshpande et al. [6] discussed two methods of benchmarking (statistical and model-based energy benchmarking).

A theoretical benchmarking target (minimum value) is determined by model-based approach using mathematical models and thermodynamic relations. Theoretical minimum value based on model-based approach is always less than the statistical benchmarking target as it is based on the actual (measured) operating parameters. The most commonly used performance indices used for benchmarking energy consumption are specific energy consumption (SEC) and specific fuel consumption (SFC) (Eqs. 9.1 and 9.2 respectively).

$$\mathrm{SEC} = E/Q \tag{9.1}$$

$$\mathrm{SFC} = M_\mathrm{f}/Q \tag{9.2}$$

E, M_f are energy consumption and mass of fuel consumed, respectively; and Q represents the output—material handled or production of the process or industry. The benchmarking techniques can be applied to mining processes like truck haulage and loading as well as to total mine operations to calculate the energy consumption targets.

Haul trucks consume 32% of the total energy usage in surface mines [1, 7]. The energy consumed by haul trucks depend on engine parameters, like *engine speed, vehicle speed*, and mine parameters such as *topography*, *distance of haul road,* and other external parameters like *rainfall, wind speed.* These parameters affect the specific fuel consumption of haul trucks operating in a mine. The diesel consumption pattern of different heavy earth moving machines (HEMMs) for a typical surface mine is given in Fig. 9.1. Diesel consumption in haul trucks accounts for 56% of the total diesel consumption [8]. Further, the quantity of fuel consumed in haul trucks affects the cost of production per ton of material handled. Hence, benchmarking is important to assess the energy performance of haul trucks.

Table 9.1 summarizes a review of the literature on haul truck energy efficiency and the models used in such work. Researchers have used analytical models, artificial neural networks, and stochastic models to study fuel consumption and

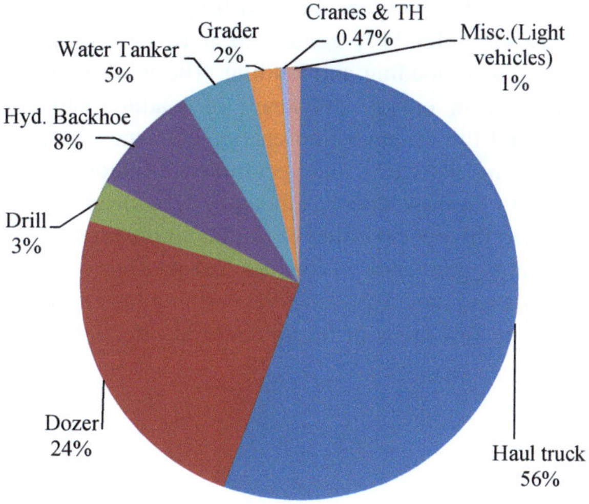

Fig. 9.1 Diesel consumption pattern for an opencast mine in India

Table 9.1 Summary of research work done in haul truck and vehicle operation

Work	Objective of research	Model type	Method/remarks
Sahoo et al. [7]	Benchmarking energy consumption for dump trucks in mine	Analytical	Constrained optimisation
Irdemoosa et al. [9]	Prediction of fuel consumption for mining dump trucks: A neural network approach	Neural network	ANN model
Awuah-Offei et al. [10]	Modelling truck/shovel energy efficiency under uncertainty	Statistical	Stochastic simulation
Saboohi et al. [11]	Optimisation of fuel consumption of passenger vehicle	Analytical	Constrained optimisation
Saerens et al. [12]	Develop model for minimizing fuel consumption	Analytical	Dynamic optimisation
Vemba [5]	Maximize throughput by optimizing haulage cycle	Analytical	Multivariate computation
Vasil'ev et al. [13]	Maximize throughput of motor transport system	Analytical	Multivariate computation
Chang et al. [14]	Minimise fuel consumption of vehicle	Analytical	Langrangian method
Chung et al. [15]	Optimize mine truck allocation	Probabilistic	Stochastic optimisation
Tolouei et al. [16]	Effect of mass of vehicle on fuel consumption	Statistical	Cross-sectional data analysis

energy efficiency of haul truck operations [7, 9, 10]. Past research has mostly used analytical models to maximize the throughput by optimizing the haulage cycle, vehicle speed and minimize fuel consumption [5, 9–14]. Chung et al. [15] used a stochastic model to optimize the mine truck allocation and increase the material handling rate. Tolouei et al. [16] also used a statistical model to study the effect of the mass of vehicle on fuel consumption.

Awuah-Offei [17] conducted a review of energy efficiency in mining with emphasis on loading and hauling. The work shows that the role of the operator is important in energy efficiency of loading and hauling operations. The author classified the research into energy efficiency initiatives in mining, processing and supporting activities and recommended industrial best practices for the mining sector to improve energy efficiency. The need for disruptive technology to achieve closer to theoretical efficiency limits in material handling, the influence of operators on energy efficiency in loading and hauling operations, models for life cycle cost analysis and the effectiveness of training program in achieving the energy efficiency goals are few areas of future research directions.

9.2 Benchmarking Techniques

Though there are several techniques available for benchmarking industrial processes, statistical benchmarking has been mostly applied by many researchers to benchmark the performance parameters due to its simplicity. Model-based benchmarking is used to calculate minimum energy consumption based on the mathematical models.

9.2.1 *Statistical Benchmarking*

In statistical benchmarking, the best performance of any process or industry is estimated by comparing the performance parameters within the process using time series data or cross-sectional data for similar processes or industries. Accurately estimating the benchmarking target requires a large quantity of data. Statistical benchmarking uses regression analysis for calculating best performance. This chapter presents a few examples of benchmarking using this statistical approach.

Example 9.1 Boyd et al. [18] proposed an energy performance indicator (EPI) for benchmarking industrial plant energy use. EPI is a statistical benchmarking tool, which relates the level of energy use in a plant to the level of production along with external factors like climate and material quality. The EPI approach uses linear regression to estimate the ***lowest possible plant energy use*** based on these factors. EPI is percentile ranking of energy efficiency of the plant and is a dimensionless performance parameter. Boyd et al. used the automobile industry as an example to illustrate how to estimate EPI. The EPI of an automobile plant was compared with *EPI of baseline plant for previous to the current year* as well as with the *efficient plant*. The EPI for baseline is calculated to be 40 and that of the current year is 53. The EPI of the efficient plant was estimated at 75. Hence, the benchmarking target for best performing plant was proposed to be 75.

Example 9.2 Phylipsen et al. [19] benchmarked energy efficiency of Dutch industries by comparing their current level of energy efficiency with that of most efficient plant of the countries and regions. They used an energy efficiency index (EEI) as the performance indicator to benchmark the energy efficiency.

They defined the EEI_a of an industrial sector (a) as:

$$\begin{aligned}\mathrm{EEI}_a &= 100\frac{\mathrm{SEC}_a}{\mathrm{SEC}_{\mathrm{ref},a}} \\ &= 100\frac{\sum_i E_i/\sum_i Y_i}{(\sum_i Y_i\mathrm{SEC}_{\mathrm{ref},i})/\sum_i Y_i} \\ &= 100\frac{E_a}{\sum_i Y_i\mathrm{SEC}_{\mathrm{ref},i}}\end{aligned} \tag{9.3}$$

where E_a and SEC_a are energy consumption and specific energy consumption for sector a; $SEC_{ref,a}$ is the reference specific energy consumption of best-operating plant in sector a; and E_i and Y_i indicate the energy consumption and production for product i. Chauhan et al. [20], Chung et al. [21], and Omid et al. [22] have all used regression analysis for benchmarking energy consumption for industries. However, only a few studies have carried out energy benchmarking in the mining sector based on statistical methods. Cooke and Randal [23] discussed energy use benchmarking for opencast coal mining. They established range of energy efficiency benchmarks and targets based on studies undertaken at a number of coal mines in Hunter valley [23]. They estimated a benchmark target of 19.6 MJ/m^3 for opencast mines without draglines and a target of 14 MJ/m^3 for opencast mines with draglines. Reddy et al. [24] studied Indian opencast coal mines using data envelope analysis.

Authors have used a statistical technique for benchmarking energy consumption of haul trucks operating in Dipka opencast mine of M/s South Eastern Coalfields Ltd, Bilaspur, India [8, 25]. We used total diesel consumption data of the haul trucks operating in the mine. We have taken the minimum specific fuel consumption (SFC) from past 3 years' progressive data to benchmark diesel consumption. However, this approach can not accurately assess the benchmark value as the progressive SFC of previous years does not consider the present energy performance of haul trucks. Hence, we formulated the statistical benchmarking method using disaggregate diesel consumption of the mine equipment for the year of the study.

Benchmarking based on actual field measurements can be done using disaggregate data of fuel consumption from individual equipment. This method requires instrumentation to measure the required parameters during the trial period to calculate the benchmarking index. Since, different capacities of dump trucks operate in surface mines to transport material from shovels to crushers or dumping points, the specific fuel consumption (SFC) is a good performance index for benchmarking. The mathematical formulation [25] for benchmarking using a statistical benchmarking approach based on disaggregate data for individual haul trucks operating in a mine is given in Eqs. 9.4–9.8. This formulation is only useful for the haul trucks operating in mines in which measurement of the payload is not possible. Few manufacturers have real-time equipment monitoring software (e.g. Caterpillar's vital information management systems (VIMS) software) that are capable of measuring the payload. The global positioning system (GPS)-based vehicle/fleet management software was used to monitor the number of trips, payload, cycle time, wait time to maximize the material handling rate of haul trucks.

The specific fuel consumption (SFC) of a single dump truck is defined as the ratio of diesel consumed $m_f(t)$ during field trial period (t) to the material handled. The material handled is determined as the product of shovel bucket capacity (C_b), actual numbers of buckets $n_b(t)$ filled in dump trucks during the trial period (t), and fill factor (C_f) $m_f(t)$ and $n_b(t)$ are measured parameters whereas C_b is design capacity and fill factor $(C_f = 0.8)$ is assumed.

$$\mathrm{SFC}_1 = \frac{m_\mathrm{f}(t)}{C_\mathrm{b} n_\mathrm{b}(t) C_\mathrm{f}} \tag{9.4}$$

If 'Q_i' is the material handling rate in cubic meters per year by different capacity dump trucks ($i = 1, 2, 3\ldots$), an analyst can obtain the benchmark SFC ($\mathrm{SFC_{BM}}$) of multiple dump trucks using Eq. 9.5 for rainy and dry seasons. The minimum SFC is considered as the benchmark for the same capacity dump trucks operating on the same routes.

$$\mathrm{SFC_{BM}} = \frac{\mathrm{Min} \sum_{i=1}^{n} \mathrm{SFC}_i Q_i}{Q} \tag{9.5}$$

where,

$$Q = \sum_{i=1}^{n} Q_i \tag{9.6}$$

Average SFC of dump trucks for the rainy or dry season is estimated as:

$$\mathrm{SFC_{Avg}} = \frac{\mathrm{Avg} \sum_{i=1}^{n} \mathrm{SFC}_i Q_i}{Q} \tag{9.7}$$

Equations 9.4–9.7 are used for calculating benchmark SFC and average SFC for rainy and dry seasons. Equation 9.8 estimates the practical minimum SFC for the whole year. T_or and T_r are a number of months of a year for dry and rainy seasons, respectively.

$$\mathrm{SFC_{practical\ minimum}} = \frac{\mathrm{SFC_{BM,or}} T_\mathrm{or} + \mathrm{SFC}_{\mathrm{BM},r} T_r}{12} \tag{9.8}$$

The trial period assumptions and conditions include:

- The trial period for field measurement of each haul truck is 2 h.
- Diesel consumption is measured by tank fill method in diesel pump.
- Shovels should be engaged in loading operation only.
- Number of trips is counted or taken from trip report of GPS data.
- Annual material handling of haul trucks (Q) is taken from previous year data.

The above method is based on the number of buckets loaded by shovel into the haul truck and the bucket capacity. If the sophisticated VIMS/GPS based vehicle/ fleet management software are installed in haul trucks, specific fuel consumption for each haul truck shown in Eq. 9.4 is calculated as the ratio of fuel consumption to actual payload, measured directly.

9.2.2 Model-Based Benchmarking

Model-based benchmarking estimates the benchmarking target for a process based on mathematical models that use thermodynamic relationships, vehicle dynamics, and other theoretical relationships. Also, this approach correlates model variables to the process requirements [6, 7].

Example 9.3 Sardeshpande et al. [6] proposed an energy benchmarking model for glass furnaces using mass and energy balance including heat loss equations for different zones and empirical equations based on operating practices. The model was verified for field data of end fired glass furnace. The minimum energy consumption of 3830 kJ/kg is estimated for a 100 TPD glass furnace. The potential of energy saving of about 20–25% is possible by comparing the similar glass furnaces of similar production scale. The model can be applied to estimate the target energy performance for a new design as well as an operating furnace.

Sahoo et al. [7] developed a mathematical model for calculating specific fuel consumption (SFC) of truck haulage systems for surface mines using vehicle dynamics and engine and mine characteristics. Figure 9.2 shows the model input parameters in the modelling framework of the energy benchmarking of haul trucks in [2].

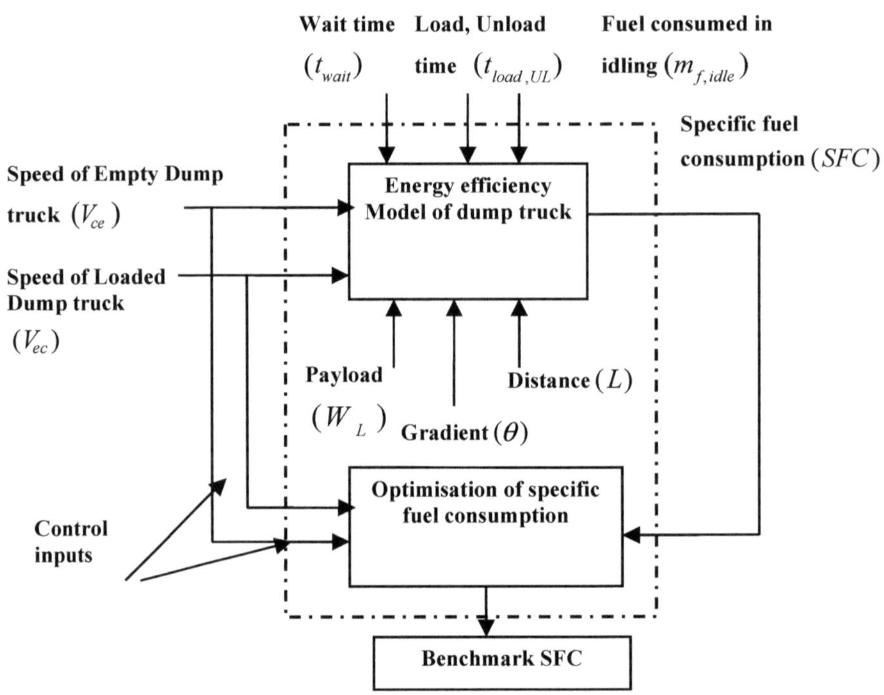

Fig. 9.2 Schematic of energy benchmarking for dump trucks in a mine [2]

Sahoo and co-workers modelled SFC of haul truck as:

$$\mathrm{SFC} = f(V_{\mathrm{ce}}, V_{\mathrm{ec}}, t_{\mathrm{load,UL}}, W_L, W_E, L, t_{\mathrm{wait}}, \theta, m_{f,\mathrm{idle}}) \quad (9.9)$$

From Fig. 9.2, the specific fuel consumption varies with speed of haul trucks, distance and payload. The speed of empty haul trucks and loaded haul trucks are the two decision variables for the optimization problem. The fuel consumption is dependent on the speed of the truck and torque. The torque depends on the payload of the truck. For a constant payload, speed of empty haul truck and speed of loaded haul are only two variables which affect specific fuel consumption. Hence, these two variables are taken as decision variables for optimization. The optimal speed of operation for empty and loaded haul trucks are calculated by solving the optimization problem using Excel solver.

9.2.2.1 Optimization of Specific Fuel Consumption

Haul trucks are allocated in each shift at the loading points (excavators) for loading and transportation of ore/coal to crushers or stockpiles and overburden to waste dumps. A generic surface mine transportation problem with M dumping points (crushers, stockpiles and waste dumps), N excavators and K dump trucks are shown in Fig. 9.3 [7]. The loaded dump trucks move from excavators to dumping points and the empty dump trucks return back to excavators to complete the cycle. The dump trucks consume fuel during loading, unloading, waiting and travel period. The fuel consumption in dump trucks varies with many variables including distance, speed, payload, and material handling demand. The objective of the modelling framework is to determine the operating strategy for the haul trucks in order to minimize the specific fuel consumption. The modelling framework [7] is presented in Eqs. 9.10–9.47.

The gross mass of dump truck (W_{G}) is the sum of the weight of empty dump truck (W_{E}) and the payload (W_{L}). The payload is defined as the material carried by a dump truck in a single trip.

The mass balance of a dump truck is given as:

$$W_{\mathrm{G}} = W_{\mathrm{L}} + W_{\mathrm{E}} \quad (9.10)$$

The payload is restricted by the maximum capacity of the dump truck $(W_{\mathrm{L,max}})$ as given in Eq. 9.11:

$$W_L \leq W_{\mathrm{L,max}} \quad (9.11)$$

The power required by the empty dump truck moving from dumping point to excavator (P_{ce}) is the sum of power required at constant speed profile (drag, frictional, and gradient resistance) and the power required due to acceleration [26] as shown in Eq. 9.12. The power required by the loaded dump truck moving from

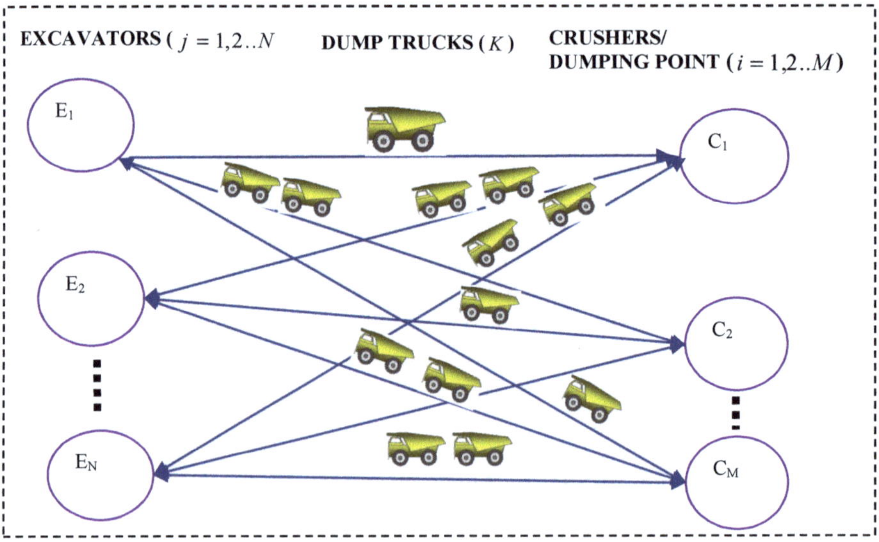

Fig. 9.3 Schematic of a transportation network of opencast mine [7]

excavator to the crusher (P_{ec}) is obtained in a similar fashion and given in Eq. 9.13. a is constant for drag resistance b, c are constants for rolling, friction and gradient resistances.

$$P_{ce} = V_{ce}(aV_{r,ce^2} + bW_E) + \frac{0.5W_E V_{ce^3}}{L} \quad (9.12)$$

$$P_{ec} = V_{ec}(aV_{r,ec^2} + cW_G) + \frac{0.5W_G V_{ec^3}}{L} \quad (9.13)$$

V_r, the velocity of the wind relative to ground is given as:

$$V_r = V \pm U \cos\beta \quad (9.14)$$

In Eq. 9.14, U speed of wind and direction of wind is indicated by a positive sign (direct headwind) or a negative sign (direct tailwind). The effect of crosswind is neglected $(\beta = 0)$ in both the cases of direct head and tailwind. The relative velocity is equal to average speed of the dump truck for still air $(U = 0)$.

The power supplied to the dump truck is restricted by the maximum engine power (P_{max}):

$$P_{ce,ec} \leq P_{max} \quad (9.15)$$

The vehicle speeds for the empty and loaded truck are restricted by their maximum speed ($V_{\max}$):

$$V_{\mathrm{ce,ec}} \leq V_{\max} \tag{9.16}$$

In Eqs. 9.12 and 9.13, constants a, b, c for negative gradient mine are given as:

$$a = \frac{1}{2} C_{\mathrm{d}} \rho_{\mathrm{air}} A_{\mathrm{F}} \tag{9.17}$$

$$b = g \cos\theta(f + Crr) + g \sin\theta \tag{9.18}$$

$$c = g \cos\theta(f + Crr) - g \sin\theta \tag{9.19}$$

where, Cd is drag coefficient, Crr is the coefficient of rolling resistance, A_{F} is frontal area of haul truck, and θ is mine gradient. The negative gradient mine implies that the loaded dump truck is assisted by the gravity and hence negative sign is taken for $g \sin\theta$ in Eq. 9.19 for gradient resistance. The positive sign is taken for $g \sin\theta$ in case of empty dump truck as shown in Eq. 9.18.

The engine power of an internal combustion engine is a function of the angular velocity of the engine [27] shown in Eq. 9.20.

$$P(w) = \sum_{i=1}^{3} Piw^{i} \tag{9.20}$$

P_1, P_2,P_3 are constants for direct injection diesel engine and are obtained from engine characteristics curve supplied by the manufacturer.

The engine power available (P_{avail}) is given as Eqs. 9.21 and 9.22 by converting angular velocity to linear velocity using Eq. 9.23.

$$P_{\mathrm{avail,ce}} = a_1 V_{\mathrm{ce}} + a_2 V_{\mathrm{ce^2}} - a_3 V_{ce^3} \tag{9.21}$$

$$P_{\mathrm{avail,ec}} = a_1 V_{\mathrm{ec}} + a_2 V_{\mathrm{ec^2}} - a_3 V_{\mathrm{ec^3}} \tag{9.22}$$

a_1, a_2, a_3 are constants. These are constants of direct injection diesel engine defined in Eq. 9.20 and obtained after conversion from angular velocity to linear velocity.

The speed of dump truck (V_{ce}) and (V_{ec}) are given as:

$$V_{\mathrm{ce,ec}} = \frac{R_{\mathrm{w}} w}{G_{\mathrm{R}}} \tag{9.23}$$

where R_{w} is wheel radius, w is angular velocity and G_{R} is gear ratio.

The maximum speeds of dump truck $V_{\mathrm{max,ce}}$ and $V_{\mathrm{max,ec}}$ are obtained by equating Eqs. 9.12 and 9.13 with Eqs. 9.21 and 9.22 at the condition of the maximum power that engine can deliver.

The brake specific fuel consumption of dump truck $B_{F,ce}$ and $B_{F,ec}$ are given as:

$$B_{F,ce} = b_1 V_{ce}^2 - b_2 V_{ce} + b_3 \quad (9.24)$$

$$B_{F,ec} = b_1 V_{ec}^2 - b_2 V_{ec} + b_3 \quad (9.25)$$

b_1, b_2, b_3 are constants obtained from the engine characteristics.

The fuel consumption of the dump truck (m_f) moving between a dumping point and an excavator and return journey are given as:

$$m_{f,ce} = P_{ce} B_{F,ce} \quad (9.26)$$

$$m_{f,ec} = P_{ec} B_{F,ec} \quad (9.27)$$

The travel times from *i*th dumping point to *j*th excavator and return journey are calculated as:

$$tij = \frac{Lij}{Vce} \quad (\mathrm{i} = 1, 2, 3\ldots M;\ \mathrm{j} = 1, 2, 3\ldots N) \quad (9.28)$$

$$tji = \frac{Lji}{Vec} \quad (\mathrm{i} = 1, 2, 3\ldots M;\ \mathrm{j} = 1, 2, 3\ldots N) \quad (9.29)$$

The generalized equation for the fuel consumption per trip ($M_{f,ij}$) of each dump truck moving between *i*th dumping point and *j*th excavator is given as Eq. 9.30.

$$M_{f,ij} = m_{f,ij} t_{ij} + m_{f,ji} t_{ji} + m_{f,idle}(t_{load,UL} + t_{wait}) \quad (9.30)$$

where

$$t_{load,UL} = t_{load} + t_{UL} \quad (9.31)$$

In the present model, $t_{load,UL}$ is assumed to be constant.

For optimisation,

$$t_{wait} = 0 \quad (9.32)$$

The travel time (t_{travel}) and cycle time of the dump truck ($t_{d,cycle}$) are given as follows:

$$t_{travel} = t_{ce} + t_{ec} \quad (9.33)$$

$$t_{d,cycle} = t_{load,UL} + t_{travel} \quad (9.34)$$

The waiting time is the time the dump truck has to wait at the excavator before it is loaded for its subsequent trip. The waiting time of the dump truck is given in Eq. 9.36.

$$\text{If, } t_{\text{d,cycle}} < n_d t_{\text{load}} \tag{9.35}$$

$$t_{\text{wait}} = n_d t_{\text{load}} - t_{\text{d,cycle}} \tag{9.36}$$

The cycle time of the dump truck given in Eq. 9.34 is modified as Eq. 9.37 considering waiting time:

$$t_{\text{d,cycle}} = t_{\text{load,UL}} + t_{\text{travel}} + t_{\text{wait}} \tag{9.37}$$

The number of trips (x_{ce}) per hour is calculated using Eq. (9.38):

$$x_{\text{ce}} = \frac{60}{t_{\text{d,cycle}}} \tag{9.38}$$

The theoretical output of a dump truck (q_d) per hour is calculated using Eq. 9.39:

$$q_{\text{d}} = W_{\text{L}} x_{\text{ce}} \tag{9.39}$$

The specific fuel consumption of a dump truck ($\text{SFC}_{\text{dump truck}}$) operating between a dumping point and an excavator is defined as the ratio of the fuel consumed by dump truck to the material handled in the same time period and is given in Eq. 9.40.

$$\text{SFC}_{\text{dump truck}} = \frac{x_{\text{ce}} M_{\text{f,ce}}}{q_{\text{d}}} \tag{9.40}$$

The formulation of the optimisation is given as:
Minimize

$$\text{SFC}_{\text{dump truck}} = \frac{\sum_{i=1}^{M} \sum_{j=1}^{N} H X_{\text{ij}} M_{\text{f,ij}}}{\sum_{i=1}^{M} D_{\text{i}}} \tag{9.41}$$

Subject to the following constraints:
Material transported from all excavators is greater than the material handling demand of each crusher (D_{i}):

$$\sum_{j=1}^{\text{N}} X_{\text{ij}} W_L H \geq D_i \quad (i = 1, 2, 3 \ldots M) \tag{9.42}$$

If there is no stockpile, then material handled by dump trucks are directly fed to the crusher hopper. In that case, material handled by the dump trucks is less than the crusher capacity $(Q_{c,i})$:

$$\sum_{j=1}^{N} X_{ij} W_L \leq Q_{c,i} \quad (i = 1, 2, 3 \ldots M) \tag{9.43}$$

Material handled by the dump trucks is less than the excavator capacity $(Q_{e,j})$:

$$\sum_{j=1}^{M} X_{ij} W_L \leq Q_{e,j} \quad (j = 1, 2, 3 \ldots N) \tag{9.44}$$

The total number of dump trucks transporting material from all excavators is equal to the total number of dump trucks received at all dumping sites:

$$\sum_{j=1}^{N} n_{dj} = \sum_{i=1}^{M} n_{di} \tag{9.45}$$

where

$$n_{dj} = \frac{t_{d,\,cycle,\,ij}}{t_{load}} \quad (i = 1, 2, 3 \ldots M; \, j = 1, 2, 3 \ldots N) \tag{9.46}$$

And, $n_d = \text{round}(\text{nd}, 0)$

Similar dump trucks with same engine characteristics and capacity are assumed to operate in each route for multiple dump trucks model.

Total number of trips made between ith dumping site and jth the excavator is given as:

$$X_{ij} = n_{dj} x_{ij} \tag{9.47}$$

All variables are positive,

n_d, is an integer variable For mine transportation problems with M crushers or dumping sites, N excavators and K dump trucks, the number of variables and constraints depend on the value of M and N.

9.2.2.2 Solution Procedure

The model for multiple dump trucks operating between multiple crushers and multiple excavators is solved by linear programming using the simplex method. The fuel consumption for each route is minimized by finding the optimal speed profile. The minimum fuel consumption per trip $(M_{f,ij})$ calculated for each route between *i*th dumping site and *j*th excavator is taken as input data for the LP problem to minimize the specific fuel consumption of the transport network for given material handling demand. That LP problem is solved in excel solver with an objective function to minimize the specific fuel consumption given in Eq. 9.41 subject to constraints Eqs. 9.42–9.47 including Eqs. 9.10–9.40. This research work uses analytical and statistical models to optimize the fuel consumption and theoretical material output of haul trucks. The modelling approach minimizes the objective function (SFC) using optimization techniques though this estimate is difficult for the industry to achieve as a benchmarking target. A case study of a large surface coal mine in India is used to illustrate statistical benchmarking. The work calculates practical minimum SFC based on actual field measurement of sample trucks. Following this, the case study of a limestone mine is used to illustrate the model-based approach.

9.3 Case Studies on Benchmarking of Haul Trucks

9.3.1 Statistical Benchmarking

A study conducted by Central Institute of Mining and Fuel Research (CIMFR), India at one of the largest opencast mines of Coal India Ltd (CIL) in India is used as the case study for statistical benchmarking. In this case study, CIMFR proposes a benchmarking target for specific fuel consumption (SFC) by comparing annual progressive SFC of the previous 3 years. This resulted in a benchmark SFC of 0.81 l/cu.m. for the mine and 0.47 l/cu.m. for haul trucks [8]. The specific energy consumption is calculated to be 31 MJ/cu.m. for the coal mine and 17.98 MJ/cu.m. for haul trucks, assuming a heating value of diesel of 46 MJ/kg and specific gravity of diesel of 0.832. Diesel consumption and SFC analysis for the 3 years in question are given in Table 9.2. The annual progressive SFC was considered for benchmarking as the monthly SFC varies significantly due to seasonal effects on the performance of haul trucks. The monthly SFC varies from a minimum value of 0.39 l/cu.m. in October 2014 to a maximum of 0.93 l/cu.m. in August 2012.

The authors also used disaggregated data of individual haul trucks of different capacity (240t, 120t, 100t) and model (BEML/Caterpillar/Terex) used in the mine for transporting overburden (OB) and coal for benchmarking. The installed capacity of the mine is 25 Mt/y, whereas the present coal production is 41 Mt/y.

CIMFR used individual data for haul trucks with a composite production of 14.1 Mcu.m. in 2014–2015 to estimate a benchmarking target of specific fuel consumption (SFC), at one of the largest opencast coal mines of Coal India Ltd (CIL) in India, to be 0.45 l/cu.m. Only departmental coal and department overburden were used for calculating the composite production. The specific energy consumption benchmarking of the haul truck is calculated to be 17.22 MJ/cu.m.

Table 9.2 Progressive SDC analysis of mine and haul trucks

Year	Composite OB (Mcu. m./y)	Total diesel consumption in Mine (Ml/y)	Diesel consumption in dump trucks (Ml/y)	Progressive mine SFC (l/cu.m.)	Progressive haul truck SFC (l/cu. m.)
2012–13	10.4	11.2	7.27	1.072	0.69
2013–14	12.0	11.18	6.27	0.93	0.52
2014–15	14.1	11.50	6.60	[a]0.81	[a]0.47

[a]Minimum SFC

Table 9.3 Input data for estimating minimum SFC for off rainy season

Capacity of dump truck	SFC (l/cu. m.)	Distance (km)	Effective SFC (l/cum/km)	Q = Material handled (2014–15) in (Mcu.m.)
240t	0.44	1.76	0.25	7.48
120t	0.44	0.65	0.67	2.44
100t	0.39	0.65	0.60	4.10
				Total Q = 14.1

The input data for calculating the benchmarking target of SFC for rainy and off rainy seasons was taken from the Tables 9.3 and 9.4.

The authors estimated the best-operating SFC for off rainy, using Eq. 9.5 and data in Table 9.3, to be 0.42 l/cu.m. Similarly, the authors estimated the best-operating SFC for rainy season, using Eq. 9.5 and data in Table 9.4, to be 0.52 l/cu.m. The results are given in Table 9.5.

Now, using Eqs. 9.7 and 9.8, the practical minimum and average SFC is 0.45 and is 0.50 l/cu.m., respectively. The fuel saving potential (*S*) in a mine is calculated by comparing the practical minimum SFC with the average SFC and calculated using Eq. 9.48. The diesel saving potential in the mine is estimated as 10% using disaggregated data.

$$S = \frac{\mathrm{SFC}_{\mathrm{Avg}} - \mathrm{SFC}_{\mathrm{Practical\ minimum}}}{\mathrm{SFC}_{\mathrm{Avg}}} \times 100 \tag{9.48}$$

The benchmark SFC obtained by the statistical benchmarking approach using disaggregate data is 0.45 l/cu.m. and aggregate data is 0.47 l/cu.m. Though the difference in the results obtained by the two approaches is negligible, statistical benchmarking using disaggregated data is more pragmatic as results are based on field measurement.

9.3.2 *Model-Based Benchmarking*

A case study of a limestone surface mine located in Rajasthan, India is used to illustrate the model-based approach of benchmarking [7]. For a single dump truck

Table 9.4 Input data for estimating minimum SFC for rainy season

Capacity of dump trucks	SFC (l/cu. m.)	Distance (km)	Effective SFC (l/cum/km)	Q = Material handled (2014–15) in (Mcu. m.)
240t	0.43	2.2	0.195	7.48
120t	0.73	2.9	0.25	2.44
100t	0.59	2.97	0.198	4.10
				Total Q = 14.1

Table 9.5 Average and minimum SFC for rainy and off rainy season

Capacity of dump trucks	Average SFC (l/cu.m.)		Minimum SFC (l/cu.m.)	
	Off rainy	Rainy	Off rainy	Rainy
240t	0.44	0.45	0.44	0.43
120t	0.61	0.87	0.44	0.73
100t	0.43	0.71	0.39	0.59
Aggregate SFC (Using Eqs. 9.5 and 9.7)	0.46	0.59	0.42	0.52

of 65t capacity, operating between a crusher and an excavator 1.5 km apart, the minimum SFC is 84.8 g/t. Three trucks are required to serve the loading point to achieve optimal SFC for multiple dump trucks moving between a crusher and excavator for a distance of 1.5 km by solving optimization model. Haul trucks of two different capacities (52t, 65t) operate between three excavators and two crushers to fulfill the material handling demand of 2600 t/h. The model estimates the optimal SFC to be 89 g/t for multiple dump trucks by solving Eqs. 9.10–9.47. The model estimates the optimal speeds to be 28 and 25 km/h for empty and loaded dump trucks, respectively, and an optimal number of trucks to achieve minimum fuel consumption to be three. SFC varies with payload, distance and operating speed of the haul trucks.

The variation of SFC with payload is shown in Fig. 9.4. SFC decreases with payload and minimum SFC is 84.8 g/t. The specific energy consumption calculated for multiple dump trucks is 10.48 MJ/cu.m. (89 g/t). The lower value of specific energy consumption is due to the mine topography. The limestone mine is a down gradient mine and the material is transported by loaded truck from higher to lower elevation. A potential fuel saving of 17% is estimated by comparing with monthly SFC from 2008 to 2009 (a range of 106 g/t to 152 g/t) using the model-based approach. For the same mine, the average SFC for haul trucks, calculated using the statistical approach, is 120 g/t and is equivalent to 14.46 MJ/cu.m. The model-based approach considers the engine and mine parameters as well as external parameters to obtain minimum SFC and is applicable for haul trucks operating in any mine topography. The statistical approach is mine specific and required more

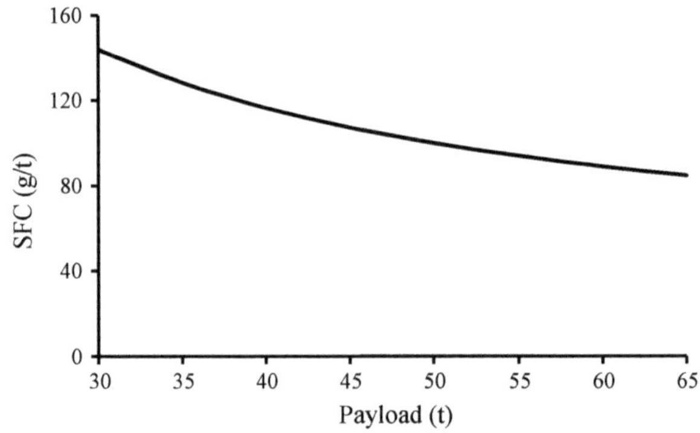

Fig. 9.4 Variation of SFC with payload [7]

data for accuracy. The minimum SFC in the statistical approach is calculated based on the average energy performance of a sample of haul trucks.

The authors also conducted a sensitivity analysis of the model for a 65t capacity dump truck. The sensitivity index [7] is defined in Eq. 9.49 as the ratio of change in SFC computed to the expected change in input base value.

$$\text{Sensitivity Index} = \frac{\dfrac{(\text{SFC, Change from base input} - \text{SFC, base input})}{\text{SFC, base input}}}{\dfrac{(\text{Change from base input} - \text{Base input})}{\text{Base input}}} \tag{9.49}$$

A positive sensitivity index indicates that an increase in base value increases the SFC and vice versa. A negative sensitivity index indicates that an increase in base value decreases the SFC and vice versa. The sensitivity analysis of the SFC for a 65t capacity dump truck operating in down gradient and up gradient mines is analyzed by Sahoo et al. [7]. An increase in the base value of the payload (65t) by 5% decreases SFC by 2.7% whereas the same decrease in base value of payload increases SFC by 3.18% for down gradient mines. The SFC decreases by 2.83% with an increase in base value of the payload by 5% for up gradient mine.

The SFC of the dump truck decreases with increasing payload so long as the rated payload is not exceeded significantly and the minimum SFC of 84.8 g/t is estimated at the maximum payload of 65t as shown in Fig. 9.4. The material handling rate of a dump truck increases with operating speed. However, the SFC also increases with speed and subsequently with handling rate. The SFC of the dump truck decreases with gradient for a down gradient (negative gradient) mine and vice versa for an up gradient (positive gradient) mine. The SFC of the haul trucks also varies linearly with haulage distance.

9.4 Discussions

Benchmarking haul truck operations in surface mines can help to set rational targets for the energy consumption by haul trucks and result in more efficient mining operations. Though the statistical approach based on aggregate data is more simple and easy to apply for benchmarking, it may not give the appropriate benchmarking target. For example, if we decide the benchmarking target will be the minimum value of annual progressive specific fuel consumption (SFC) of haul trucks from last three years and the minimum value occurs in the first of the three years, the SFC value may change depending on the haul distance, gradient and number of machines used during the period. This will not account for recent performance and operating conditions of haul truck. Hence, the statistical benchmarking approach based on disaggregate data may be better for setting the benchmarking target. However, operating parameters of haul trucks such as payload, distance and speed are not taken into account.

Model-based benchmarking for dump trucks developed by Sahoo et al. [7] is the most appropriate method because it considers engine parameter, mine parameters as well as external parameters to obtain the minimum SFC. It can be applied to haul truck operations for any surface mine with variations in mine topography whereas statistical benchmarking is applicable for a specific mine and is based on the average performance. However, due to some assumptions considered in the model, such as zero waiting time and idle engine characteristics, the model is limited in its ability to predict the theoretical value of the benchmarking target. The model-based approach can appropriately predict the minimum SFC than statistical approach.

Statistical benchmarking based on disaggregating data of haul trucks is useful for single as well as multiple haul trucks operating in a mine as it evaluates the present best-operating practice from actual field measurements. The benchmarking index is calculated using simple mathematical equations. When benchmarking a number of similar mines or processes, statistical benchmarking is better as it is simpler and determines target by comparing the performance index. The model-based approach is generic and can be applied to any surface mine regardless of the type of material (coal, manganese, limestone, iron ore, zinc, etc.). The effect of different operating relevant parameters like the speed of haul trucks, haul distance, gradient, and payload can be analyzed using model-based benchmarking. In addition, it facilitates decisions for optimal truck allocation to optimize the fuel consumption and reduce the cost of extraction.

Based on the models and discussions, the authors offer the following recommendations to optimize fuel consumption:

- Operate dump trucks at their optimal operating speed calculated using model.
- Allocate an optimal number of trucks at excavators/shovels.
- Maximize payload during loading ore in haul trucks.
- Install equipment monitoring software in haul trucks to monitor number of trips, wait time, cycle time so as to maximize the material handling rate.

The benchmarking truck haulage for energy efficiency is a new and emerging area of research in the mining sector. Only limited work has been done relating to benchmarking fuel consumption for mine transportation using truck haulage. Sahoo et al. [7] have given a direction for the future direction of benchmarking energy in mining using a model-based approach.

Some possible future research directions include:

- Standardization of methodology of benchmarking using disaggregate data for haul trucks for different topography, material and geo-climatic conditions
- Optimization of operational parameters like operating speed, payload, gradient, etc.
- Better understanding of the effect of monsoon on specific fuel consumption of haulage trucks (relevant for India)
- Selection of haul truck-shovel combination for optimization of fuel consumption
- Optimisation of fuel consumption in haul trucks considering uncertainty of material handling demand, soil density, mine and environmental conditions
- An analysis of the 3-D mine topography to plan optimal mine progress and minimum SFC
- Evaluation of the feasibility of low-carbon mining operations using electric vehicles and renewable energy.

9.5 Conclusions

This chapter presented a review of benchmarking of energy consumption in haul truck operations in surface mines. The chapter presented a methodology of benchmarking specific fuel consumption using a statistical approach based on both aggregate and disaggregate data of fuel consumption and material handling rate. The statistical approach is illustrated with the case study of a large surface coal mine in India and resulted in minimum SFC of 17.98 MJ/cu.m. using aggregate fuel consumption of haul trucks and 17.22 MJ/cu.m using disaggregated fuel consumption data. The statistical approach is useful for quick assessment of the minimum energy required per ton of throughput in haul trucks using the field data.

The chapter also presented a model-based approach based on the optimization of specific fuel consumption using a mathematical model that accounts for operating parameters. The application of the model resulted in SFC benchmark of 89 g/t (10.48 MJ/cu.m.) for a limestone mine in India. The model-based approach is recommended for benchmarking energy consumption in haul trucks as it is based on a generic model that accounts for operating parameters. The benchmarking methods discussed in this chapter can be applied to other heavy earth moving machines like hydraulic excavators and dozers to establish the specific fuel consumption norm. Hopefully, the contents of the chapter will encourage further research work on benchmarking in truck haulage. This will also help practicing mine engineers (mining and mechanical), who want to implement benchmarking techniques to save fuel and minimize the cost of extraction in mines.

Acknowledgements We are thankful to M/s South Eastern Coalfields Ltd, Chhattisgarh, India for providing data of studies conducted for benchmarking diesel consumption during 2015 and Director, CSIR-Central Institute of Mining & Fuel Research for giving permissions and support to this publication.

References

1. Sahoo LK, Bandyopadhyay S, Banerjee R (2010) Energy performance of dump trucks in opencast mine. In: Proceedings of ECOS 2010, Lausanne, Switzerland 14–17 June 2010, pp 1899–1906
2. Sahoo LK, Bandyopadhyay S, Banerjee R (2013) Benchmarking energy efficiency of opencast mining in India. In: Proceedings of ICAER 2013, Mumbai, India. 10–12 Dec 2013, pp 1684–1692
3. Shuzhao C et al (2011) Energy consumption comparison and energy saving measures of open pit development. In: Proceedings of international conference on computer distributed control and intelligent environmental monitoring (CDCIEM), China
4. Benchmarking the energy consumption of Canadian openpit mine (2005) Report No. ISBN 0-662-39538-7, *Mining association of Canada*
5. Vemba MMDS (2004) The loading and transport system at SMC-Optimisation. J S Afr Inst Min Metall, 139–147
6. Sardeshpande V, Gaitonde UN, Banerjee R (2007) Model-based energy benchmarking for glass furnace. Energy Convers Manage 48:2718–2738
7. Sahoo LK, Bandyopadhyay S, Banerjee R (2014) Benchmarking energy consumption for dump trucks in mines. Appl Energy 113:1382–1396
8. CIMFR studies (2015) conducted by first authour at Dipka opencast mine of South Eastern Coalfields Ltd (SECL), Bilaspur, India of Coal India Ltd
9. Irdemoosa ES, Dindarloo SR (2015) Prediction of fuel consumption for mining dump trucks: A neural network approach. Appl Energy 151:77–84
10. Awuah-Offei et al (2011) Modelling truck/shovel energy efficiency under uncertainty. Trans Soc Min Metall Explor 330:573–584
11. Saboohi et al (2009) Model for developing an eco-driving strategy of passenger vehicle based on least fuel consumption. Appl Energy 86:1925–1932
12. Saerens et al (2009) Minimization of fuel consumption of gasoline engine using dynamic optimisation. Appl Energy 86:1582–1588
13. Vasil'ev MV et al (1974) Optimisation of the journey schedules of high capacity Quarry dump trucks. In: The Institute of mining: Ministry of Ferrous metallurgy of USSR 1:63–69
14. Chang et al (2005) Vehicle speed profiles to minimise work and fuel consumption. J Transp Eng 3:173–182
15. Chung et al (2005) A stochastic optimisation approach to mine truck allocation. Int J Surf Min Reclam Environ 19(3):162–175
16. Tolouei et al (2009) Vehicle mass as a determinant of fuel consumption and secondary safety performance. Trans Res D-Trans Environ 14(6):385–399
17. Awuah-Offei K (2016) Energy efficiency in mining: a review with emphasis on the role of operators in loading and hauling operations. J Clean Prod 117:89–97
18. Boyd G, Dutrow E, Tunnesssen W (2008) The evolution of energy star energy performance indicator for benchmarking industrial plant energy use. J Clean Prod 16:709–715

19. Phylipsen D, Blok K, Worrell E et al (2002) Benchmarking the energy efficiency of the Dutch industry: an assessment of the expected effect on energy consumption CO_2 emissions. Energy Policy 30:663–679
20. Chauhan NS, Mohapatra PKJ, Pandey KP (2006) Improving energy productivity in paddy production through benchmarking: an application of data envelopment analysis. Energy Convers Manag 47:1063–1085
21. Chung W, Hui YV, Lam YM (2006) Benchmarking the energy efficiency of commercial buildings. Appl Energy 83:1–14
22. Omid M, Ghojabeige F, Delshod M et al (2011) Energy use pattern and benchmarking of selected green houses in Iran using data envelopment analysis. Energy Convers Manag 52 (1):153–162
23. Cooke D, Randall C (2005) Energy use benchmarks for opencast coal mines. The AusIMM cannual conference, New Castle, pp 1–8
24. Reddy TG, Sudhakar K, Krishna JS (2013) Benchmarking of coal mine using data envelope analysis. Int J Adv Trends Comput Sci Eng 2(1):159–164
25. Sahoo LK, Kumar R, Topno S (2017) Benchmarking diesel consumption for dump trucks in opencast mines. In: International conference DEEP 16, IIT Kharagpur, 24–26 January 2017, pp 1–8
26. Mackay JCD (2009) Sustainable energy –without the hot air: UIT Cambridge Ltd, Cambridge
27. Jajar RN (2009) Vehicle dynamics: theory and application. Springer, New York

Chapter 10
Role of the Operator in Dragline Energy Efficiency

Maryam Abdi-Oskouei and Kwame Awuah-Offei

Abstract Dragline operators, as controllers of one the most energy-intensive equipments in surface coal mines, play a significant role in dragline energy efficiency and thus mine profitability. The literature lacks work that explores monitoring system data and applies data-driven methods to gain a better understanding of dragline operation and develop more effective training approaches. This chapter provides a framework for assessing dragline energy efficiency performance using monitoring data and using such work to improve operator training. The first step in improving dragline performance is the assessment using data from dragline monitoring systems to estimate an overall performance indicator. Next, the analyst should apply a comprehensive algorithm to quantify the relationship between different operating parameters and the overall performance indicator. Finally, operators' performance can be improved by using the results to optimize operator training.

Keywords Energy efficiency · Operator's skills · Data-driven analysis
Mining equipment performance · Dragline

10.1 Introduction

Coal plays an important role in electricity generation and steel production. Coal-fired power plants generated about 40% of total electricity worldwide in 2012. U.S. Energy Information Administration (EIA) predicts that by 2040, control policies over power plant air pollution and greenhouse gas emission will reduce the role of coal as a dominant fuel for electricity generation in many countries.

M. Abdi-Oskouei (✉)
University of Iowa, 417 IATL, Iowa City, IA 52246, USA
e-mail: maryam-abdioskouei@uiowa.edu

K. Awuah-Offei
Missouri University of Science & Technology, 326 McNutt Hall, Rolla, MO 65409, USA
e-mail: kwamea@mst.edu

K. Awuah-Offei (ed.), *Energy Efficiency in the Minerals Industry*, Green Energy and Technology, https://doi.org/10.1007/978-3-319-54199-0_10

However, in China and India, coal-fired power plants will continue to generate a great portion (69% combined by 2040) of electricity [1].

More stringent regulation on the carbon footprint of coal's full life cycle (from cradle-to-grave) forces the mining sector to improve operational efficiency to remain environmentally and economically viable. Based on the U.S. Department of Energy (DOE) energy bandwidth study, there is a high potential for improving energy efficiency in the mining sector [2]. Improving energy efficiency of the mining sector will benefit it by decreasing energy cost, carbon footprint, and environmental costs [3].

The dragline is one of the most energy-intensive equipments in surface coal mines. With high production rate and low cost of operation, draglines are commonly used for removing large amounts of overburden to expose coal seams for extraction. Draglines consume about 15–30% of total mine energy consumption. Improving dragline energy efficiency has a significant impact on the mine energy efficiency and profitability [4]. Dragline operation, not including the walking process, is a cyclic process and consist of filling the empty bucket, hoisting the bucket, swinging out or away to the dumping pile, dumping, returning to the digging spot, and positioning the bucket to start the next cycle. Dragline operation performance is influenced by operating conditions, equipment characteristics, mine design and planning, and operator skills.

Studies have shown that operators have a significant effect on dragline energy efficiency [5–8]. This represents a significant energy efficiency improvement opportunity, given that most of the other factors (perhaps with the exception of mine planning and design) are not as easy to control as operator effects. Consequently, a focus on the role of operators in dragline energy efficiency is warranted.

As pointed by Levesque et al. [9], higher emphasize on production rather than energy conservation is one of the challenges in energy efficiency programs of the mining sector. Lumley [10] shows that a switch in the strategies of mining companies, from a cost-conscious to a volume-based (maximizing production) one, has resulted in a decline in productivity in the mining industry since the mid-2000s. As a result of these pressures, dragline operator performance assessment studies have stressed production and failed to adequately incorporate energy consumption and cost into the performance equation. There is a lack of peer-reviewed publications on dragline energy efficiency research and development in the mining sector [9]. As a result, the sector lacks consistency in terminology and method regarding dragline energy efficiency.

This chapter addresses some of these limitations by presenting a summary of previous studies to establish the current knowledge on dragline energy efficiency, the role of operators, and improvement opportunities. It also provides a framework for assessing dragline energy efficiency performance using monitoring data and using such work to improve operator training. This work can be used as a guideline to improve dragline performance and energy efficiency based on the dragline monitoring setup and mine priorities and policy.

10.2 Assessing, Identifying Key Factors, and Improving Dragline Performance

As pointed out earlier, dragline performance is influenced by operating conditions, equipment characteristics, mine design and planning, and operator skills. These four categories encompass a variety of factors which individually or in combination affect the productivity and energy efficiency of dragline operations.

Operating condition includes geology, material properties, groundwater level, and weather conditions, which vary in different mines, and impact dragline performance [11–13]. It is not possible to change these conditions in a particular mine. These conditions can only be managed by optimizing the mine design to be most compatible with the conditions. An appropriate bucket size, sufficient motor power, and proper gear ratios are some of the equipment characteristics that can increase dragline productivity and reduce energy consumption [14–17]. Mines frequently purchase used draglines and, in some of those cases, the bucket size and drive system may not be completely compatible with the operating condition. Modifying drive system or bucket size can be costly, in practice.

Mine design and planning such as dragline operating task assignment, dragline positioning, maintenance strategies, and blast performance have a significant impact on dragline performance. An optimum mine design should assign tasks to the dragline in proper sequence to maximize mine productivity and keep energy consumption, maintenance cost, and idle time minimum. Assigning inappropriate tasks, such as deep cuts, to draglines can increase cycle time and energy consumption, and therefore make the operation inefficient [12, 18, 19]. Proper maintenance strategies and scheduling will increase dragline availability and utilization, and hence improve dragline performance [20, 21]. Blast performance determines materials diggability, bucket fill factor, and consequently dragline performance [11, 22].

Operators' practice, skills, and habits have been observed to be an important factor affecting dragline performance. Nicholes et al. [23] study is one of the first studies, to the best knowledge of the author, to observe variations in operators' performance using statistical and signal analysis. They observed differences in amplitude histograms, average cycle time, and mean power demand between three operators [23]. Later, Lumley [24] observed a 35% difference in productivity between the best and worst operator in a database containing 150 million cycles of dragline operations from Australia, the U.S., South Africa, and Canada [24, 25]. Other studies have observed differences up to 86% in energy efficiency during the digging phase and up to 15.7% in the complete operational cycle [6–8]. Among the four categories of governing factors, mine design and planning and operator skills and practice are the easiest and least costly to improve.

Operators' habits during the dragline cycle can affect dragline performance and efficiency. Energy consumption and production during the filling process can be influenced by operators' bucket filling practices. Operators can control the engagement and disengagement position of the bucket during the filling process (bucket positioning), fill factor, and time spent on positioning (spot time) and filling the bucket which can influence dragline energy consumption and production [5, 11, 18, 19]. As an example, Bogunovic and Kecojevic [11] investigated the impact of different filling practices on dragline production and digging energy consumption. They showed that, contrary to the standard operator practice, filling the bucket as full as possible regardless of filling time and digging energy does not result in highest production and energy efficiency. In the case study presented by Oskouei and Awuah-Offei [5], dump height, which is controlled by the operator, and stockpile design had a significant impact on dragline energy efficiency.

In this section, we address how to assess, identify key factors, and improve dragline performance, with particular emphasis on the role of the operator. We refer to dragline performance often to simplify the discussion. By this, however, we mean the performance of the dragline when operated by particular operators. This section answers three questions regarding the process of improving dragline performance:

1. How to assess dragline performance?
2. How to identify factors that impact dragline performance?
3. How to improve dragline performance?

Figure 10.1 summarizes the procedure to improve the dragline performance. To assess dragline performance, an analyst needs information on different dragline operating variables measured by a dragline monitoring system (DMS). Next, the analyst can use a well-defined performance indicator to assess and compare dragline performance quantitatively using the measured data to identify the key factors leading to disparities in performance. DMS datasets are excellent sources of information and can be explored with different methods to identify the operating factors that control dragline performance. Feeding operators with information regarding their operation in real time are likely to improve dragline performance and, consequently, mine profitability [14]. Besides, management can further investigate such datasets to personalize and optimize operator training for better results. Each section of this process (outlined by the three questions above) is discussed in more detail in the following subsections.

10.2.1 Assessing Dragline Performance

The process of improving dragline performance starts with measuring relevant operating parameters. Data collected by DMS is the cornerstone of dragline performance analysis. Statistical analysis of DMS data is used to investigate dragline efficiency and quantify the impact of governing factors such as operator practices and mine design on dragline performance. Displaying operating parameters and operator performance relative to other operators, in real time, and notifying the operator in case of dragline overload improves dragline performance and reduces unnecessary stresses on the dragline [14, 26].

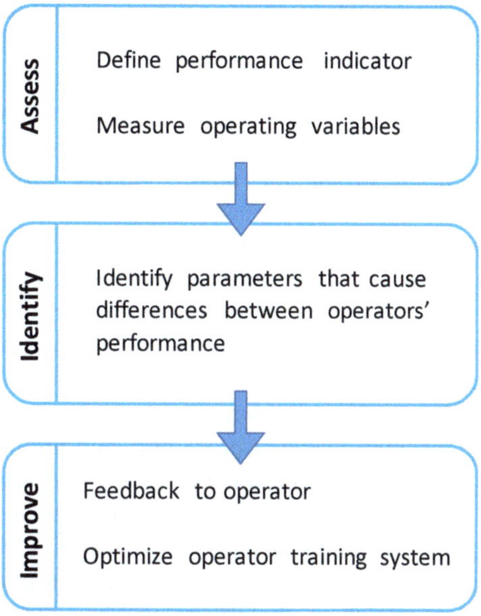

Fig. 10.1 Procedure to improve dragline performance

Before the 1980s, swing charts were mostly used to collect operating data manually. Engineers used these charts and the amount of material moved to assess dragline performance. Later, several different data loggers were developed with the ability to record, store, and report parameters such as total operating time, productive operating time, machine motion performance, average swing angle, vertical hoist to dump, average and maximum drag force, average bucket load, average maximum lowering, and payout speeds [27]. The Tritronics 9000 Monitor, first developed in 1983, is one of the oldest and most popular monitoring systems. In this system, an onboard computer and a radio telemetry were used to store, analyze, and transfer the information to the mine office computer. This system was capable of measuring production in each cycle and determining the bucket position in real time [28, 29]. AccuWiegh™ by Drives & Controls Services (DCS) and Virtual Information Management System (VIMS) by Caterpillar® are other examples of dragline monitoring systems [7, 8, 30]. Advances in computational power and sensor technology have lowered the cost and improved the accuracy of DMS measurements. Today, most draglines are equipped with DMS to measure various metrics during the operation.

To identify best practices and compare the performance of different operators, one needs to define a measure of performance (we call this the overall performance indicator). It must incorporate proxies for the cost and benefit of dragline operation to rank operators' performance in a useful manner. DMS setup, energy price, mine priorities, and policies are some of the constraints and considerations in defining an overall performance indicator. Application of overall performance indicator is not just limited to investigating the role of operators' practice. It can also be used to evaluate and compare dragline operation performance in different mine plans, designs, and maintenance strategies.

Productivity is a common overall performance indicator. Some of the reasons for its popularity, especially when evaluating the impact of mine design and planning on dragline performance, are that it is simple to measure, tangible, and easily related to profit. Production, production rate, production efficiency, and productivity have been used interchangeably in different studies all indicating a measure of production (e.g., material moved) over a measure of time (e.g., cycle time). Rai et al. [12] provided detailed definitions of different measures of production and their application. In this work, production (function of bucket size, bucket fill factor, hours available, swell percentage, average cycle time, rehandle percentage) is used to compare different operating methods and conditions, production index (bank measure of overburden volume moved per period per rated bucket volume) is used in machine selection procedure, and production rate (bank measure of overburden per period) is used for production forecasting and scheduling.

In addition to productivity and similar measures, various studies have investigated the impact of different factors such as blast performance, cut type, and dragline positioning on cycle time (a component of productivity) [12, 18, 31]. Mohammadi et al. [20] recommend overall equipment effectiveness (OEE) as a comprehensive index to measure the performance of loading and hauling equipment (including draglines). OEE is a function of equipment availability, utilization, and production. Since it incorporates time loss (availability and utilization) along with production, it can be used to compare different mine plans and schedules.

While maximizing production rate (increase material moved and decrease cycle time) is critical in evaluating dragline performance, it does not capture the entire picture. With increasing awareness of energy efficiency programs in the mining sector and energy prices, overall performance indicators must include a measure of energy consumption to represent the total cost of the operation. Operators with the highest production rate may consume higher energy for the same amount of work compared to other operators with lower production rate. An operator ranking system based on production rate alone fails to comprehensively capture the production efficiency.

Energy efficiency addresses this issue by incorporating both energy consumption (cost or input) and production rate (benefit or output). It describes how well draglines (or other equipment) converts energy input to useful work. Energy efficiency is defined as the ratio of energy output (work done) to energy input. Due to the limitation in DMS, it is not always possible to measure the exact energy input or output. Different proxy parameters are employed instead of actual values in performance evaluation of different equipments. In hauling and loading operations, the amount material handled (production) and fuel or electricity consumption are used as proxies for energy output and input, respectively [6]. Studies have used electricity consumption during the digging phase or total electricity consumption in a given period (e.g., each cycle) as the proxy parameters for dragline input energy, and total payloads during that period as a proxy parameter for energy output [5–7]. Depending on the proxy parameters, units of energy efficiency vary. Whatever indicator is used, it is important that it is a ratio of energy output to input or proxies of these.

As discussed, different overall performance indicators have been used to fulfill various objectives in different studies. The objective of the study is one of the considerations in defining the overall performance indicator. For example, OEE incorporates time losses (availability and utilization) along with the production and thus can effectively be used to evaluate dragline performance with respect to maintenance strategies. Also, when evaluating dragline performance with respect to operators' practice and skills, it is important to have a measure of energy in the overall performance indicator due to significant variations in operators' energy consumption and efficiency. Including availability, in this case, may be irrelevant since equipment availability depends on mine planning and maintenance strategies and not operators' skills (at least not a major factor). Mine management policy and energy price can also influence the choice of the overall performance indicator. In general, mining companies are moving toward adopting environmental management systems and energy conservation practices due to strict environmental regulation and higher energy prices [9, 32]. As a result, more mines are incorporating measures of energy consumption in the overall performance indicator. However, mines that do not follow energy conservation practices are less likely to consider equipment energy consumption when evaluating the equipment performance and ranking operators [24]. DMS capability and setup can impose constraints on the choice of overall performance indicator and the proxies used to determine this indicator. Improvements in DMS have enabled us to measure and record energy consumption of draglines and incorporate these measurements into performance analysis. Yet, the number of studies using the data on dragline energy consumption to evaluate dragline performance is still limited [5, 6].

10.2.2 Identifying the Factors that Impact Dragline Performance

Overall performance indicators can be used to rank and classify operators based on their performance using statistical analysis [7]. However, identifying the best operator does not provide us with any additional information regarding the specific habits and skills that have led to the superior performance. Further analysis is required to detect the factors that cause differences between operators' performance and result in superior performance. Many studies have used the term "Key Performance Indicator (KPI)" to refer to these factors. Investigating the relationship between operator skills and KPIs provides valuable information to build more effective operator training programs and increase dragline energy efficiency.

Reviewing the literature, only a few studies have explored the relationship between dragline operator skills and dragline performance [5, 11]. However, there are a number of other studies that have looked at the impacts of other governing

Table 10.1 Important KPIs identified in the reviewed literature after the year 2000

KPI	Operating condition	Equipment characteristics	Mine design and planning	Operator practice and skills	Reference
Payload	X	X		X	Lumley [25], Vynne [14], Kizil [33]
Cycle time			X	X	Erdem and Duzgun [18], Rai et al. [22], Mohammadi et al. [31]
Dig time			X	X	Rai et al. [12], Rai [21], Lumley [25], Erdem and Duzgun [18], Williams [4], Bogunovic and Kecojevic [11]
Bucket filling positioning				X	Bogunovic and Kecojevic [11], Oskouei and Awuah-Offei [5], Mohammadi et al. [19]
Fill factor	X		X	X	Bogunovic and Kecojevic [11],

factors on KPIs and dragline performance. Table 10.1 provides a summary of well-known KPIs identified in multiple studies (from the year 2000 to 2016) along with the governing factors that influence them.

Most of these studies use a classic hypothesis-based approach to identify KPIs. First, the authors develop a hypothesis, based on the available prior knowledge or observation, that an operating parameter, which is controlled by governing factors, has an influence on equipment performance (i.e., is a KPI). Then, they design experiments and collect data to test the hypothesis. An example of this method is the study by Bogunovic and Kecojevic [11]. They hypothesized that dragline bucket fill factor (KPI), which is affected by operator skill and practices, has an impact on dragline production rate and digging energy consumption (overall performance indicators). They then designed experiments where operators performed with different targeted fill factors and different descriptive filling practices. By comparing measured production rates and digging energy consumption with the bucket fill factors, they observed differences among tested filling practices and identified the optimal practice.

Dragline operation is a complicated system to model. There are many factors that affect dragline performance in each cycle. Due to the large scale of production and energy consumption in each cycle, a slight variation in the operational condition can result in significant changes in the performance. Data collected for a short duration of the experiment might not capture the high variability of the operation. Besides, the number of operating parameters that can be tested in each experiment is limited. Therefore, it is not possible to explore many different conditions, simultaneously. This limits the range of possible applications of the hypothesis-based approach.

Modern monitoring systems on loading and hauling equipment have provided us with abundant high-quality data. Hence, data-driven approaches that analyze acquired data to draw the same inferences as the hypothesis-based approaches are viable alternatives. This approach addresses the high variability in the system by taking advantage of the abundance of data in DMS datasets and high computational power available in modern computers. It uses statistical analysis and data mining tools to explore and quantify the relationship between operating parameters and overall performance indicator and detects KPIs based on the output of this analysis.

An example of this method is the study by Oskouei and Awuah-Offei [5]. They retrieved the data collected by a DMS for a 1-month period. After removing outliers, they selected operators with sufficient amount of data to study. Next, they used correlation analysis to identify operating parameters that are correlated with energy efficiency (overall performance indicator). Finally, they used regression analysis to quantify the relationship between correlated parameters and the overall performance indicator. Based on probabilistic results of the regression analysis, they identified correlated parameters with a higher probability of being KPIs than the desired threshold.

The limitation of the data-driven approach is that it requires large amounts of high-quality data. To get valid inferences, it is critical to confirm that the amount of data investigated is adequate. Small datasets fail to capture the high variability of the problem. A related caution is that large datasets may result in losing valuable details in the datasets [34]. Due to the nature of this problem, one should expect significant amounts of outliers in the dataset. Engineers and researchers should use appropriate methods to handle outliers.

An important advantage of the data-driven approach is that the results (quantified relation between KPI and dragline performance) can be displayed to the operator in near real time to improve the performance of the operation. In spite of the advantages of this approach, the number of studies that used data-driven approaches in mining equipment performance evaluations is very limited in the literature.

10.2.3 Improving Dragline Performance

The critical effect of operator practice on dragline performance and energy efficiency has been demonstrated in the literature [5, 6, 11]. Various operator training programs are available to improve operators' performance. In a traditional operator training method (sometimes called crew coaching), a leading operator (expert operator) spends time watching and evaluating the less-experienced operator and provides him/her with feedback to increase his/her performance, based on observed suboptimal practices [35]. The feedback is intuitive, based on the operating parameters that have impacts on dragline performance (KPIs). This method of training might be more psychologically acceptable by operators but may result in production loss and dragline damage during the training [36].

Using training simulators to train new or experienced operators helps mines to improve dragline performance. Training simulators build a near-reality condition and give operators the chance to experience different operating conditions without losing production or putting the dragline under stress. Dorey and Knights [37] show a significant improvement in the performance of three out of the four subject operators after training with dragline simulator. This performance improvement was sustained for the month-long duration of the experiment in the case of two of these operators.

Continuous monitoring and training during the actual operation are required to maintain the performance improvement. Data-driven-based monitoring systems provide operators with personalized information on their performance (KPIs) relative to other operators in real time through in-cabin displays. Ranking systems based on comparison to a peer-generated benchmark are more acceptable by dragline operators than a threshold set by mine management [37]. The advantage of data-driven approaches for training is that the feedback is based on real data from the operator and his/her peers. When this data is well communicated, it is a powerful tool to convince the operator of his/her weaknesses and is likely to cause a change in behaviors.

Rapid increase in computational power has enabled us to perform more sophisticated analysis on the data collected by DMS in real time. Displaying the results of this analysis in a user-friendly format to the operators during the operation can help them to adjust their operating practices to match the maximum performance. The detailed information regarding the technology and algorithms used in modern monitoring systems are not published by the developers (proprietary information). However, the information provided by these systems can be used to further quantify the impact of different training systems on performance improvement.

10.3 Recommendations for Future Research

A DMS collects and stores different sets of parameters in each cycle depending on the system. Monitoring dragline operation for even a short period will result in a large dataset that can be overwhelming. This data is a valuable source of information to better understand the role of the operator in the complicated dragline operation. Statistical and data mining methods can be used to explore DMS datasets to derive meaningful information regarding dragline performance in real time, identify less-efficient practices during operation, and develop the best strategies to improve performance and energy efficiency of the operation. In the past, only a small portion of the collected information contributed to useful results, because of data overload and the absence of post-processing software [38, 39]. With increasing concerns about the efficiency of the mining operation, more mines are utilizing advanced monitoring and training systems to extract the information on loading equipment operation and improve the operators' performance. Despite the invaluable outputs of data-driven analysis, not enough studies have employed these methods in dragline studies (only one [5]). Data-driven methods such as cluster analysis and regression analysis have been recently used to investigate the role of operators' practice on electric rope shovel energy efficiency [40, 41]. Similar approaches can be used in the future to better understand dragline operation and design a more effective operators ranking system. The positive impact of operator training simulators on operators' performance is undeniable. However, these improvements may last for a short period [37]. Continuous or real-time training system can be a solution to this problem. Evaluating operators' performance in real time and conveying the information through a user-friendly platform to operators can help them to adjust their operating practices. To develop a real-time training system, one needs a computationally efficient algorithm to convert the real-time data to practical information. Further research should explore how to develop and evaluate the performance and speed of such algorithms in different mines and conditions training system. The result of such studies will lay the groundwork for developing algorithms for automating dragline operation.

10.4 Summary

This chapter summarizes the current knowledge on the role of the operator in dragline energy efficiency, highlights the gaps in the literature, and discusses the possible future paths of research. It also provides a framework for assessing dragline energy efficiency performance using monitoring data and using such work to improve operator training. In general, studies targeting dragline energy efficiency are very limited. Most papers focus on increasing dragline production by maximizing payload and minimizing cycle time and maintenance time. Given the high rate of dragline energy consumption and the urge to improve the energy efficiency of the mining operation, mines have to shift their goal from just maximizing production rate to optimizing production rate and maximizing energy efficiency.

The chapter provides an overview of how operators affect the energy efficiency of dragline operations. The reviewed studies have been summarized in the form of a framework for improving dragline performance. This framework can act as a guide for mine managers and engineers interested in improving dragline performance. The framework is a three-step process that includes assessing, identifying key factors that affect, and improving dragline performance.

The first step in improving dragline performance is to assess dragline performance based on data collected by dragline monitoring systems and a well-defined overall performance indicator. We suggest that mines upgrade their monitoring systems to record energy consumption during the operation and include a measure of energy consumption in the overall performance indicator. The second step is to develop a comprehensive algorithm to explore and quantify the relation between different operating parameters and dragline performance. To capture the high variability and complexity in the dragline operating system, we suggest that mine managers and engineers perform statistical and data mining analysis on the data collected by the monitoring system. Finally, displaying the results (in a user-friendly format) in real time to operators through in-cabin displays provide operators with valuable information regarding their performance. Based on this information, operators can adjust their performance to achieve better efficiency.

In spite of the large volume of operating data collected by monitoring systems, the literature is lacking work that explores this data specifically to improve the energy efficiency of dragline operations. Applying data-driven analysis on DMS data can help to identify less well-known KPIs and better understand the role of operators on dragline performance. The computationally efficient data-driven algorithm can convert real-time operating measurements to information on operators' operating practices and help operators to improve their performance through a real-time training system. This can result in long-lasting improvement in operators' performance, dragline energy efficiency, and thus mine profitability.

References

1. U.S. Energy Information Administration (2016) International Energy Outlook 2016-Electricity. In: International Energy Outlook. pp 81–100
2. U.S. Department of Energy (DOE) (2007) Mining Industry Energy Bandwidth Study
3. Steele R, Sterling D (2011) Identifying Opportunities to reduce the consumption of energy across mining and mineral processing plants. In: SME Annual Meeting, pp 1–5
4. Williams G (2005) Achievements through the dragline improvement group (DIG) in Anglo coal. Inst Quarr South Africa
5. Abdi Oskouei M, Awuah-Offei K (2015) A method for data-driven evaluation of operator impact on energy efficiency of digging machines. Energy Effic. doi:10.1007/s12053-015-9353-3
6. Abdi Oskouei M, Awuah-Offei K (2014) Statistical methods for evaluating the effect of operators on energy efficiency of mining machines. Min Technol 123:175–182. doi:10.1179/1743286314Y.0000000067
7. Komljenovic D, Bogunovic D, Kecojevic V (2010) Dragline operator performance indicator. Int J Mining Reclam Environ 24:34–42. doi:10.1080/17480930902778191
8. Bogunovic D, Kecojevic V, Lund V et al (2009) Analysis and control of energy consumption in surface coal mining. SME Annual Meeting 1–7
9. Levesque M, Millar D, Paraszczak J (2014) Energy and mining-The home truths. J Clean Prod 84:233–255. doi:10.1016/j.jclepro.2013.12.088
10. Lumley G (2014) Mining for efficiency
11. Bogunovic D, Kecojevic V (2011) Impact of fill factor on dragline production rate and energy consumption. Min Eng
12. Rai P, Trivedi R, Nath R (2000) Cycle time and idle time analysis of draglines for increased productivity–a case study. Indian J Eng Mater Sci 7:77–81
13. Bogunovic D (2008) Integrated data environment for analysis and control of energy consumption (Ide-Ace) in surface coal mining. The Pennsylvania State University

14. Vynne JF (2008) Innovative dragline monitoring systems and technologies. In CIM Conference
15. Rowlands C, Just GD (1992) Performance characteristics of dragline buckets. In: Third Large Open Pit Mineral Conference, pp 89–92
16. Pippenger JG (1995) Competing with the big boys: productivity and innovation at the Freedom lignite mineral, pp 3–6
17. Isokangas E (1997) Measuring dragline performance improvement initiatives. Int Congr Autom Technol
18. Erdem B, Düzgün HŞB (2005) Dragline cycle time analysis. J Sci Ind Res 64:19–29
19. Mohammadi M, Rai P, Oraee SK (2015) A critical investigation of digging time segment of draglines in a large surface mine. Geotech Geol Eng 33:763–771. doi:10.1007/s10706-015-9857-9
20. Mohammadi M, Rai P, Gupta S (2015) Performance measurement of mining equipment. Int J Emerg Technol Adv Eng 5:240–248
21. Rai P (2004) Performance assessment of draglines in opencast mines. Indian J Eng Mater Sci 11:493–498
22. Rai P, Yadav U, Kumar A (2011) Productivity analysis of draglines operating in horizontal and vertical tandem mode of operation in a coal mine-a case study. Geotech Geol Eng 29:493–504. doi:10.1007/s10706-011-9398-9
23. Nichols ST, Barton TH, Gunthrope G (1981) Load model of a dragline. IEEE Trans Ind Appl 356–361
24. Lumley G (2004) How to increase dragline productivity without spending too much money. In: Sixth Annual AJM Open Cut Coal Mining Conference
25. Lumley G (2005) Reducing the variability in dragline operator performance. In: Coal Conference, pp 97–106
26. Mirabediny H, Baafi E V (1998) Statistical analysis of dragline monitoring data. In: Third Reg APCOM Symposium, pp 7–9
27. Matuszak RA (1982) How do you measure your dragline output?. First Int, SME-AIME Fall Meet
28. Torrance A, Baldwin G (1990) Blast performance assessment using a dragline monitor. In: Third International Symposium Rock Fragmentation by Blasting, pp 219–224
29. Hawkes PJ, Spathis T, Sengstock GW (1995) Monitoring equipment productivity improvements in coal mines. In: EXPLO conference, pp 4–7
30. Drives & Controls Services (2003) AccuWeigh production monitoring systems
31. Mohammadi M, Rai P, Singh U, Singh SK (2016) Investigation of cycle time segments of dragline operation in surface coal mine: a statistical approach. Geotech Geol Eng. doi:10.1007/s10706-016-9987-8
32. Kusi-Sarpong S, Sarkis J, Wang X (2016) Assessing green supply chain practices in the Ghanaian mining industry: a framework and evaluation. Int J Prod Econ. doi:10.1016/j.ijpe.2016.04.002
33. Kizil M (2010) Improving dragline productivity using a diggability index as an indicator. SME Annu, Meet
34. Sandelowski M (1995) Focus on qualitative methods sample size in qualitative. 179–183
35. Norman S (2011) Variability reduction in dragline operator performance. In: SME Annual Meeting pp 1–2
36. Bernold L, Lloyd J, Vouk M (2003) Equipment operator training in the age of internet2. Nist special publication, pp 505–510
37. Dorey F, Knights PF (2015) Quantifying the benefits of simulator training for dragline operators. Min Technol 124:97–106. doi:10.1179/1743286315Y.0000000007
38. Hettinger D, Lumley G (1999) Using data analysis to improve dragline productivity. Coal Age

39. Morrison R, Scott A (2002) Maximising the value of "A Wealth of Information." Value Track symposium, pp 7–8
40. Babaei Khorzoughi M, Hall R (2016) A study of digging productivity of an electric rope shovel for different operators. Minerals 6:48. doi:10.3390/min6020048
41. Vukotic I, Kecojevic V (2014) Evaluation of rope shovel operators in surface coal mining using a Multi-attribute Decision-making model. Int J Min Sci Technol 24:259–268. doi:10.1016/j.ijmst.2014.01.019

Part III
Mineral Processing and Extractive Metallurgy

Chapter 11
Energy-Efficient Comminution: Best Practices and Future Research Needs

Bern Klein, Chengtie Wang and Stefan Nadolski

Abstract The mining energy value chain starts at the face and extends to smelting and refining. System designs that conserve energy and apply energy-efficient technologies can result in significant reductions in overall energy usage. A main component of the energy value chain involves comminution, which accounts for about 50% of all energy used by mines. This chapter summarizes innovations of the state of the art with respect to energy-efficient comminution technologies and process circuit designs. The chapter will also present practices to measure and benchmark energy efficiency.

Keywords Comminution · Comminution energy value chain · Technology Circuit design · Measuring energy efficiency

11.1 Introduction

Mining is an energy-intensive activity and is vulnerable to risks associated with fluctuating energy costs as well as to government regulations aimed at reducing greenhouse gas (GHG) emissions. Therefore, energy conservation and efficiency are motivated by the need to both reduce operating costs and to meet the demands of society to reduce the carbon footprint. Several international and national mining associations, as well as utility companies, have developed initiatives aimed at improving energy conservation and improving energy efficiency [1–5]. The Mining Association of Canada initiated a program called Towards Sustainable Mining (TSM) that produced a reference guide for energy and greenhouse gas emission

B. Klein (✉) · C. Wang · S. Nadolski
Mining Engineering, University of British Columbia, Vancouver V6T 1Z4, Canada
e-mail: bklein@mining.ubc.ca

C. Wang
e-mail: chengtie@mail.ubc.ca

S. Nadolski
e-mail: snadolski@mining.ubc.ca

K. Awuah-Offei (ed.), *Energy Efficiency in the Minerals Industry*, Green Energy and Technology, https://doi.org/10.1007/978-3-319-54199-0_11

management [6]. The program led to substantial advances related to energy use and greenhouse gas emissions such that by 2014, 75% of Canadian mining operations had management systems, 87% had implemented reporting systems and 60% had established and met targets [7]. Additionally, the Coalition for Energy-Efficient Comminution (CEEC) was established to promote knowledge transfer leading to improved energy practices in comminution [5].

System designs that conserve energy and apply energy-efficient technologies can result in significant reductions in overall energy usage. Improvements in energy performance can be gauged by comparison to benchmarks for comminution energy. This chapter summarizes innovations of the state of the art with respect to energy-efficient comminution technologies and process circuit designs. Also, the chapter discusses benchmarking of comminution energy. We use a review of the pertinent literature to summarize the state of the art and to determine the best practices to measure and benchmark energy efficiency. This chapter relies mainly on the most recent (mainly work published since 2007) information about energy-efficient technologies. The chapter ends with recommendations for further research to further improve the energy efficiency of comminution.

11.2 Comminution Energy Value Chain

The mining energy value chain starts at the face and extends to smelting and refining. The main component of this value chain is comminution, which accounts for 53% of all energy used in the mining industry [5]. For large open-pit mines, comminution has been reported to account for up to 70% of all electrical energy consumed at a mine [2]. Therefore, any improvements to the energy performance of comminution processes represent significant overall energy savings.

Figure 11.1 presents the scale of size reduction that takes place within the comminution component of the energy value chain. While the main comminution methods are blasting, crushing, and grinding, other important technologies that can affect energy usage include continuous mining, ore sorting, rock weakening and size classification technologies. Systems engineering approaches that integrate mining and processing activities have led to concepts of Mine to Mill and Geometallurgy that can be used to develop strategies to optimize the overall energy use.

While the focus of this chapter is on comminution technologies, operational improvements and circuit designs, there are associated technologies that should be considered for improving the overall efficiency of the comminution energy value chain.

For example, sorting systems can reduce the amount of material crushed and the associated energy consumed. Figure 11.2 shows a sensor-based sorting system. Energy savings are directly proportional to the mass of rejected material that does not require crushing and grinding [9, 10]. One study on seven nickel operations in Ontario, Canada showed substantial reductions in operating costs and overall

Fig. 11.1 Comminution component of energy value chain [8]

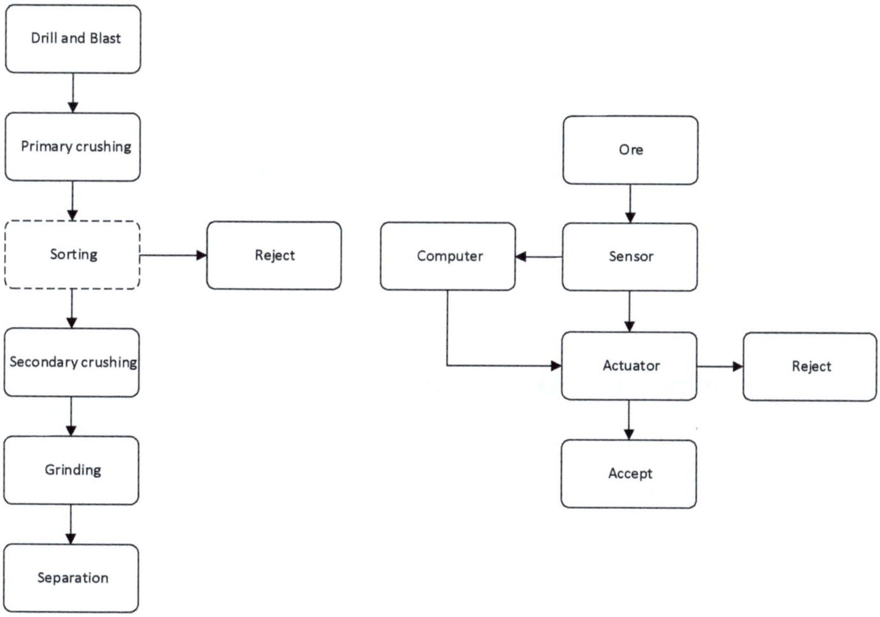

Fig. 11.2 Sensor-based sorting system

energy savings of 20% [10]. Researchers have developed sorting systems for rocks as well as bulk materials [9, 11].

Several technologies have been developed at the conceptual and pilot evaluation stages that if commercialized stand to improve overall energy efficiency. Some advances have been achieved in the area of rock/particle weakening. A significant breakthrough has been made using microwaves that fragment or weakens rock [12]. A very recent study published details of the design, commissioning, and operation of a pilot-scale high-power microwave treatment process with a continuous throughput capacity of up to 150 tph [13]. Similarly, researchers have developed the Selfrag technology that uses a high voltage pulse to break or weaken rocks by inducing electrical explosions causing shockwaves [14–16]. The weakening results in lower grinding energy and breakage is reported to be preferential along grain boundaries enhancing liberation for improved mineral separation.

In open-pit mines, improvements in mill performance have been achieved by increasing the energy intensity of blast patterns, thereby reducing the size of feed to the mill. A net improvement in energy efficiency results due to a reduction in crushing and grinding requirements [17]. The approach has been primarily aimed at increasing mill production rates, which are often in the order of 20% for the same grinding input energy levels. Hydraulic fracturing is carried out to improve cave-ability and reduce hang-ups by improving fragmentation [18]. While no specific studies have assessed the impacts on downstream comminution, hydraulic fracturing induces fractures resulting in smaller fragments, which is similar to increasing blast fragmentation, and therefore should reduce crushing and grinding energy usage.

The implementation of geo-metallurgical programs to support design and mine planning can also lead to improvements in operational performance as well as significant energy savings. Studies conducted by Amelunxen et al. [19] and Bueno et al. [20] applied to geometallurgy and Monte Carlo simulation to support trade-off studies and design decisions for high pressure grinding rolls (HPGR) and semi-autogenous comminution circuits. When considering ore variability, the effect of specific energy requirements on operating costs supported such design decisions.

11.3 Comminution Technologies

The principal purpose of comminution is to increase the degree of liberation between valuable and non-valuable constituents of the ore. The basic components of the comminution taking place at mining operations are blasting, crushing and grinding.

Conventional crushing and grinding technologies such as jaw crushers, gyratory crushers, cone crushers, autogenous/semi-autogenous grinding mills (AG/SAG mills), high pressure grinding rolls (HPGRs), rod mills, and ball mills are described in the well-known mineral processing textbooks [21, 22]. Apart from improvements in machine material and capacity, process control, modeling, and simulation, there have not been many breakthroughs in these comminution technologies. Although not new technologies, two energy-efficient technologies that represent best available technologies, with respect to energy usage, are high pressure grinding rolls, and high-speed stirred mills.

11.3.1 Crushing Technologies

Professor Schönert [23] developed high pressure grinding technology following his work which showed that improvements in energy efficiency could be achieved by breaking particles in compressive beds. He applied the concept to the development of the HPGR in which material is choke fed between two counter-rotating rollers, one of which is fixed and the other is floating in connection with a hydraulic cylinder (Fig. 11.3) [24].

In comparison to SAG mill circuits, the HPGR has resulted in electrical energy savings that range from 20 to 30% over SAG mill circuits [25, 26]. Despite these energy benefits, the uptake of the technology was slow. While the HPGR found application in the cement industry and for iron ore mines, the first large-scale installation in a base metal operation was not until 2007, almost 30 years later after the technology was conceived in 1979. The slow uptake was attributed to issues that had to be overcome in relation to roller wear rates and capital costs that are higher than the more conventional SAG-based circuits.

To advance the HPGR technology, a piston press test method was developed [27–29] to enable HPGR assessment for early stage projects, testing ore variability and modeling HPGR circuits to support operational improvements. Compared to conventional HPGR pilot testing, piston press tests only require a small amount of sample and provide process data that is suitable for scoping level assessment.

Another technology that breaks particles under compression is the Loesche Vertical Roller Mill. The VRM can achieve high reduction ratios, reducing particle sizes to below 45 µm. A study conducted on Platreef ore compared the energy requirements of the VRM to ball milling from size reduction from a P_{80} of 12 mm to a P_{80} of 45 µm [30]. The VRM energy requirement was about 6 kW-h/t as compared to 13 kW-h/t for the ball mill, representing an energy savings of about 55%.

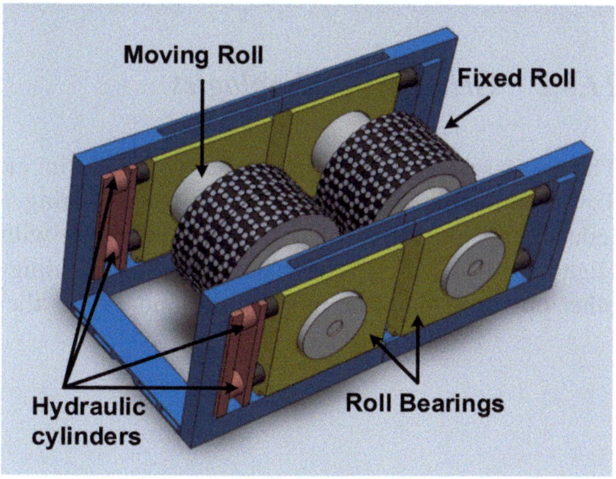

Fig. 11.3 High pressure grinding rolls [27]

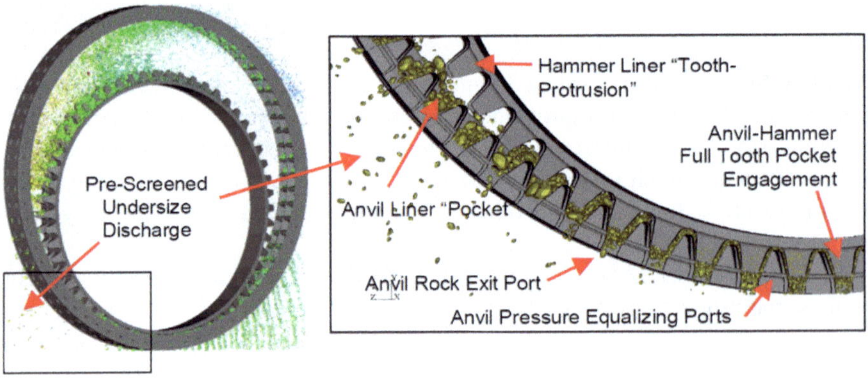

Fig. 11.4 CAHM Mill—Hammer/Anvil Mesh [32]

Conjugate Anvil Hammer Mill (CAHM) is a novel technology that breaks particles under compression. Compression loading is applied to the particles; however, the broken particles pass through a slotted opening preventing wasting energy on bed compression. Based on DEM modeling, the CAHM shows significant potential to be more energy-efficient than the HPGR [31, 32] (Fig. 11.4).

There are several emerging and novel crushing technologies that are aimed at fine or superfine crushing that is more energy-efficient than grinding. These technologies can replace conventional grinding and are considered more energy-efficient. These include the Loesche Vertical Roller Mill as well as the Vibrocone, Horomill, HICOM mill and IMPTEC crusher [33–36]. IMPTEC have developed a technology that extends the limits to "Superfine Crushing" generating products with P_{80}'s below 20 μm. The crushing is achieved in a rotating crushing chamber with a gyrating mandrel [36]. Challenges related to these technologies relate to maintenance and scale-up to production levels needed in large-scale mining.

11.3.2 Grinding Technologies

The most widely used grinding technologies include SAG, rod, and ball mills. Tower mills were introduced to the metal mining industry in the 1970s and the industry applied them, mostly, to applications with target grinds below 20 μm. High-speed stirred mills have been used for grinding clays, calcium carbonates and other powders for some time. In the mid-1990s, they were introduced to fine and ultrafine grinding in metal mines. Mount Isa Mines worked with Netzsch to scale-up a high-speed stirred mill as a means of liberating fine-grained polymetallic ores [37]. Similarly, the Stirred Media Detritor (SMD) which was originally used for kaolinite grinding started to be used for metal mining. Since the introduction of these two high-speed stirred mills, others like the VXP and HIG mills have been developed for metal mining [38, 39]. The technology represents a breakthrough due to its capability to produce particle sizes finer than 10 μm with relatively low energy input compared to tower mills or ball mills. For example, comparison of the ISA Mill to the Vertimills for fine regrinding to 80% passing 18 μm, showed energy savings of 50% [40]. For grinding below 10 μm, energy savings are even greater. Research has also shown that selective comminution can be achieved using the speed control on a high-speed stirred mill. Where valuable minerals are harder than the gangue, energy utilization can be improved by operating at a shaft speed that promotes liberation of valuable minerals and minimizes the grinding of gangue [41–43].

11.4 Comminution Circuit Designs

It is well known that the energy efficiency of crushing technologies is better than downstream grinding. There is, therefore, a strong motivation to design comminution processes that maximize the effort in the crushing circuit. In conventional circuit design, secondary cone crusher product is fed to tertiary HPGRs, which operate in closed-circuit with wet fine screens followed by ball mills.

Advances in cone crusher technology and the introduction of the Vertical Shaft Impactor (VSI) crusher have led to the extension of the feed and product size ranges of crushers to finer particle sizes. Research has shown that mines can save energy by using quaternary crushing to feed stirred mills such as tower mills.

Figure 11.5 shows a fine crushing flowsheet with possible combinations of tertiary and quaternary crushers including cone crushers for all stages, HPGRs for tertiary and quaternary crushing and the VSI for quaternary crushing. The objective of the circuit designs is to maximize energy utilization in the crushing circuit to prepare fine feed for downstream processing.

A study to assess combining HPGR and stirred mill into a single flowsheet without tumbling mills for several copper operations in British Columbia, Canada showed a reduction in energy consumption of 15–36% compared to existing circuit configuration [25, 26] (Fig. 11.6).

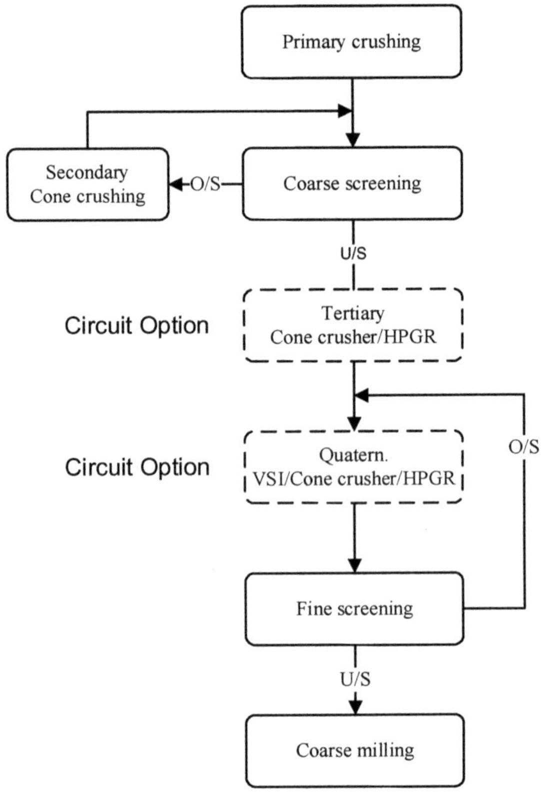

Fig. 11.5 Proposed fine crushing circuits arrangement for comparison

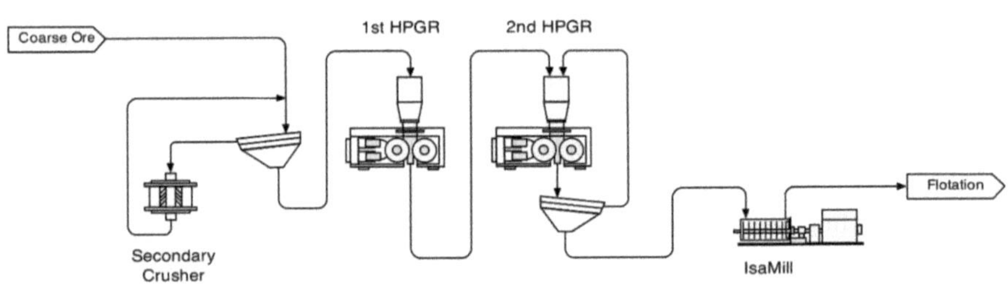

Fig. 11.6 Proposed HPGR—stirred mill circuit

11.5 Operational Improvements

The primary means for improving energy efficiency has been to make incremental changes to the design and operation of mines using well-established and mature technologies. There are numerous examples demonstrating the benefits of process control and automation that represent best practices and can help to reduce energy usage by about 25% [44].

Energ-focused operational improvements are realized through iterative tuning and optimizing of the comminution circuit and individual comminution devices. One of the keys to maximizing comminution energy efficiency is to optimize and balance power used for breakage in all installed comminution equipment, to control the transfer size within the comminution circuit. A recent study at the Detour Lake Mine showed that choke feeding a gyratory crusher has a positive impact on overall throughput, and reduces the specific energy consumption of downstream processes [45]. Another common practice is to adjust the transfer size between SAG mill and ball mill circuit by changing the screen opening size, which improves overall throughput capacity and consequently the energy performance of the circuit.

For any comminution equipment, the specific energy consumption is defined as the energy consumed per unit mass of throughput and determined by a range of operational parameters such as ore properties, equipment parameters, and operating conditions. There are optimum conditions of these operational parameters that lead to efficient comminution operation. For example, In an HPGR operation, key parameters are feed size and moisture content, pressing force, and the rotational speed of the rollers. It is known that there is diminishing size reduction due to the phenomenon of energy saturation, which suggests an optimum energy input for size reduction in the HPGR. Similarly, in SAG milling, there is an optimum point for rock/ball filling, mill speed, and product size to achieve optimum throughput and specific energy consumption [46, 47]. However, operational strategies at many mine sites are based on a simple philosophy—maximizing installed power (such as running maximum HPGR force pressure or maximum SAG/Ball mill rotational speed), which may not result in optimum grinding performance and certainly will not result in improved energy efficiency.

Mill liners and rotational speed have a significant impact on the charge trajectory in the mill and consequently affect the milling efficiency. Various studies demonstrated the approaches for the optimization of liner profile and configuration for operational performance improvement [47, 48]. Large mining corporations such as Barrick Gold [49] and Newmont Mining [50] have also recognized the significance of energy efficiency. They have conducted numerous evaluation studies in several of their worldwide operations. Both reported that significant energy savings and CO_2 emission reductions were achieved by implementing best practices, and some of their successes resulted directly from the improved liner and lifter design and selection.

Advances in speed controls have enabled development of control systems that result in more responsive and improved process operation and lower energy consumptions. Flexible speed control can support process optimization and eliminate design risk for a process [51]. When ore specifications such as ore hardness and particle size distribution vary, operational parameters can be continuously adapted to the load to prevent high energy loss. In addition, variable speed drives allow the tumbling mills to be slowed down during periods when upstream or downstream constraints occur [52].

11.6 Energy Efficiency and Energy Benchmarking

Efforts by mining operations to reduce energy consumption requires benchmarks to compare against. Natural Resources Canada [53, 54] and the US Department of Energy [44] surveyed operations to benchmark energy consumption. According to the DOE report, in 2007 the average overall energy consumption in metal mining was almost 49.5 kW-h/t, 45.8 kW-h/t for coal and 9.7 kW-h/t for industrial minerals. The main energy consumers are material handling (17%), ventilation (10%) for underground mines and grinding (40%) [44]. The report also states that applying best practices and conducting research and development on energy-efficient technologies would result in a 54% savings of total energy consumed by the mining industry. Furthermore, innovations that significantly reduce energy consumption in any of these three main areas will have an impact on overall energy consumption.

Defining energy efficiency as the energy required for breakage divided by the energy used by the mechanical system, the energy efficiency for comminution is reported to range from 0.1 to 2% [55, 56]. Based on fundamentals of fracture mechanics, Tromans [57] hypothesized that there is a maximum limiting energy efficiency for comminution of between 5 and 10%.

To achieve best practice at an operating plant, there are three key areas of focus.

1. Implement an energy management system (e.g., ISO 50001) and enhance training both at management and operation level [58]. The human factor is always considered as one of the major contributors to success.
2. Optimize the comminution circuit. This is achieved through modeling and process simulation to improve production rates and energy efficiency. Due to rapid increases in computing capacity, it is now possible to conduct highly sophisticated and realistic simulations of comminution processes using techniques such as Discrete Element Modeling (DEM), Computational Fluid Dynamics (CFD), and Finite Element Methods (FEM). Advanced process control systems with high-tech instruments allow a plant to better tackle variable ore types and process fluctuations, thus improving the plant performance and reducing overall specific energy.
3. Measure and benchmark energy efficiency so that operational changes aimed at improving energy efficiency can be assessed.

The CEEC supports an Energy Curve Program that compiles comminution energy usage information from mining operations [5, 59]. The energy curve tool can be used to assess equipment performance by comparing to similar operations in the database. A similar approach was developed by Doll [60].

While benchmarking to similar operations allows a broad assessment of energy efficiency, the assessment is complicated by variations in ore types within a deposit. Therefore, an operation may not be able to assess the effect of mine-specific changes on energy efficiency. One approach is to compare operational energy usage to energy usage as predicted by the Bond Work Index, which is recommended by the Global Mining Standards and Guidelines [61]. The approach assumes 100% efficiency for an operation with comminution energy consumption equal to the Bond predicted energy consumption. Another approach is to use a single passing screen size to identify the energy efficiency of comminution equipment [62], which is referred to the size-specific energy (SSE) method. In this method, the specific energy is plotted against the generation of new material finer than a marker size, typically 75 μm.

One issue with the Bond method is that for some new energy-efficient technologies such as the high pressure grinding roll and high-speed stirred mills, the method can indicate efficiencies exceeding 100%. Also, the size-specific energy method is not effective to assess fine grinding process due to the limitation of marker size. The Benchmark Energy Factor (BEF) is an alternative approach and is defined as the ratio of operational energy usage to the minimum practical energy required to carry out the comminution duty of the crushing and grinding circuit that is being assessed [63–65].

$$\text{BEF} = \frac{\text{Measured energy used for comminution at the mine } (\text{kWh/t})}{\text{Minimum practical comminution energy } (\text{kWh/t})}$$

Using single-particle compression testing, the energy-breakage relationship is determined for a wide particle size range of approximately 2–65 mm. The minimum practical energy is determined from the plant feed and product sizes, and the hardness of the ore as determined by the test regimen. Quantifying improvements in energy efficiency due to implementing operational changes can be difficult due to the variation in feed size and ore hardness that takes place before and after plant modifications have been made. The BEF method accounts for this variation in plant feed properties and can be used as an effective tool for gauging the energy performance of an operation.

11.7 Future Research Needs

There is significant opportunity to improve energy efficiency for comminution. To develop new energy-efficient technology, we need improved understanding of fracture mechanics applied to rocks. In mineral processing, mineral particles are broken by abrasion, attrition and massive fracture and fragmentation depend on the applied shearing loads, off-center loading, slow compression or impact. There is a volume of knowledge outside of the mineral industry describing breakage mechanisms as a function of parameters such as rates of loading and the role of microstructures for brittle materials [66–68]. Such knowledge is fundamental to the development of technologies that will use less energy and break particles along grain boundaries rather than across them to enhance liberation.

Advances in the following areas will lead to improved comminution energy efficiency:

- Standard and accurate testing procedures to support energy-efficient plant design,
- Energy-efficient comminution machines,
- Selective comminution technologies that promote breakage along grain boundaries to enhance liberation without needlessly fracturing particles across grains,
- Enhanced instrumentation devices for advanced process control, and advanced modeling and simulation tools for plant design and optimization,
- More efficient high capacity size classification technologies, particularly for size range below which screening rates and efficiencies are too slow and above which classifying hydrocyclones can be applied,
- Advance innovative particle weakening technologies such as microwave, electric pulse and hydrofracking that can be applied in-situ or during material handling.

There is also a need for research to recover waste energy during comminution. Comminution processes generate heat that may be recoverable for secondary usage such lighting and heating buildings.

11.8 Summary

Based on the DOE Report [44], application of best available technologies will reduce energy usage by 30% as compared to current practices. Based on potential energy-saving technologies, the potential energy savings are about 60% and by assessing the theoretical minimum, the energy savings can be up to 80%. In the 10 years since 2007, with the advancement and increased application of energy-efficient technologies such as the HPGR, high-speed stirred mills and the development of sensor-based sorting to reject rock ahead of comminution, savings that sum up to 50% have been reported [10, 26, 40]. Research into the areas listed above will allow even greater energy savings in the future.

References

1. MAC (2004) Towards sustainable mine guiding principles. http://mining.ca/towards-sustainable-mining/tsm-guiding-principles
2. BC Hydro (2016) Power smart. https://www.bchydro.com/powersmart/business/types-of-business-customers/mining.html
3. ICMM (2015) International council for mining and metals statement on climate change. http://www.icmm.com/en-gb/news/2015/icmm-issues-statement-on-climate-change. Accessed 20 July 2017
4. ICMM (2015) International council for mining and metals principles. http://www.icmm.com/en-gb/about-us/member-commitments/icmm-10-principles/the-principles
5. CEEC (2017) Coalition for energy efficient comminution. http://www.ceecthefuture.org/energy-curve-program/
6. MAC (2014) Energy and greenhouse gas emissions management reference. http://mining.ca/sites/default/files/documents/EnergyandGreenhouseGasEmissionsManagementReferenceGuide2014.pdf
7. MAC (2015) Towards sustainable mining progress report 2015. http://mining.ca/towards-sustainable-mining/tsm-progress-report-2015/energy-efficiency-aggregate-performance
8. Klein B, Altun NE, Scoble MJ (2015) Comminution: technology, energy efficiency and innovation. In Proceedings of 24th International Mining Congress and Exhibition of Turkey (IMCET2015), Antalya, Turkey
9. Wotruba H (2006) Sensor sorting technology—is the minerals industry missing a chance? In: XXIII International mineral processing congress
10. Bamber AS, Klein B, Pakalnis RC, Scoble MJ (2008) Integrated mining, processing and waste disposal systems for reduced energy and operating costs at Xstrata Nickel's Sudbury Operations Institute of Materials. Min Technol 117:142–153
11. Klein B, Bamber AS (2017) Sensor based sorting. In: SME Mineral processing handbook
12. Charikinya E, Bradshaw S (2014) Use of X-ray computed tomography to investigate microwave induced cracks in sphalerite ore particles. In: International mineral processing congress
13. Buttress AJ, Katrib J, Jones DA, Batchelor AR, Craig DA, Royal TA, Dodds C, Kingman SW (2017) Towards large scale microwave treatment of ores: Part 1—basis of design, construction and commissioning. Miner Eng 109:169–183. doi:10.1016/j.mineng.2017.03.006
14. Cabri LJ, Rudashevsky NS, Rudashensky VN, Oberthür T (2008) Electric-pulse disaggregation (EPD), hydrosearation (Hs) and their use in combination for mineral processing and advanced characterization of ores. In: 40th Annual Canadian mineral processors conference, pp 211–235
15. Zou W, Shi F, Manlapig E (2014) The effect of metalliferous grains on electrical comminution of ore. In: International mineral processing congress
16. van der Wielen K, Weh A, Giese H, Käppeler J (2014) High voltage breakage: a review of theory and applications. In: International mineral processing congress
17. Esen S, La Rosa D, Dance A, Valery W, Jancovic A (2007) Integration and optimisation of blasting and comminution processes. In: EXPLO conference
18. He Q, Suorineni FT, Oh J (2016) Review of hydraulic fracturing for preconditioning in cave mining. Rock Mech Rock Eng
19. Amelunxen P, Mular MA, Vanderbeek J, Hill L, Herrera E (2011) The effects of ore variability on HPGR trade-off economics. In: Fifth international conference on autogenous semiautogenous grinding
20. Bueno M, Foggiatto B, Lane G (2015) Geometallurgy applied in comminution to minimize design risks. In: Sixth international conference on semi-autogenous high press. Grinding technology

21. Wills BA, Finch J (2015) Wills' mineral processing technology: an introduction to the practical aspects of ore treatment and mineral recovery. Butterworth-Heinemann
22. Napier-Munn TJ, Centre JKMR (1996) Mineral comminution circuits: their operation and optimisation. Julius Kruttschnitt Mineral Research Centre
23. Schönert K (1979) Aspects of the physics of breakage relevant to comminution. In: Tewksbury 4th symposium on fracture. Fracture at work
24. Schönert K (1988) A first survey of grinding with high-compression roller mills. Int J Miner Process 22:401–412
25. Wang C (2013) Comparison of HPGR—ball mill and HPGR—stirred mill circuits to the existing AG/SAG mill—ball mill circuits. University of British Columbia
26. Wang C, Nadolski S, Mejia O, Drozdiak J, Klein B (2013) Energy and cost comparisons of HPGR based circuits with the SABC circuit installed at the Huckleberry mine. In: 45th Annual meeting Canadian mineral processors
27. Davaanyam Z, Klein B, Nadolski S (2015) Using piston press tests for determining optimal energy input for an HPGR operation. In: Sixth international conference on semi-autogenous high press. Grinding technology
28. Davaanyam Z, Klein B, Nadolski S, Kumar A (2013) A new bench scale test for determining energy requirement of an HPGR. In: Materials science and technology. Montreal, Quebec, Canada, pp 1917–1925
29. Davaanyam Z (2015) Piston press test procedures for predicting energy—size reduction of high pressure grinding. Doctoral dissertation, The University of British Columbia
30. Little WM, Mainza AN, Becker M, Gerold C, Langel J, Naik S (2015) Assessing the performance of the Vertical Roller Mill for grinding Platreef ore. In: Sixth international conference on semi-autogenous high press. Grinding technology
31. Nordell L (2012) Conjugate anvil-hammer mill
32. Nordell L, Potapov A (2015) Novel comminution machine may vastly improve crushing-grinding efficiency. In: Sixth international conference on semi-autogenous high press. Grinding technology
33. Genç Ö, Benzer AH (2009) Horizontal roller mill (Horomill) application versus hybrid HPGR/ball milling in finish grinding of cement. Miner Eng 22:1344–1349. doi:10.1016/j.mineng.2009.08.010
34. Altun D, Gerold C, Benzer H, Altun O, Aydogan N (2015) Copper ore grinding in a mobile vertical roller mill pilot plant. Int J Miner Process 136:32–36. doi:10.1016/j.minpro.2014.10.002
35. Cleary PW, Owen PJ (2016) Using DEM to understand scale-up for a HICOM mill. Miner Eng 92:86–109. doi:10.1016/j.mineng.2016.03.004
36. Kelsey CG, Kelly JR (2014) Super-fine crushing to ultra-fine size the "IMP" super-fine crusher. In: International mineral processing congress
37. Gao M, Weller KR, Allum P (1999) Scaling-up horizontal stirred mills from a 4-litre test mill to a 4000-litre "IsaMill." In: Powder technology symposium
38. Giyani BF, Klein B, Rahal D (2017) Analysis of variables governing the operation of a vertical stirred mill. In: Conference on metallurgy 2017
39. Erb H, van de Vijfeijken M, Hanuman Y, Rule CM, Swart WCE, Lehto H, Keikkala V (2015) An Initial review of the metallurgical performance of the HIGmill in a primary milling application in the hard rock mining industry. In: Sixth international conference on semi-autogenous high press. Grinding technology
40. Parry J, Klein B, Lin D (2006) Ultrafine grinding for improved mineral liberation in regrind concentrates. Mineral Engineering Ultrafine Grinding 2006
41. Parry J (2006) Ultrafine grinding for improved liberation in flotation concentrate. MASc thesis, University of British Columbia
42. Tong L, Klein B, Zanin M, Quast K, Skinner W, Addai-Mensah J, Robinson D (2013) Stirred milling kinetics of siliceous geothitic nickel laterite for selective comminution. Miner Eng 109–115

43. Roufail R, Klein B (2016) Effect of mineral type on specific breakage entering in a stirred mill. In: Proceedings of XXVIII international mineral processing congress
44. DOE (2007) Mining industry energy bandwidth study
45. Dupont JF, McMullen J, Rose D (2017) The effect of choke feeding a gyratory crusher on throughput and product size. In: 49th Annual Canadian minerals processors conference
46. Powell MS, Weerasekara NS, Cole S, Laroche RD, Favier J (2011) DEM modelling of liner evolution and its influence on grinding rate in ball mills. Miner Eng 24:341–351. doi:10.1016/j.mineng.2010.12.012
47. Powell MS, Smit I, Radziszewski P, Cleary P, Rattray B (2006) Selection and design of mill liners. B Adv Comminution 331–376
48. Radziszewski P, Orford I, Strah L (2003) Lifter design using a DOE approach with a DEM charge motion model. In: 35th Annual meeting Canadian mineral processors, pp 527–542
49. Buckingham L, Dupont J-F, Stieger J, Blain B, Brits C (2011) Improving energy efficiency in Barrick grinding circuits. In: Fifth international conference on autogenous semiautogenous grinding
50. Giblett A, Hart S (2016) Grinding circuit practices at Newmont. 13th AusIMM Mill Oper. Conf
51. Atutxa I, Legarra I (2015) Stepping forward: using variable speed drives for optimizing the grinding process in SAG and ball mills. In: Sixth international conference on semi-autogenous high press. Grinding technology
52. Cebeci T, Klein B, Wang C, Atutxa I, Legarra I (2016) Speed is the key factor: optimizing grinding process in comminution circuits by using variable speed drives. Comminuton'16
53. NRCan (2005) Benchmarking the energy consumption of Canadian Open Pit Mines
54. NRCan (2005) Benchmarking the energy consumption of Canadian Underground Bulk Mines
55. Fuerstenau DW, Abouzeid A-Z (2002) The energy efficiency of ball milling in comminution. Int J Miner Process 67:161–185
56. Tromans D, Meech JA (2002) Fracture toughness and surface energies of covalent minerals: theoretical estimates for oxides, sulphides, silicates and halides. Miner Eng 15:1027–1041
57. Tromans D (2008) Mineral comminution: energy efficiency considerations. Miner Eng 21:613–620
58. NRCan (2014) Global energy management system implementation: case study
59. Ballantyne GR, Powell MS (2016) Using the comminution energy curves to assess equipment performance. Comminution'16
60. Doll AG (2016) An updated data set for SAG mill-power model calibration. In: Proceedings of XXVII international mineral processing congress
61. McIvor R (2016) The GMSG guideline for determine the bond efficiency of industrial grinding circuits. In: Proceedings of XXVIII international mineral processing congress
62. Levin J (1992) Indicators of grindability and grinding efficiency. J South African Inst Min Metall 92:283–290
63. Nadolski S, Davaanyam Z, Klein B, Zeller M, Bamber A (2013) A new method for energy benchmarking of mineral comminution. In: 23rd World mineral congress, Montreal
64. Nadolski S, Klein B, Kumar A, Davaanyam Z (2014) An energy benchmarking model for mineral comminution. Miner Eng 65:178–186. doi:10.1016/j.mineng.2014.05.026
65. Nadolski S, Klein B, Gong D, Davaanyam Z, Cooper A (2015) Development and application of an energy benchmarking model for mineral comminution. In: Sixth international conference on semi-autogenous high press. Grinding technology
66. Ramesh KT, Hogan JD, Kimberley J, Stickle A (2015) A review of mechanisms and models for dynamic failure, strength and fragmentation. Planet Space Sci 107:10–23
67. Hogan JD, Farbaniec L, Daphalapurkar N, Ramesh KT (2016) On compressive brittle fragmentation. J Am Ceram Soc 1–11
68. Leissner T, Duing HH, Rudolph M, Heinig T, Bachmann K, Gutzmer J, Schubert H, Peuker UA (2016) Investigation of mineral liberation by transgranular and intergranular frac-ture after milling. In: XXVII International mineral processing congress

Chapter 12
Energy Efficiency of Electrowinning

Michael S. Moats

Abstract The winning of high purity metal from aqueous solutions through electrodeposition is the final processing recovery step for many nonferrous metals. Direct electrical current/voltage provides the necessary driving force to promote the necessary reactions at an industrially relevant rate. Energy, especially electrical, is often the highest cost for electrowinning operations. Therefore, energy efficiency is a paramount concern for modern facilities. This chapter discusses electrical energy consumption in aqueous electrowinning with a specific focus on cell voltage and current efficiency. It also presents potential improvements.

Keywords Energy · Voltage · Current · Efficiency · Electrowinning

12.1 Introduction

Electrowinning is the primary method for producing high purity metal from aqueous solutions. High purity copper, zinc, nickel, cobalt, manganese, gold, silver, indium, and cadmium are all produced commercially by electrowinning. Other metals, such as aluminum, rare earths, magnesium, and sodium are produced electrolytically as well from molten salt electrolytes. This chapter discusses the energy efficiency of electrowinning from aqueous solutions, as it is more likely to be practiced by the mining industry.

Electrowinning can be defined as the "winning" of metal from an aqueous solution using electrical current. That is, a solid metal product is produced by applying a direct current between two electrodes. The two types of electrodes are anodes and cathodes. Electrochemical oxidation occurs at the anodes; this is most often the decomposition of water to create protons and oxygen gas. Reduction occurs at the cathodes; the solid metal is produced here.

M. S. Moats (✉)
Materials Research Center, Missouri University of Science and Technology, 220 McNutt Hall, Rolla, MO 65409, USA
e-mail: moatsm@mst.edu

K. Awuah-Offei (ed.), *Energy Efficiency in the Minerals Industry*, Green Energy and Technology, https://doi.org/10.1007/978-3-319-54199-0_12

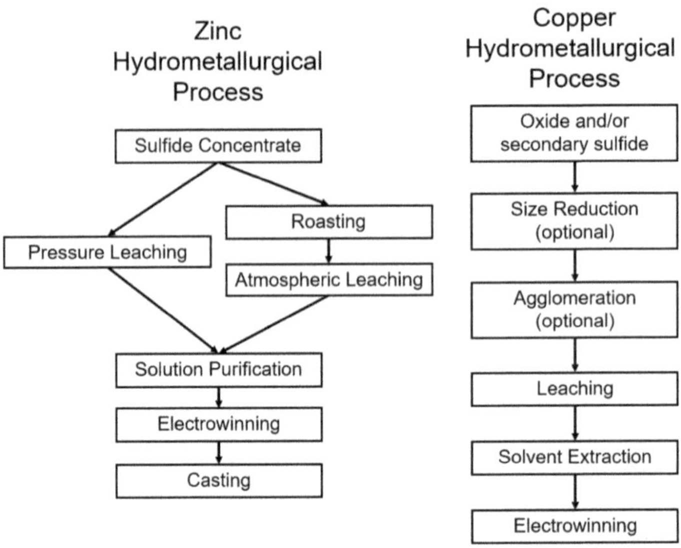

Fig. 12.1 Schematic flow diagrams for hydrometallurgical processing of zinc and copper adopted from Free and Moats [1]

Prior to electrowinning, the metal is placed into solution by leaching of ore or concentrate. The leachate is then, often, processed to concentrate the target metal ion and reduce deleterious impurities. It is important to recognize that electrowinning is often the end of a series of chemical processing steps.

Figure 12.1 shows schematic box flow diagrams for the hydrometallurgical production of zinc and copper using electrowinning. In zinc processing, sphalerite containing ore is processed through flotation to produce a concentrate. The concentrate is either leached directly using elevated temperature and pressure in an autoclave or roasted and then leached at atmospheric pressure. The leach solution is treated to remove impurities by pH adjustment and cementation using zinc dust. The purified electrolyte is then fed to a cellhouse, where zinc is electrodeposited. The zinc cathode is recovered and melted to produce ingots of customer specific alloys.

In the hydrometallurgical production of copper, oxide, or secondary sulfide ores are heap leached [2]. Pretreatment of the ore prior to heap leaching may include crushing and/or agglomeration. The decision to crush and/or agglomerate is based on the economic evaluation to determine if increased extraction can justify the added cost of the pretreatment. Heap leaching produces a pregnant leach solution (PLS), which is processed by solvent extraction (SX). Solvent extraction uses a highly selective reagent dispersed in an organic phase to extract copper from the PLS, while leaving the vast majority of the impurities in the leach solution. The copper is then stripped from the organic phase by mixing with electrolyte from the electrowinning plant. Finally, copper is recovered as a cathode in the electrowinning tankhouse. Copper is not typically melted at the electrowinning facility and cathodes are either sold or shipped to a rod plant.

12.2 Energy Consumption

In electrowinning, most of the energy consumed is related to the electrical energy needed to drive the electrochemical reactions. For industrial copper and zinc electrowinning, the median specific electrical energy consumptions are 2.0 and 3.0 kWh kg^{-1}, respectively [3, 4]. This electrical energy consumption accounts for 80% of the total energy consumed in copper electrowinning [5] and 94% in zinc electrowinning [6]. The remaining energy consumption is electrical energy to harvest metal, pump solution and run ventilation or cooling fans and thermal energy to heat electrolyte. Therefore, this chapter will focus on efficiency related to electrical energy consumption for electrochemical reactions.

From basic physics of electricity, electrical energy (E) is simply the product of power (P) and time (t), Eq. 12.1. Electrical power is the product of current (I) and voltage (V), Eq. 12.2.

$$E = P * t \tag{12.1}$$

$$P = I * V \tag{12.2}$$

where the units most often used for industrial operations are kW-h for E, kW for P, hours for t, kiloamperes for I, and volts for V.

The theoretical mass of metal deposited by an electrowinning facility is calculated by Faraday's law (Eq. 12.3)

$$m_{\text{theoretical}} = I * N * t * \text{AW}/(n * F) \tag{12.3}$$

where $m_{\text{theoretical}}$ is the theoretical mass of metal electrodeposited in kg, I is current in kA, N is the number of cells in the facility, t is the deposition time in seconds, AW is the atomic weight of the metal being deposited, n is the number of electrons used to reduce the metal ion to a solid, and F is Faraday's constant (96,485 C per mole of electrons).

The actual mass of metal deposited is determined by weighing the product from the facility. Using the actual mass and theoretical mass, every electrowinning operation calculates and monitors current efficiency using Eq. 12.4.

$$\text{CE}(\%) = m_{\text{actual}}/m_{\text{theoretical}} * 100 \tag{12.4}$$

where CE(%) is current efficiency as a percentage and m_{actual} is actual mass produced in kg.

All facilities also track energy consumption per mass of product. This value is measured and estimated by calculation. For an individual cell, the specific electrical energy consumption can be calculated by combining Eqs. 12.1–12.4 to produce Eq. 12.5.

$$EC = 2680 * V * n/[CE(\%) * AW] \tag{12.5}$$

where EC is the specific electrical energy consumption in kWh per kg and all other variables are as previously defined. From Eq. 12.5, it is clear that specific energy consumption for electrowinning is a function of cell voltage and current efficiency.

12.3 Cell Voltage

Electrowinning cells are operated in electrical series as shown schematically in Fig. 12.2. Each cell is subjected to the same current as supplied by a rectifier. The voltage of the system is the sum of the series of cells and resistances of the electrical distribution system (e.g., bus bars). The voltage of the electrowinning system is measured at the rectifier or calculated using Eq. 12.6.

$$V_{\text{system}} = \sum V_{IR\,\text{of busbars}} + \sum V_{\text{cell}} \tag{12.6}$$

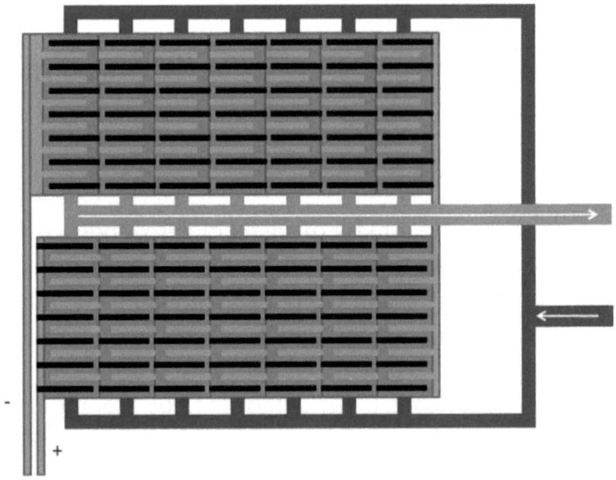

Fig. 12.2 Schematic diagram for electrowinning facility showing electrical connections and electrolyte flow. Each cell contains a multiple of anodes (dark) and cathodes (light). Each electrode type is electrically in parallel. Cathodes are in connect with the negative busbar from the rectifier. Anodes are connected to the positive busbar. Copper busbar are placed between cells to allow electron flow from the anodes in one cell to the cathodes in the adjacent cell

Table 12.1 Approximate cell voltage parameters for commercial copper [2] and zinc electrowinning [7] in sulfate based electrolytes

Parameter	Copper in sulfate electrolyte	Zinc in sulfate electrolyte
E_{anodic}	1.23	1.23
$E_{cathodic}$	0.34	−0.77
anodic	0.50	0.6
cathodic	0.06	0.1
$IR_{solution}$	0.25	0.5
IR_{other}	0.30	0.3
V_{cell}	2.0	3.5

The voltage drop associated with the electrical distribution is a function of the thickness of the copper bus bars and distance between the rectifier and electrowinning circuit. This is usually between 3 and 5 V.

The voltage of an electrowinning cell is also the sum of a series of electrical components which include the electrochemical reactions. Equation 12.7 is a mathematical formula for the cell voltage.

$$V_{cell} = E_{anodic} - E_{cathodic} + \eta_{anodic} + \eta_{cathodic} + IR_{solution} + IR_{other} \tag{12.7}$$

where V_{cell} is the cell voltage, E_{anodic} is the thermodynamic potential of the anode reaction, $E_{cathodic}$ is the thermodynamic potential of the cathode reaction, η_{anodic} and $\eta_{cathodic}$ are the overpotentials that occur as current density is applied, $IR_{solution}$ is the voltage caused by current flow through the electrolyte and IR_{other} is the voltage caused by other resistances in the cell, such as current flow through metal electrodes and the contact between the electrodes and electrical bus system. Table 12.1 displays typical values for each of these parameters and the total cell voltage in commercial copper and zinc cells.

12.3.1 Thermodynamic Potentials

The thermodynamic potentials can be calculated using the Nernst equation (Eq. 12.8):

$$E = E^{o} + RT/nF \ln Q \tag{12.8}$$

where E is the electrode potential versus the standard hydrogen electrode (SHE), E^{o} is the standard reduction potential versus the standard hydrogen electrode, R is the gas constant, T is temperature, n is the moles of electrons transferred, F is Faraday's constant, and Q is the reaction quotient. The reaction quotient is related to the activities of the reactants and products. More details about calculating thermodynamic potentials for electrochemical reactions can be found in numerous textbooks [8].

For copper and zinc electrowinning in sulfate electrolytes, the main anodic reaction is water decomposition that produces acid and oxygen gas (Eq. 12.9). The main cathodic reactions are the depositions of the metal (Eqs. 12.10 and 12.11).

$$H_2O_{(l)} = 2\,H^+{}_{(aq)} + 1/2\,O_{2(g)} + 2\,e^- \tag{12.9}$$

$$Cu^{2+}{}_{(aq)} + 2\,e^- = Cu_{(s)} \tag{12.10}$$

$$Zn^{2+}{}_{(aq)} + 2\,e^- = Zn_{(s)} \tag{12.11}$$

In commercial systems, these reactions are near equilibrium at their standard reduction potentials as shown in Table 12.1.

12.3.2 Overpotential

The overpotential, η, is the energy needed to move a reaction away from equilibrium and achieve the necessary current density. Thus, the overpotential is the potential applied to the equilibrium potential or $E - E_{eq}$.

Electrochemical reactions such as metal deposition and oxygen evolution occur at the surface of an electrode. The current density (current divided by area) dictates the energy required. The current density within a cell is calculated by dividing the applied current by the total anode and/or cathode area in the cell. As anodes and cathodes are often slightly different in size, anodic and cathodic current densities do usually differ slightly. Anodes are usually slightly smaller than cathodes to reduce the amount of metal deposited at the edges of the electrode which may interfere with deposit harvesting.

The overpotential created by the applied current density is described by the Butler–Volmer equation (Eq. 12.12). The equation can be derived from first principles of the activated complex theory. Free provides a derivation in his Hydrometallurgy textbook [8].

$$i = i_o\left(e^{\frac{\alpha_{Ox} nF(\eta)}{RT}} - e^{\frac{-\alpha_{Red} nF(\eta)}{RT}}\right) \tag{12.12}$$

where i is the applied current density, i_o is the exchange current density of the reaction on the electrode being used, η is the overpotential ($E - E_{eq}$), α is the transfer coefficient which indicates how reversible the reaction is in the forward and backward direction. The exchange current density indicates how reactive the surface is for the reaction in question and is really the net exchange rate when the reaction is at equilibrium.

At high values of overpotential (>50 mV), the Butler–Volmer equation can be simplified into an empirical formula called the Tafel equation (Eq. 12.13).

$$\eta = a + b \log i \tag{12.13}$$

where a and b are constants with b being the so-called "Tafel" slope. By comparing the Butler–Volmer and Tafel equations, the reader can see that

$$a = 2.3\,RT/\alpha nF \log i_o \tag{12.14}$$

$$b = 2.3\,RT/\alpha nF \tag{12.15}$$

The largest voltage component for electrowinning other than the voltage needed to overcome thermodynamics is that for anode overpotential. Traditional base metal electrowinning uses lead alloy anodes in sulfate based electrolytes. When new lead anodes are inserted in sulfuric acid-based electrolytes, they dissolve and quickly become covered in precipitated lead sulfate. Upon energizing of the cell, the lead sulfate oxidizes to lead oxide by the chemical reaction shown in Eq. 12.16.

$$PbSO_{4(s)} + 2\,H_2O_{(l)} = PbO_{2(s)} + 4\,H^+{}_{(aq)} + SO_4{}^{2-}{}_{(aq)} + 2\,e^- \tag{12.16}$$

As the standard reduction potential for this reaction is 1.69 V versus SHE, ~460 mV is needed above the thermodynamic oxygen evolution potential to stabilize PbO_2 from $PbSO_4$. This value is added to the overpotential needed to decompose water on PbO_2 resulting in a large anode overpotential for lead alloy anodes.

12.3.2.1 Anode Overpotential

To reduce anode potential and corrosion of the lead alloy plate, additives are added to the lead or the electrolyte. In zinc electrowinning, the anodes contain 0.5–1.0 wt % silver. Silver acts as a catalyst for oxygen evolution and reduces the anode overpotential [9].

The use of silver in anodes is expensive. To overcome this cost, copper electrowinning uses cobalt in the electrolyte to catalyze oxygen evolution and lower the anode potential [10]. The anode overpotential decreases with increasing cobalt concentration. Recently, Abbey, and Moats [11] proposed an empirical equation (Eq. 12.17) for the anode potential for oxygen evolution as a function of cobalt concentration for a rolled Pb–Ca–Sn anode in 170 g L^{-1} H_2SO_4, 0.6 g L^{-1} Mn, 0.6 g L^{-1} Fe at 50 °C. Research is ongoing to further refine this equation and add other parameters and interactions.

$$\text{Anode Potential (V vs. SHE)} = 1.928 - 0.135[\text{Co}]^{0.32} \tag{12.17}$$

where [Co] is cobalt concentration in g L^{-1} over the range of 0.0–0.6 g L^{-1}.

While additives can reduce the anode overpotential of lead anodes, there is still a fundamental problem with Pb anodes, because they need lead oxide on their surface to allow for current flow (lead sulfate is non-conductive). This has resulted in a multiple decade search to find a replacement for lead anodes. This search has resulted in the developed of alternative anodes, such as MMO anodes or coated titanium anodes.

Moats [12] reviewed the economics and challenges of replacing lead anodes with coated titanium anodes. The electrocatalytic coating typically consists of IrO_2 and Ta_2O_5. This use of iridium and titanium makes MMO anodes more expensive than lead alloy anodes. Thus, the main challenge in using these anodes is justifying the additional anode cost relative to the operational savings when compared to lead alloy anodes.

Recently three copper electrowinning tankhouses operated by Freeport McMoRan (Chino, New Mexico, USA; Bagdad, Arizona, USA and El Alba, Antofagasta, Chile) have converted from lead anodes to MMO anodes [13]. The MMO anodes are reported to reduce the cell voltage by 15%, produce a current efficiency improvement of 1–2% and eliminate the need for cobalt in the electrolyte. It thus appears that further improvement in anode technology and/or increases in electrical energy costs could result in replacement of lead alloy anodes with MMO anodes. This topic will be discussed in greater detail later in the chapter.

12.3.2.2 Cathode Overpotential

Fundamental electrochemistry dictates that flowing current in a preferred direction (i.e., electrodepositing metal) will cause overpotential. Thus, the cathode overpotential is directly related to the current density used in the electrowinning cell. Using data from Brown et al. [14], the author developed a formula for cathode overpotential for copper electrodeposition in a laboratory cell at current densities up to 200 A m^{-2} from a solution of 0.65 M Cu and 2 M H_2SO_4 at 60 °C (Eq. 12.18).

$$\eta_{\text{cathodic}} = 0.473 + 0.070 \log(i) \qquad (12.18)$$

where η_{cathodic} is cathodic overpotential in volts and i is cathode current density in A m^{-2}.

While this formula shows the expected relationship between cathode overpotential and the logarithm of current density, it does not completely describe what is occurring in a commercial cell.

In industrial electrowinning, organic additives are added to foster the growth of smooth dense metal electrodeposits and to operate cells at higher current efficiency. In zinc electrowinning, cellhouses use gelatin and/or licorice [3, 6]. In copper electrowinning, guar, polysaccharides, and/or polyacrylamides are utilized [4]. These smoothing agents or deposit modifiers can increase the overpotential [15–17], though some polysaccharides cause minimal increases in cathode overpotential [18]. While the increase in cathode overpotential increases the cell voltage slightly, the ability to produce smooth deposits with higher current efficiency overcomes this deficiency.

12.3.3 Solution Resistance

Solution resistance occurs because ions must transport the charge between electrodes in an electrowinning cell. The voltage is simply related to Ohm's law (Eq. 12.19).

$$V = IR_{\text{soln}} \tag{12.19}$$

where I is the applied current and R_{soln} is the resistance of the solution or electrolyte. The resistance of the electrolyte is a function of the distance between the anodes and cathodes, the electrolyte conductivity and the area of the electrode (Eq. 12.20).

$$R_{\text{soln}} = d/(kA) \tag{12.20}$$

where d is the distance between electrodes, k is the specific conductance and A is the electrode area.

The specific resistances ($1/k$) of copper and zinc electrolytes have been reported in the literature. For copper sulfate electrolytes, an empirical formula derived by [19] is given in Eq. 12.21. For zinc sulfate electrolytes, a formula was derived by Guerra and Bestetti [20] is provided in Eq. 12.22.

$$\begin{aligned} 1/k = 2.87 + 10^{-3} * (9.0 * [\text{Cu}] + 10.9 * [\text{Ni}] \\ + 1.1 * [\text{As}] + 11.6 * [\text{Fe}] - 5.8 * [\text{H}_2\text{SO}_4] - 11.6 * T) \end{aligned} \tag{12.21}$$

where concentrations of ions [ ] are in g L^{-1} and temperature is in °C and $1/k$ is in ohm cm.

$$\begin{aligned} 1/k = 1/[-10.79[\text{Zn}^{2+}] - 53.90[\text{H}_2\text{SO}_4] - 1.814[\text{Zn}^{2+}]^2 - 6.380[\text{H}_2\text{SO}_4]^2 \\ -1.027[\text{Zn}^{2+}][\text{H}_2\text{SO}_4] + 16.73[\text{Zn}^{2+}]^2[\text{H}_2\text{SO}_4] + 0.0585[\text{Zn}^{2+}]T \\ + 12.46[\text{Zn}^{2+}][\text{H}_2\text{SO}_4]^2 - 6.762[\text{Zn}^{2+}]^2[\text{H}_2\text{SO}_4]^2 + 0.3345[\text{H}_2\text{SO}_4]T \\ - 0.1341[\text{Zn2}+][\text{H}_2\text{SO}_4]T \end{aligned} \tag{12.22}$$

where T is temperature (K) and concentrations in Eq. 12.22 are in M.

12.3.4 Other Resistances

Other voltage drops caused by resistance are associated with the passage of current through the electrode body into the header bar and then into the bus system. Boon et al. [21] calculated the resistance between various styles of intercell contact bars and different types of electrodes assuming clean contacts. The reported resistances were between 15 and 42 μΩ depending on the material and mass of the electrode. For a copper electrowinning cell, each electrode will generally carry 500–600 A indicating a contact voltage drop of 7–25 mV for a clean contact on both the anode and cathode. Modern jumbo electrodes used in zinc electrowinning will carry 1200–1900 A and produce a slightly larger voltage drop at the contacts.

With that said, voltage drop at the contact between the bus bar and electrode is often larger than these values as salt formation occurs. This requires routine cleaning of the bus system to mitigate the voltage drop which causes increased power consumption and current maldistribution within the cell.

To mitigate current maldistribution, some facilities are using "equipotential" bars [22]. This results in a double contact on each electrode which creates a better opportunity for current to flow evenly to all electrodes.

Another source of resistance is the contact between the electrode plate and the electrode header bar. Wiechmann et al. [23] discussed the voltage drop associated with different electrode designs used in copper electrowinning. Using finite element analysis, they calculated voltage drops of 16 mV for an anode and 31–43 mV for different styles of stainless steel cathodes. They concluded that a stainless steel sheathed solid copper core header bar welded to a stainless steel plate offered the best performance for the cathode.

12.3.5 Analysis of Cell Voltage

In terms of electrical energy efficiency, electrowinning operations are limited in their abilities to affect the cell voltage on a day to day basis. Operations can diligently clean bus bars to reduce the voltage drop between the header bar and bus. Major reductions in cell voltage, however, will only occur if the anode reaction is changed (decrease E_{anode}) or the facility can achieve a reduction in anodic overpotential (η_{anodic}). Minor reductions in cell voltage might be achieved by minimizing electrode spacing (decrease IR_{soln}) or optimizing the electrolyte conductivity (decrease IR_{soln}).

12.4 Current Efficiency

Besides cell voltage, current efficiency is the other major variable affecting the specific energy consumption of an electrowinning cell. Current efficiency describes how much of the current is used to produce metal. Well-run zinc cellhouses report current efficiencies of 90–92% [3, 6]. Highly efficiency copper tankhouses report 92–94% [4]. The major causes for the 6–10% inefficiencies are side reactions and short circuiting.

12.4.1 *Side Reactions*

The predominant side reaction in electrowinning depends on the reduction potential of the metal being recovered. For zinc, the reduction potential of the metal is less than hydrogen (e.g., $E^{o}_{Zn(II)/Zn} < E^{o}_{H(+)/H_2}$). This results in hydrogen evolution being a major source of current inefficiency in zinc. Since copper has a relatively noble reduction potential as compared to hydrogen (e.g., $E^{o}_{Cu(II)/Cu} > E^{o}_{H(+)/H_2}$), the loss of current to hydrogen gas evolution is very small unless the concentration of copper in the cell becomes very low. In copper electrowinning, the main side reaction is ferric reduction to ferrous (e.g. $E^{o}_{Cu(II)/Cu} < E^{o}_{Fe(III)/Fe(II)}$). Iron is present in the copper electrolyte due to incomplete solution purification prior to electrowinning.

12.4.1.1 Hydrogen Evolution in Zinc

Reduction of hydrogen ion to hydrogen gas is much more thermodynamically favorable as compared to reduction of zinc ion to zinc metal. Thus, to produce zinc at a high current efficiency, hydrogen gas evolution needs to be kinetically inhibited. Fortuitously, the rate of hydrogen evolution on zinc is low (e.g., the exchange current density for hydrogen evolution on zinc is low). Additionally, the use of aluminum mother blanks also suppresses hydrogen evolution.

Other practical methods of reducing hydrogen evolution, thus increasing current efficiency, are to (1) use extremely pure electrolyte with (2) a high concentration of zinc ions at (3) an appropriate sulfuric acid concentration, and (4) optimal temperature. Most cellhouses deposit zinc from a solution of 50–75 g L^{-1} Zn [3]. A higher zinc concentration favors zinc electrodeposition by shifting the cathode equilibrium potential in the positive direction based on the Nernst equation. Since current efficiency and cell voltage are both affected in the same manner by increasing zinc concentration (both increase), increasing sulfuric acid concentration [both decrease] and increasing temperature (both decrease), zinc operations strive to find optimal conditions which serve to minimize energy consumption.

Hydrogen evolution is kinetically hindered in zinc electrowinning. Therefore, impurities that electrodeposit and become preferred sites for hydrogen evolution are problematic. Table 12.2 lists the effects of common electrolyte impurities on current efficiency in single variable experiments. The impact of impurities on current efficiency, however, is complex. There are synergies between impurities and between impurities and gelatin [24, 25]. A detailed review of impurities and their effects on current efficiency and cathode morphology should be the topic of future study.

Current efficiency is maximized by stringent electrolyte purification and/or by the optimal addition of gelatin. Gelatin is added to inhibit the diffusion of impurities to the zinc surface, specifically antimony [26]. However, if too much gelatin is

Table 12.2 Impurity effects on current efficiency from Parada and Asselin [27]

Impurity	Ault and Frazier [25]	Mackinnon et al. [24]
Ge	NA	Decrease
Sb	Decrease	Decrease
Te	NA	Decrease
Se	NA	Decrease
Sn	NA	Decrease
Bi	NA	Decrease
Ga	NA	Decrease
As(III)	NA	No Effect
As(V)	NA	Decrease
Tl	NA	Increase
In	NA	Increase
Cu	Decrease	Increase
Cd	NA	Increase
Pb	Increase	Increase
Ni	Decrease	Decrease
Co	Decrease	NA

added then zinc deposition is hindered and this promotes hydrogen evolution due to a shift in electrode potential. Gelatin also changes the crystal structure of the zinc deposit, which further affects current efficiency.

As indicated by Parada and Asselin [27], improving current efficiency by optimization while useful for zinc operations "will not result in a major breakthrough in energy savings."

12.4.1.2 Ferric Reduction in Copper

In copper electrowinning, the reduction of ferric ion is more thermodynamically favorable than copper deposition and is not kinetically hindered. Therefore, ferric reduction will occur at its limiting current density. The limiting current density is controlled by diffusion of ferric ion across the Nernst boundary layer. Operating factors that will affect the rate of ferric reduction are ferric concentration, solution viscosity, cell hydrodynamics (stirring, flow, etc.) and temperature.

The effect of ferric ion on copper electrowinning current efficiency has been known for some time [28–30]. The qualitative results of operating parameters on current efficiency caused by ferric reduction are summarized in Table 12.3 [31].

Khouraibchia and Moats [32] used surface response methodology to develop an empirical model (Eq. 12.23) that could predict current efficiency for copper electrowinning where solution agitation is minimal. The model uses as inputs: ferric concentration, copper concentration and current density. The model was constructed at 40 °C and is most accurate for the ranges of 1.27–5.0 g Fe^{3+} L^{-1}, 30–50 g Cu L^{-1}, and 300–400 A m^{-2}.

Table 12.3 Qualitative effect of electrowinning parameters on current efficiency related to ferric reduction

Parameter	Effect on current efficiency
Ferric Iron	Significant decrease with increasing concentration
Copper	Increase with increasing concentration
Acid	Increase slightly with increasing concentration
Manganese	No effect
Chloride	No effect
Cobalt	No effect
Current density	Increase with increasing current density
Temperature	Decrease with increasing temperature

$$\begin{aligned}\mathrm{CE}(\%) = {} & 88.19 - 4.91 \times \mathrm{Fe(III)}\left(\mathrm{g\,L^{-1}}\right) + 0.52 \times \mathrm{Cu(II)}\left(\mathrm{g\,L^{-1}}\right) + 1.81 \times 10^{-3} \times \mathrm{C.D.}\left(\mathrm{A/m^2}\right) \\ & - 6.83 \times 10^{-3} \times \mathrm{Cu(II)}^2\left(\mathrm{g\,L^{-1}}\right) + 0.028 \times \mathrm{Fe(III)}\left(\mathrm{g\,L^{-1}}\right) \times \mathrm{Cu(II)}\left(\mathrm{g\,L^{-1}}\right) \\ & + 4.015 \times 10^{-3} \times \mathrm{Fe(III)}\left(\mathrm{g\,L^{-1}}\right) \times \mathrm{C.D.}\left(\mathrm{A/m^2}\right)\end{aligned} \tag{12.23}$$

The great utility of this formula is that it uses data that most tankhouses generate on a regular basis. It can be used to determine a best case scenario current efficiency quickly. While the model seems to be effective over the ranges listed, it has been found to under-predict current efficiencies at current densities less than 300 A m^{-2}.

While iron reduces current efficiency, it has a positive benefit in solvent extraction/electrowinning operations. Some iron is needed to control the formation of permanganate in electrowinning cells, which can adversely affect solvent extraction [33].

Optimizing current efficiency by controlling the effect of ferric reduction will help industrial operations, but as with the case of hydrogen evolution in zinc, it will not result in a major breakthrough in energy consumption.

12.4.2 Short Circuits

Short circuits result from direct contact between an anode and a cathode within a cell. There are many possible causes for short circuits. Joy et al. [34] listed the main reasons for short circuits at a starter-sheet copper electrowinning facility as: "… bent sheets, bowed sheets, slippers (e.g., electrodes that have fallen into the cell), top roping, bottom roping, ear nodulation, body nodulation, bad anode, insulators, double contact, and no contact." In copper facilities that use stainless steel blanks, common reasons for short circuits include electrode misalignment, bent electrodes, not removing copper nodules from previous harvests, broken edge stripping and body nodulation. In zinc electrowinning, common reasons for short circuits include electrode misalignment, bent electrodes, and anode scale (manganese and lead oxide film) touching the zinc cathode.

Practically all tankhouses monitor for short circuits on at least a daily basis. Robinson et al. [4] indicated that a majority of copper tankhouses surveyed used infrared detectors to monitor for shorts; while others use Gauss detectors or physical inspection. This monitoring is a potential source of data. Most well-run operations collect this data and use it to track the frequency of short circuiting.

Short circuiting is minimized by using straight electrodes, spacing the electrodes evenly, regular anode cleaning to remove loose scale, and producing smooth metal electrodeposits. Smooth metal deposits are produced by adding organic polarizing/smoothing agents, keeping the solution metal concentration and temperature in normal operating ranges, operating at typical current densities for the metal in question and maintaining a consistent flow of electrolyte.

Eliminating short circuits is, of course, necessary to achieve operational excellence for modern electrowinning facilities. A poorly run facility can lose 10–15% current efficiency due to short circuiting if attention is not given to details. Even so, controlling short circuits will only improve the energy consumption of electrowinning so much.

12.5 Future Opportunities

This review has examined the underlying fundamentals related to energy consumption of aqueous electrowinning for copper and zinc. The two parameters that affect energy consumption are cell voltage and current efficiency. Operations which are not operating optimally may use the information provided to track a course to improve their energy efficiency by 5–15%. Well-run operations have limited opportunity to further reduce their energy efficiency without a major change in electrowinning technology.

The biggest opportunity for a reduction in energy efficiency is by changing the anode reaction to one with a lower potential (E_{anodic}), changing the cathode reaction to reduce the number of electrons needed or changing anode materials to reduce anode overpotential (η_{anodic}). These will be discussed in further detail below.

12.5.1 Alternative Anode Reactions

The use of an alternative anode reaction instead of water oxidation has received some attention in the past. In zinc electrowinning, the oxidation of organic specie and hydrogen oxidation have been examined. Parada and Asselin [27] conducted a review of the literature on these topics. For methanol and formic acid oxidation, the poisoning of platinum catalysts by CO adsorption is a major hurdle to overcome. However, research focused on organic fuel cell catalyst development may lead to a breakthrough [35, 36]. For hydrogen oxidation, the development of an economic gas diffusion electrode is required. Some effort is continuing in this area based on recent patents and patent applications [37, 38].

In copper electrowinning, researchers have explored ferrous oxidation [39, 40], oxidation of sulfurous acid [41, 42], and the oxidation of sulfur dioxide [43, 44] as possible routes to reduce the energy consumption of the process. The most advanced demonstrations of these reactions were performed using ferrous oxidation [45–49].

Ferrous oxidation can reduce the cell voltage for copper electrowinning by 46% at an iron concentration of 30 g L^{-1} [49]. Ferrous oxidation also eliminates the generation of acid mist caused by oxygen bubbles bursting at the air/electrolyte interface. Stainless steel anodes can be used instead of lead-alloy anodes which eliminates the production of hazardous by-products caused by lead anode corrosion. Finally, the removal of lead anodes eliminates the need for cobalt addition to the electrolyte.

To promote ferrous oxidation, high concentrations (20–50 g L^{-1}) of iron are needed in the electrolyte [5]. As previously stated, ferric reduction is preferred over copper reduction, so the presence of significant quantities of ferric ion in the electrolyte would be very detrimental to energy consumption. To avoid ferric reduction at the cathode, ferrous oxidation is coupled with the reduction of ferric by sulfur dioxide. The use of sulfur dioxide adds complexity to the process flowsheet [49]. Marsden [5] estimated even with the extra process steps, conversion of copper electrowinning from anodic decomposition of water to ferrous oxidation would reduce the energy consumption from 1746 to 1011 kJ kg^{-1}, respectively. The implementation of ferrous oxidation in copper electrowinning has not yet occurred on a large commercial scale.

12.5.2 Alternative Cathode Reaction

Copper ions have two stable oxidation states in aqueous solutions, Cu^{+} and Cu^{2+}. Since the amount of copper produced per ampere depends on the number of electrons required per mole of metal, changing from cupric to cuprous ion could reduce needed amperes by 50%. Hence, the energy consumed could be cut in half. Since zinc is only stable as Zn^{2+}, no reduction in energy consumption is possible by changing oxidation states.

Researchers and companies have examined the coupling of chloride leaching of chalcopyrite with electrowinning directly from the chloride solution fairly extensively [50, 51]. In fact, a facility operated using the CLEAR process produced 30,000 tonnes of copper powder in 1981 using copper chloride electrometallurgy [52]. However, today the electrodeposition of copper from chloride solutions is limited to by-product recovery at a few nickel refineries [53]. Some of the reasons for the low popularity of chloride electrometallurgy for copper are low product quality, materials costs, and production of a powder.

The electrowinning of copper from cuprous chloride solutions is also challenging because copper powder or dendrites are produced at the cathode due to the high exchange current density of copper electrodeposition from a cuprous chloride solution. Smooth, dense copper electrodeposit can be produced if applied current density, electrolyte agitation, and smoothing agent are controlled [51].

The combination of chloride leaching of low grade chalcopyrite ores with energy-efficient copper electrowinning seems to be a topic that should be reexamined.

12.5.3 Alternative Anodes

Besides alternative reactions, the largest source of energy inefficiency in base metal electrowinning is the overpotential of oxygen evolution from the decomposition of water on lead alloy anodes. There have been many attempts over the past 40+ years to exchange lead alloys with more energy-efficient materials. The most common material examined is ceramic coated titanium. These anodes are called dimensionally stable anodes (DSA), mixed metal oxide (MMO) anodes, and coated titanium anodes (CTA). In this chapter, they will simply be referred to as MMO anodes.

MMO anodes use grade 1 or grade 2 commercial pure titanium as a substrate. To this substrate, a coating of mixed metal oxides is applied. The coating composition for oxygen evolution application is typically IrO_2-Ta_2O_5 [12], but researchers have evaluated numerous other compositions. Besides coating composition, manufacturing parameters can have an effect on coating performance [54, 55].

Some copper electrowinning facilities have converted from lead alloy anodes to MMO anodes [56]. The benefits of MMO anodes are lower energy consumption (10–15%), elimination of cobalt addition to the electrolyte and elimination of lead-bearing hazardous by-products. The disadvantages of MMO anodes in copper are materials cost and damage caused by short circuiting [12]. Research should be focused on minimizing the lifetime cost of the coating and suppressing undesired side reactions along with mitigation of short circuits [57]. Morimitsu's group has been very active in this area and his work should be reviewed prior to embarking on investigations [58–60].

In zinc electrowinning, the presence of 2–5 g L^{-1} Mn in commercial electrolyte can be problematic for MMO anodes. Manganese oxide can electrodeposit on the coating surface [61], which can reduce the anode's life and/or raise its overpotential. Innovation to suppress the formation of manganese oxidation has occurred [54], but more work is likely warranted.

12.6 Summary

In summary, metal electrowinning as practiced by the mining industry is a mature process. At the operations level, being energy efficient is related to maximizing current efficiency and optimizing electrolysis parameters. At the process level, significant energy savings can only be achieved by changing one of the underlying electrochemical reactions or reducing the anode overpotential. Significant changes will not likely occur in electrowinning without a major technological breakthrough or a drastic escalation in energy prices.

References

1. Free M, Moats M (2013) Hydrometallurgical production. In: Seetharaman S (ed) Treatise on process metallurgy volume 3: industrial processes part A. Elsevier, Oxford, pp 949–982
2. Schlesinger ME, King MJ, Sole KC et al (2011) Extractive metallurgy of copper, 5th edn. Elsevier, Oxford
3. Moats M, Guerra E, Gonzalez JA (2008) Zinc electrowinning—operating data. In: Centomo L, Collins M, Harlamovs J et al (eds) Zinc and lead metallurgy. CIM, Montreal, pp 307–314
4. Robinson TG, Sole KC, Sandoval S et al (2013) Copper electrowinning: 2013 world tankhouse operating data. In: Abel R, Delgado C (eds) Proceedings of copper 2013. vol V, pp 3–14
5. Marsden JO (2008) Energy efficiency and copper hydrometallurgy. In: Young C, Taylor P, Anderson C et al (eds) Hydrometallurgy 2008 proceedings of the sixth international symposium. SME, Littleton, pp 29–42
6. Moats M, Guerra E, Siegmund A et al (2010) Primary zinc smelter operating data survey. In: Pb Zn 2010—lead-zinc 2010 symposium, held in conjunction with COM 2010, CIM, Montreal, pp 263–282
7. Ettel VA (1977) Energy requirements in electrolytic winning and refining of metals. CIM Bull 70(782):179–187
8. Free ML (2013) Hydrometallurgy: fundamentals and applications. Wiley, Hoboken
9. Pavlov D, Rogachev T (1986) Mechanism of the action of Ag and As on the anodic corrosion of lead and oxygen evolution at the $Pb/PbO_{(2-x)}/H_2O/O_2/H_2SO_4$ electrode system. Electrochim Acta 31(2):241–249
10. Nikoloski AN, Nicol MJ (2007) Effect of cobalt ions on the performance of lead anodes used for the electrowinning of copper—a literature review. Min Process Extr Metall Rev 29 (2):143–172
11. Abbey CE, Moats MS (2017) Effect of cobalt and iron concentration on the potential for oxygen evolution from Pb-Ca-Sn anodes in synthetic copper electrowinning electrolytes. In: Wang S, Free ML, Alam S et al (eds) Applications of process engineering principles in materials processing, energy and environmental technologies. Springer, Berlin, pp 89–95
12. Moats MS (2008) Will lead-based anodes ever be replaced in aqueous electrowinning? JOM 60(10):46–49
13. Sandoval S, Clayton C, Dominguez S et al (2010) Development and commercialization of an alternative anode for copper electrowinning. In: Copper 2010 proceedings, vol 4, GDMB, Clausthal-Zellerfeld, pp 1635–1648

14. Brown AP, Loutfy RO, Cook GM (1980) The electrorefining of copper from a cuprous ion complexing electrolyte: II. Experimental comparison of possible alternative electrolytes and preliminary cost engineering analysis. ANL/OEPM-80-2
15. Kerby RC (1984) Application of polarization measurements to the control of zinc electrolyte quality for electrowinning. In: Warren IH (ed) Application of polarization measurements in the control of metal deposition. Elsevier, Amsterdam, pp 84–132
16. Adcock PA, Adeloju SB, Newman OM (2002) Measurement of polarization parameters impacting on electrodeposit morphology I: theory and development of technique. J Appl Electrochem 32(10):1101–1107
17. Stantke P (1999) Guar concentration measurement with the CollaMat system. In: Dutrizac JE, Li J, Ramachandran V (eds) Electrorefining and electrowinning of copper: proceedings of the Copper 99—Cobre 99 international conference, vol III, TMS, Warrendale, pp 643–651
18. Luyima A, Moats MS, Cui W, Heckman C (2016) Examination of copper electrowinning smoothing agents. Part II: fundamental electrochemical examination of DXG-F7. Miner Metall Process 33(1):14–22
19. Biswas AK, Davenport WG (1980) Extractive metallurgy of copper, 2nd edn. Pergammon Press, Oxford
20. Guerra E, Bestetti M (2006) Physicochemical properties of $ZnSO_4$-H_2SO_4-H_2O electrolytes of relevance to zinc electrowinning. J Chem Eng Data 51(5):1491–1497
21. Boon C, Fraser R, Johnston T, Robinson D (2013) Comparison of intercell contact bars for electrowinning plants. In: Battle T, Moats M, Cocalia V (eds) Ni-Co 2013. Wiley, Hoboken, pp 177–189
22. Ashford B, Ebert WA, Vega FDM et al (2011) Double contact bar insulator assembly for electrowinning of a metal and methods of use thereof. U.S. Patent 7,993,501, 9 Aug 2011
23. Wiechmann EP, Morales AS, Aqueveque P et al (2015) On the design robustness and long term performance of the most used electrodes in the copper electrowining industry. In: Industry applications society annual meeting, 2015 IEEE, pp 1–8
24. Mackinnon DJ, Brannen JM, Fenn PL (1987) Characterization of impurity effects in zinc electrowinning from industrial acid sulphate electrolyte. J Appl Electrochem 17:1129–1143
25. Ault R, Frazer EJ (1988) Effects of certain impurities on zinc electrowinning in high-purity synthetic solutions. J Appl Electrochem 18:583–589
26. Robinson DJ, O'Keefe TJ (1976) On the effects of antimony and glue on zinc electrocrystallization behavior. J Appl Electrochem 6:1–7
27. Parada TF, Asselin E (2009) Reducing power consumption in zinc electrowinning. JOM 61:54–58
28. Anderson TN, Wright CN, Richards KJ (1973) Important electro-chemical aspects of electrowinning copper from acid leach solutions. In: Evans DJI, Shoemaker RS (eds) International symposium on hydrometallurgy, New York, pp 171–202
29. Grunenfelder JG (1960) The hydrometallurgy of copper. In: Butts A (ed) Copper: the science and technology of the metal, its alloys and compounds. ACS, Reinhold, pp 300–337
30. Mantell CL (1960) Electrochemical engineering. McGraw-Hills, New York, p 198
31. Khouraibchia Y, Moats M (2009) Effective diffusivity of ferric ions and current efficiency in stagnant synthetic copper electrowinning solutions. Miner Metall Process 26:176–190
32. Khouraibchia Y, Moats M (2010) Evaluation of copper electrowinning parameters on current efficiency and energy consumption using surface response methodology. In: Doyle FM, Woods R, Kesall GH (eds) Electrochemistry in mineral and metal processing VIII: ESC Trans, vol 28 No. 6, pp 295–306
33. Miller G (2011) Methods of managing manganese effects on copper solvent extraction plant operations. Solvent Extr Ion Exch 29(5–6):837–853
34. Joy S, Staley A, Moats M et al (2010) Understanding and improvement of electrowinning current efficiency at Freeport-McMoRan Bagdad. In: Copper 2010 proceedings, vol 4, GDMB, Clausthal-Zellerfeld, pp 1379–1392

35. El-Nagar GA, Mohammad AM, El-Deab MS et al (2013) Electrocatalysis by design: enhanced electrooxidation of formic acid at platinum nanoparticles–nickel oxide nanoparticles binary catalysts. Electrochim Acta 94:62–71
36. Tang Y, Chen Y, Zhou P et al (2010) Electro-catalytic performance of PdCo bimetallic hollow nano-spheres for the oxidation of formic acid. J Solid State Electrochem 14(11):2077–2082
37. Gulla AF, Krasovic J (2012) Gas-diffusion electrode. US Patent Application 14/342,887, 14 Aug 2014
38. Izawa Y, Ogata S, Uno M et al (2015) Oxygen gas diffusion cathode, electrolytic cell employing same, method of producing chlorine gas and method of producing sodium hydroxide. US Patent 9,1754,10, 3 Nov 2015
39. Mishra K, Cooper WC (1984) Electrochemical aspects of the direct electrowinning copper from sulfuric acid leach solutions in the presence of iron using gas sparging. In: Robinson D, James SE (eds) Anodes for electrowinning. TMS-AIME, Warrendale, pp 13–36
40. Cooke AV, Chilton JP, Fray DJ (1985) Ferrous/ferric depolarization in copper electrowinning: mass transport and current efficiency considerations. In: Bautista RG, Wesely RJ (eds) Energy reduction techniques in metal electrochemical processes. TMS, Warrendale, pp 111–141
41. Panda B, Das SC (2001) Electrowinning of copper from sulfate electrolyte in presence of sulfurous acid. Hydrometallurgy 59(1):55–67
42. Subbaiah T, Singh P, Hefter G et al (2000) Sulphurous acid as anodic depolarizer in copper electrowinning part II. J Appl Electrochem 30(2):181–186
43. Robinson DJ (1984) SO_2 electrowinning in copper hydrometallurgy for energy conservation. JOM 36(1):43–47
44. Dawson JN, Singh P, Hefter G (1999) The effects of sulfur dioxide on the energy consumption and nature of electrowon copper. Paper presented at PACRIM '99 Congress, Bali, Indonesia, 10–13 Oct 1999
45. Sandoval SP, Lei KPV (1993) Evaluation of the ferrous/ferric-sulfur dioxide anode reaction for integration into the copper leaching-solvent extraction-electrowinning circuit. In: Hiskey JB, Warren GW (eds) Proceedings of Milton E. Wadsworth (IV) International symposium on hydrometallurgy, Salt Lake City, UT pp 1091–1105
46. Dolinar WJ, Sandoval SP (1995) Copper electrowinning in the absence of acid misting using the ferrous/ferric-sulfur dioxide anode reaction—a pilot study. Trans Soc Min Metall Explor 298:1936–1942
47. Sandoval SP, Cook PR, Hoffman, WP et al (2008) Method and apparatus for electrowinning copper using the ferrous/ferric anode reaction and a flow-through anode. US Patent 7,368,049, 6 May 2008
48. Sandoval SP, Robinson TG, Cook PR (2008) Method and apparatus for electrowinning copper using the ferrous/ferric anode reaction. US Patent 7,378,011, 27 May 2008
49. Sandoval S, Cook P, Morales C et al (2010) Demonstration of the ferrous/ferric anode reaction for copper electrowinning. In: Copper 2010 proceedings, vol 4, GDMB, Clausthal-Zellerfeld, pp 1617–1634
50. Edmiston KJ (1983) An update on chloride hydrometallurgical processes for sulfide concentrates. SME pre-print 84–114, SME, Littleton, CO
51. Rodchanarowan A, Sarswat PK, Bhide R, Free ML (2014) Production of copper from minerals through controlled and sustainable electrochemistry. Electrochim Acta 140:447–456
52. Muir D, Senanayake G (1984) Refining of clear copper powders by the Parker process: a comparison of the chemistry of copper impurities in chloride and sulfate media. In: Extractive metallurgy symposium, vol 36, 12–14 Nov 1984, Melbourne, Australia, pp 353–359
53. Crundwell F, Moats M, Ramachandran V et al (2011) Extractive metallurgy of nickel, cobalt and platinum group metals. Elsevier, Oxford
54. Morimitsu M, Oshiumi N (2009) Accelerated oxygen evolution and suppressed MnOOH deposition on amorphous IrO_2-Ta_2O_5 coatings. Chem Lett 38(8):822–823

55. Morimitsu M, Oshiumi N, Yamaguchi T (2010) Amorphous oxide coated anode for energy saving of zinc electrowinning. In: Pb Zn 2010—lead-Zinc 2010 symposium, held in conjunction with COM 2010, CIM, Montreal, pp 813–818
56. Sandoval S, Garcia R, Neff T et al (2013) Operation of alternative anodes at Chino SXEW. In: Abel R, Delgado C (eds) Proceedings of copper 2013, vol V, pp 145–152
57. Fiorucci A, Calderara A, Iacopetti L et al (2013) The De Nora solution–part I, DSA® anodes for copper electrowinning. In: Abel R, Delgado C (eds) Proceedings of copper 2013, vol V, pp 107–118
58. Morimitsu M, Yamaguchi T, Oshiumi N, et al (2011) Energy-efficient electrowinning process with smart anode comprising nanooxide catalyst. In: Proceedings of European metallurgical conference, vol 3, pp 975–984
59. Morimitsu M (2012) Performance and commercialization of the smart anode, MSA™, for environmentally friendly electrometallurgical process. In: Free M, Moats M, Houlachi G et al (eds) Electrometallurgy 2012. Wiley, New York, pp 49–54
60. Zhang T, Morimitsu M (2012) A novel oxygen evolution anode for electrowinning of non-ferrous metals. In: Free M, Moats M, Houlachi G et al (eds) Electrometallurgy 2012. Wiley, New York, pp 29–34
61. Moats MS (2010) MnO_2 deposition on coated titanium anodes in copper electrowinning solutions. ERZMETALL 63(6):286–291

Chapter 13
Plant Automation for Energy-Efficient Mineral Processing

Jocelyn Bouchard, Daniel Sbarbaro and André Desbiens

Abstract Mineral processing is one of the most energy-intensive stages of the overall mining beneficiation chain, with an increasing share of the industry footprint. This chapter examines how automation represents a practical means to significantly improve energy efficiency in mineral processing operations. It introduces the fundamentals of automation, hierarchical framework of automation systems, and how the multiple functions can be integrated into an energy management information system. The discussion also explains the rationale of process control and real-time optimization approaches that facilitates lower specific energy requirements from lower variability of key process variables, and determining more appropriate operating points. Case studies are presented to illustrate the current state of the art.

keywords Process control · Process plant automation · Mineral processing
Specific energy

13.1 Introduction

The mineral processing sector holds a significant and growing share of the energy balance of the mining industry. Tromans [1] reported striking data from the U.S. Department of Energy showing that 39% of the energy footprint of mining

J. Bouchard (✉)
Département de génie chimique, Université Laval, Pavillon Adrien-Pouliot,
1065 avenue de la Médecine, Québec City, Canada
e-mail: jocelyn.bouchard@gch.ulaval.ca

D. Sbarbaro
Departamento de ingenería eléctrica, Universidad de Concepción,
Edificio Central de Ingeniería, Concepción, Chile
e-mail: dsbarbar@udec.cl

A. Desbiens
Département de génie électrique et génie informatique, Université Laval,
Pavillon Adrien-Pouliot, 1065 avenue de la Médecine, Québec City, Canada
e-mail: desbiens@gel.ulaval.ca

K. Awuah-Offei (ed.), *Energy Efficiency in the Minerals Industry*, Green Energy and Technology, https://doi.org/10.1007/978-3-319-54199-0_13

operations originated from beneficiation and processing operations in 2004. More recently, the Coalition for Energy Efficient Comminution estimated that ore fragmentation accounts for 53% of the energy consumed on a typical mine site [2].

Expectations are that these figures are on the rise, and Norgate and Haque [3] argued that globally depleting ore grades will impact the energy consumed for mining and processing, because of the additional amount of material required to be treated in these stages, and the finer grinding product needed to liberate valuable minerals. On the other hand, the effect on downstream metal extraction and refining facilities will be minimal since concentrates are typically produced at given concentrations irrespective of the initial ore grades. The growing share of the balance going to mineral separation, thus, requires special attention to improve the overall mining industry energy utilization.

Analyzing data from the U.S. Department of Energy [4], Norgate and Haque [3] emphasized that the metal mining industry in the United States could potentially cut back about 61% of its energy consumption. Implementing best practices would enable an initial 21% reduction, and investments in research and development initiatives for energy efficiency would represent the remaining 40%. Automation, process control and real-time optimization are among the avenues showing tremendous potential for tapping into these opportunities, sometimes with minimum capital expenditure since most of the required instrumentation, control and communication equipment, software and servers are often already in place. It explains why the reduction of energy consumption is regarded as the main benefit of grinding circuit control by 50% of the grinding experts polled by Wei and Craig [5].

Initiatives for reducing the energy footprint of mineral processing plants largely target comminution stages. Essentially four different approaches have been proposed to tackle this issue:

- exploiting comminution mechanisms exhibiting lower specific energy consumption, either with the so-called "mine-to-mill" approach (i.e., consistent and fine blasting product) [6, 7], or taking advantage of compression-based processing equipment (crushers, high-pressure grinding rolls, and anvil-hammer mills) [8–10];
- reducing the amount of material processed or reprocessed in grinding equipment using ore sorting [11], coarse particle processing [12], flash separators [13], or improved particle classification [14];
- recovering waste energy [15–17];
- reducing specific energy consumption by operating at higher throughput (closer to design capacity) [18], and using process control capabilities [19].

This chapter expands this last item beyond comminution and examines how automation is essential to maximize the overall energy efficiency of mineral processing plants. Automation, process control, and real-time optimization allow engineers, not only to determine the proper operating point, but also provide a means to reach and maintain it over time, regardless of fluctuating input materials attributes, and process disturbances. The next section describes the various components of a complete automation system, and their interrelation. It also discusses the implementation of an energy management information system. The benefits of automation on the energy footprint of mineral processing plants are discussed in Sect. 13.3, while Sect. 13.4 presents case studies.

13.2 Process Automation

Automation systems provide timely information and take prompt actions to ensure a safe and profitable operation in spite of the external and internal disturbances affecting the process. In addition, modern automation systems can analyze historical data, perform process analysis and real-time process optimization considering economics as well safety aspects, and integrate information from different sources and systems.

13.2.1 Hierarchy of Automation Functions

In order to address the complexity of accomplishing these objectives, their functions are divided to tackle specific objectives and organized in a hierarchy with five levels as depicted in Fig. 13.1.

The first level considers *data acquisition and processing functions*. These functions process the data obtained from sensors and send suitable commands to actuators. Typical processing operations are signal filtering, signal sampling, and data conversion from raw values to engineering units. This information can then be used for further process analysis. These functions are required to be executed within seconds in order to deal with the time variations of process variables.

The second level carries out *regulatory and safety control functions* using a combination of discrete and continuous control functions, which aim at stabilizing the process, while respecting safe margins for the operations, equipment and workforce. Safety control actions are required to be executed within less than a second, while regulatory control operations are periodically executed within periods depending on the process dynamics. In addition, in modern automation systems, operators are located in a room away from the process. Consequently, the interaction between the automation system and the operator plays an important role in ensuring effective operation. A *Human–Machine Interface* (HMI) must present relevant information timely regarding both normal and abnormal process states, and transmit operator commands to actuators through specialized keyboards or pointing devices such as a mouse, trackballs, or touch screens.

The first two levels consider these *basic functions*, which are essential to any automation system for process monitoring and control. However, the optimal operation of a mineral processing plant requires acting simultaneously over different control variables and taking into account operational constraints. Thus, the third level with *advanced control functions* is required to keep process operations

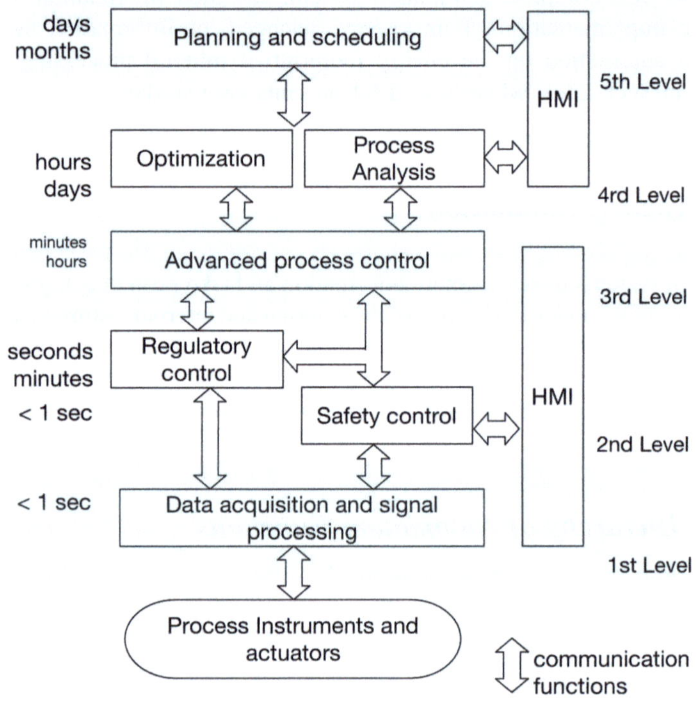

Fig. 13.1 Functions hierarchical organization

near optimum conditions expressed by a proper objective criterion. These control functions calculate the setpoints of the controllers in order to optimize an objective criterion defined over a time horizon of several minutes, typically.

In the fourth level, *process analysis and steady-state optimization functions* consider data collected during several hours or days in order to establish the best operational scenario. Process analysis functions provide modeling and analysis capabilities to characterize operations and inlet flows that can be applied for the whole process, starting with the raw materials through all the multiple stages, and down to the end-products. Steady-state optimization functions seek to improve the operational performance by taking actions according to a given performance index. These improvements can be applied to a single piece of equipment, a unit operation or a plant section. In general, steady-state optimization is always associated with process analysis functions, since it requires information concerning relevant constraints and variables affecting the operational performance.

The fifth level considers a time horizon of weeks or months to integrate corporate data, such as commercial and financial information, with operational data to improve the overall performance of the mineral processing value chain. The main functions carried out at this level are for *Planning and Scheduling*. The *Planning* functions set production goals to meet supply and logistic constraints, and address time-varying capacity and manpower utilization decisions [20]. *Scheduling* supports technical and business decisions by providing a unified view of the process by managing information and data [21]. The integration of planning and scheduling with other functions is addressed by the ANSI/ISA-95 (IEC 62264) standard. This standard consists of models and terminology to ease the exchange of information between business (sales, finance, and logistics) and operations (production, maintenance, and quality) departments. The HMI associated with the upper levels are oriented to establish a collaborative working environment and, therefore, has a different orientation compared with the process operation HMI.

Last, but not least, are the *Communication functions*, which are necessary to establish the communication across functions and levels. In many situations, control actions need to be supplemented with data located in other systems and this requires communication between different control or information systems.

Mineral processing plants are integrated operations involving multiple processes and, therefore, all automation levels are required to operate in a safe and profitable way. This means that optimizing the overall energy efficiency requires considering not only the type of ore being treated, but also the mine, concentrator, tailing, and port operations. Clearly defining the functional specifications of the automation system is a prerequisite to achieve the possible benefits offered by implementing these advanced functions [22].

13.2.2 Automation Systems

Automation functions are implemented in hardware and software systems. The organization of these systems follows a similar hierarchical structure. Field devices (sensors and actuators) are in the first level, as seen in Fig. 13.2.

The first level communicates with the second one by using field communication networks. In the second layer, there are specific microprocessor-based computer systems running control algorithms on top of special-purpose operating systems. In the mineral processing industry, distributed control systems (DCS), and programmable logic controllers (PLCs) are typically used to run regulatory control and interlocks strategies. At this level, there are also human–machine interfaces (HMI) in order to provide the operators the means to monitor and control the process. Supervisory control and data acquisition (SCADA) software provides tools to continuously monitor the process variables, manage alarms, and supervise the operation of control strategies. These systems also provide short-term process variables logs. This historical data is very useful to review trends and assess how the process and the control system evolve with time.

The fourth level has dedicated systems aiming to optimize the overall operation of the plant. Higher level optimization software can coordinate the operating conditions of control loops to meet the optimization objectives. For example, in a mineral processing plant, the objectives could be to minimize electrical energy consumption or maintain a certain production rate or quality.

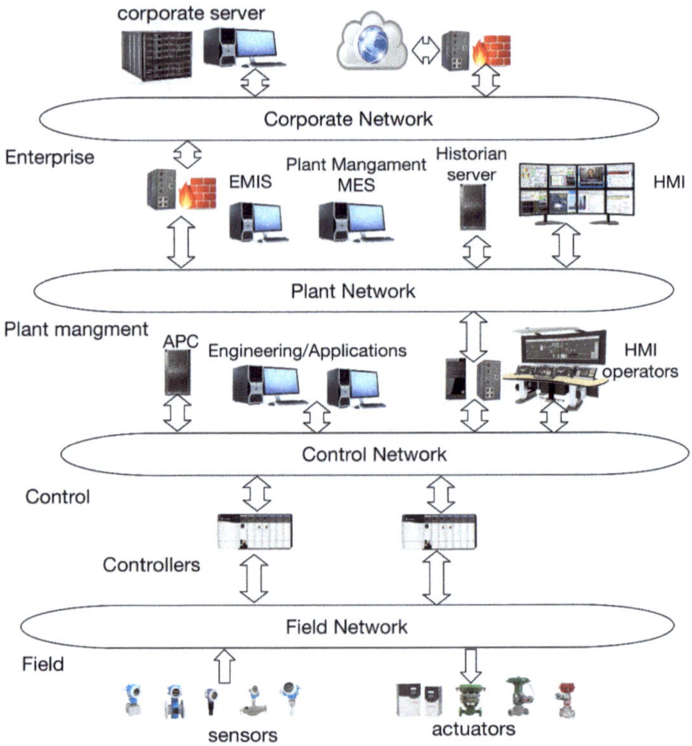

Fig. 13.2 Automation system

The systems at the second level also send data collected from the plant to a dedicated industrial data historian. The data historian is connected to a plant information system for data analysis, performance monitoring, reporting and integration with maintenance and metallurgical laboratory and accounting systems. In order to analyze and respond to real-time events, the HMI at this level enables the visualization of the current status of the plant and the enterprise in multiple media [23].

Energy management information systems (EMIS) [24, 25] are part of the fourth level and perform some specific energy management functions at a strategic level to increase the energy efficiency. The EMIS provide relevant information that makes energy performance visible to various levels of an organization, enabling them to plan, make decisions, and take effective action to manage energy [26]. The final implementation of the operational control of energy saving strategies must be carried out in real time by the lower level control loops in the control system's infrastructure.

The fifth level links the plant to the outside world. The systems at this level perform enterprise-level management function using Enterprise Resource Planning (ERP) systems. The objectives in the fourth level are decided based on the analysis performed at this level.

13.3 Energy Efficiency Improvements

This section explains how control developments can translate into improvements in energy efficiency. In order to establish this relationship, Friedmann [27] proposed an integrated approach that requires an engineer to establish the baseline case, study the impact of energy efficiency strategies on the process, analyze the constraints, and implement the actions to obtain the final reduction of energy consumption.

To estimate the impact of a control improvement on energy efficiency, the following relevant information is usually required:

- Knowledge of the current energy efficiency key performance indicators (KPIs). The International Energy Agency refers to "measures of energy efficiency performance" (MEEPs) instead of KPIs [28]. There are several KPIs representing how energy is efficiently or inefficiently used in, for example, a particular process, company, or country.
  Several MEEPs have been applied to industrial energy use including:
  1. thermal energy efficiency of equipment: energy value available for production or operation divided by input energy value,
  2. energy consumption intensity: energy value divided by certain physical value (e.g., processed mass),
  3. energy value: absolute amount of energy consumption, and
  4. diffusion rates of energy-efficient facilities/types of equipment.
- An estimate of improved control performance. Control performance improvements can be measured by the reduction of variability expressed as a percentage of original variability. For instance, in grinding circuits it is very important to control the resulting particle size distribution as the flotation process is very sensitive to this variable.
- Knowledge of applicable constraints. Most process operations are limited by constraints. These may be physical, regulatory, economical, or related to safety. Constraints can be hard or soft. A soft constraint can be violated but a hard constraint cannot. The violation of a soft constraint implies some penalty is accepted instead. The solution to a constrained steady-state optimization problem almost always lies on the boundary of the feasible region; i.e., on one or more constraints. Reduced variability allows operations closer to a limiting constraint and, therefore, near to the optimum of the deterministic optimization problem without occasionally violating the constraint. For instance, Fig. 13.3 shows the time evolution of the P80 (sieve dimension greater than 80% of ore particles, weight basis) of mill circuit product in micrometers and its upper soft constraint. It is very important for downstream processes that this variable is smaller than this limiting value. Without proper control, the mean value has to be set far away from the limit since the variations are large. With effective control, however, it is possible to move the mean value closer to the limit.

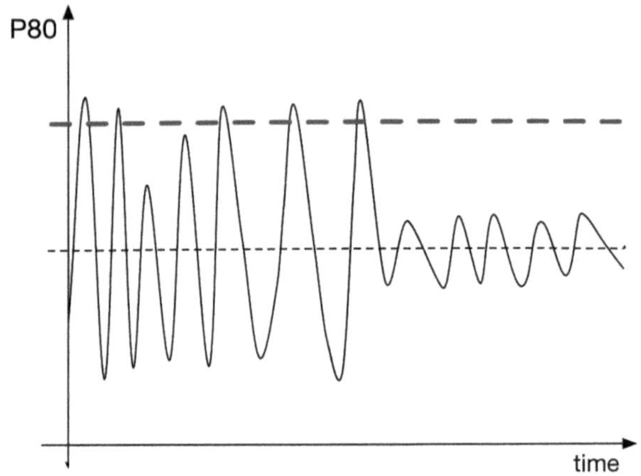

Fig. 13.3 Variability reduction

- A process model representing the relationship between control variables and MEEPs. Improved control of key process variables can increase energy efficiency because the regulation has been improved and/or new reference values for the controlled variables have been calculated in order to improve energy efficiency. In both cases, a process model is required to estimate the final improvements.

Improved control performances increase energy efficiency because one or both of the following conditions are met [29]:

1. In many situations, energy efficiency improvements are limited by operational constraints. In this case, the reduced variation permits moving the mean value of the key variable associated with energy efficiency closer to an operating target. Estimating the shift in the average operating reference depends on the particular application and the specific criteria with respect to a given operational scenario. There are several criteria for establishing the shift of the mean value such as: "best operator", same percent limit violation, same percent limit violation of the 5% limit, and limit violation x% of time [29]. Figure 13.4 shows the shift of the mean value so that the improved control will violate the limit the same percentage of time as the current operation does.
   In addition, new mean values can also be obtained by solving steady-state optimization problems considering technical as well as economic objectives of the process. These objectives are defined in quantitative terms. The improved process performance is then the result of moving the reference values to the steady-state optimal values.
2. If the relationship between energy efficiency and the key operational variable is highly nonlinear, then a reduction of the variation of the key variable will by itself change the mean value of the performance variable and, thus, energy efficiency. This is possible if the key variable is related to the process performance by a nonlinear function and the process is operating in the region of the

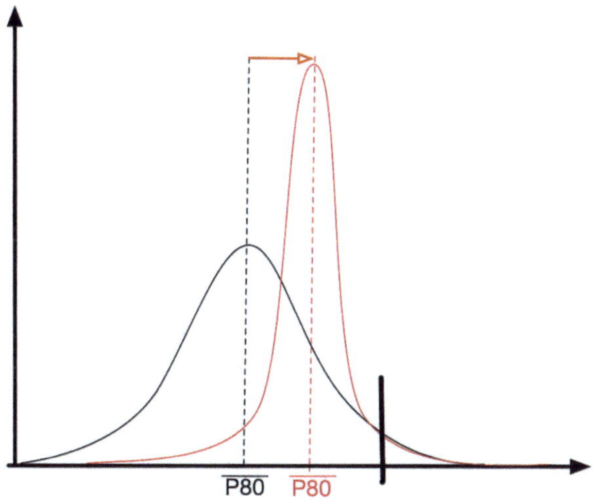

Fig. 13.4 Shifting the average operating variable

nonlinearity. Then improved regulation will change the mean value of the performance parameter, even though the mean value of the key variable is unchanged [30]. Figure 13.5 shows the relationship between the specific energy consumption in terms of the P80. Since this relationship is highly nonlinear, a reduction in the P80 variations will lead to a decrease in the specific energy consumption.

In grinding circuits, both the conditions produce a decrease in the specific energy consumption.

Solving steady-state optimization problems considering technical as well economic objectives of the process allow the calculation of new reference values. These objectives are defined in quantitative terms. The improved process performance is then the result of moving the reference values to the steady-state optimal values.

13.4 Case Studies

This section presents some selected investigations about energy consumption in the mineral processing industry. The first case studies only concentrate on a single unit even if, ideally, plant-wise assessment should be conducted. For instance, grinding mill circuits are the most cost-intensive unit operations in the mineral processing industry, but their economic and energy performance should not be evaluated locally since their product attributes highly influence downstream processes [31, 32]. In this regard, the minimum energy footprint in a plant is usually not found after locally minimizing the consumption of every individual stage [33]. This is clearly demonstrated by the last example of this section. Nonetheless, single unit/circuit studies are included because (1) it is easier to tackle smaller scale industrial

Fig. 13.5 Specific energy consumption as a function of P_{80}

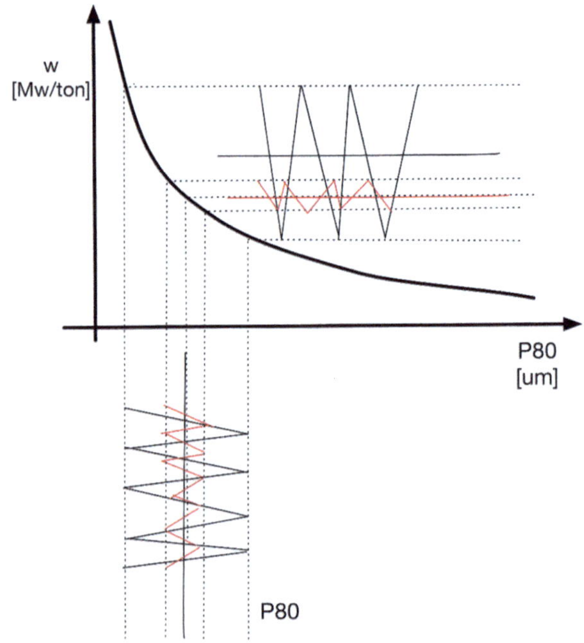

problems and show benefits, and (2) plant-wide control can only be contemplated using a tiered methodology in which controlling simple components precedes optimizing the overall system.

Some applications presented here use open-loop optimization strategies (i.e., the solution is calculated a priori, off-line, without any feedback). This notwithstanding, they reveal the potential of real-time optimization.

13.4.1 Energy Optimization of Crushing Processes

Open-loop energy management of a parallel High-Pressure Grinding Rolls (HPGR) machine is detailed by Numbi and Xia [34]. The simulated circuit has two parallel HRC™800 HPGR machines processing copper ore. The manipulated variables are the rolls rotational speed, rolls operating pressure, and feed rate. The first step of the optimization procedure is to develop models for the power consumption, total energy cost, throughput, and maximum product particle size. The manipulated variables are calculated by minimizing a cost function over a horizon subject to various constraints. The objective function is the total energy cost considering the time-of-use electricity tariff. The constraints are:

- HPGR feed bin-level limits,
- HPGR operational variables (rolls speed, pressure and gap, product rate throughput, fresh feed rate) limits,
- the fresh feed rate should be equal to throughput rate to avoid obstruction,
- a production target, i.e., the total amount of ore crushed for a given period,
- a maximum particle size of the product.

In the simulations, the horizon is 24 h and the production target is 3500 tons. Three scenarios were evaluated:

1. *Both HPGRs having identical overall drive efficiency and with a fixed rolls operating pressure*. There is potential to reduce specific energy cost by about 41.93% without any energy consumption reduction.
2. *Two parallel HPGRs having different overall drive efficiency*. With a fixed rolls operating pressure, minimizing the loading level of the less-efficient HPGR machine, and maximizing the loading level of the more-efficient one enables potential energy and energy cost savings of 1.87 and 43.17%, respectively.
3. *Both HPGRs have identical overall drive efficiency, but variable rolls operating pressure*. For any decrement of 0.2 N/mm^2 in rolls operating pressure, the potential energy saving is 4.5% without significant change in product quality.

Similar energy optimization simulations for jaw crushers and crushing processes based on a vertical shaft impactor are presented by Numbi et al. [35], and Numbi and Xia [36].

13.4.2 Control of Industrial Grinding Circuits

Example 1 Nunez et al. [19] demonstrate that better control can lead to an energy consumption reduction while improving the throughput and product quality, even if the energy is not specifically considered in the control design. One line of the grinding circuit at the Strathcona Mill (Glencore, formerly Xstrata Nickel) is composed of a 900 HP rod-mill in series with a 1750 HP ball-mill operating in closed circuit with a pump-box and hydrocyclone cluster. The main characteristics of the original control strategy consist of the following:

- PI (proportional-integral) control of the rod-mill feed rate, with the operator selecting the setpoint according to operating conditions (hard or soft ore, coarse or fine ore);
- PI control of the pump box level, the manipulated variable being the water addition flowrate setpoint; and
- no control of the cyclone overflow density.

The new features of the control design are as follows:

- PI control of the pump box level using the rod mill feed setpoint as the manipulated variable, thus automatically adjusting the throughput according to the circuit capacity to handle the different types of ore;
- PI control of the cyclone overflow density using the pump-box water addition flowrate setpoint as the manipulated variable, the density setpoint being adjusted by the operator.

Data were collected from January 2007 to May 2008 with the original strategy running and from June 2008 to September 2008 with the new approach. The following results were achieved:

- the circuit throughput was increased by 7.7% without degrading the nickel or copper recovery;
- the cyclone overflow density standard deviation was reduced from 2.0 to 0.8%;
- as a positive side effect, the specific energy efficiency (kWh/t) of the rod-mill and the ball-mill were, respectively, decreased by 7.1 and 7.5%.

For Strathcona mill engineers, the reduction of variability in the cyclone overflow density was of paramount importance as the density exhibits a very high correlation with overflow particle size (% −150 mesh). However, for purposes of this chapter, it is important to realize that the power draw for rod and ball-mills is mainly determined by steel grinding media charge, and does not vary much with throughput. Processing a higher ore feed rate for roughly the same power draw, therefore, results in lower specific energy.

Example 2 The second grinding control illustration [33, 37] also supports that an energy efficiency gain can be a positive indirect consequence of reducing variability in the process using proper control design. The Peruvian Toquepala Mine copper ore grinding circuit is composed of a rod-mill followed by three identical ball-mills in parallel. Each ball mill runs in closed circuit with a pump-box and hydrocyclone cluster. The main elements of the original control strategy are as follows:

- PI control of the ore rod-mill feed rate, the set point being determined by the operator;
- PI control of the water addition to the rod-mill ensuring the mill is operated at 80% solids;
- the rod-mill discharge is distributed to three parallel ball-mills using a three-way splitter, the distribution ratios being adjusted by the operator using two separating gates; and
- PI control of the slurry level in each pump box, the speed of the discharge pump being the manipulated variable.

Improper operation of the splitter led to an uneven distribution of the rod-mill discharge among the three ball -mills, the one receiving more feed experiencing overloads resulting in a coarser hydrocyclone overflow, which in turn disrupted the flotation circuit operation. Another common issue was increasing rod-mill hold-up resulting from increasing ore hardness. Late detection of this kind of event would require a drastic reduction of the rod-mill feed rate, and even a grind out (i.e., stopping the feed for several minutes) in some instances.

The most important changes in the new design were the following:

- ratio control of the rod-mill water addition, for fast reaction to ore feed rate fluctuations;
- PI control of the slurry distribution to the ball-mills, the position of the two gates being manipulated to maintain the same slurry level in all three pump boxes; and
- PI control of the average slurry level in the three pump boxes, the manipulated variable being the rod-mill feed setpoint, thus allowing the system to cope with ore hardness fluctuations.

The outcomes of the new control system were the following:

- improved stability of the circuit operation, particularly for the three ball-mills;
- higher circuit throughput, from 245.7 to 253.6 t/h, and a slightly finer product;
- a significant reduction in energy consumption, from 1,607,750 to 1,528,797 kWh/month, mainly explained by reduced pumping requirements resulting from a lower circulating load; and
- a reduction of specific energy consumption from 9.20 to 8.42 kWh/t.

13.4.3 Real-Time Optimization of a Grinding Circuit

The final grinding control example [38] is the simulated real-time optimization of a semi-autogenous mill operating in closed circuit with a pump box and a hydrocyclone. The hierarchical control structure corresponds to Fig. 13.1. Although not explicitly mentioned in the paper, the regulatory control layer should consist of basic PI and ratio loops for the flow rates. A 4×4 linear model predictive controller with a 10 s sampling time stands for the advanced control layer. The controlled variables are the

- product particle size,
- mill hold-up (i.e., the charge as a fraction of the total volume),
- slurry level in the pump box, and
- power drawn by the mill motor.

The manipulated variables are the percentage of critical mill speed and following flow rates (setpoints):

- slurry feed to the hydrocyclone,
- ore feed to the circuit, and
- water to the mill.

The setpoints of the four controlled variables could be calculated using the above real-time optimization layer with a sampling time of 30 min (for reasons explained in the paper, only the mill power setpoint is changed by the real-time optimization). The cost function is the difference over a 7-day horizon between the electricity consumption cost, taking into account time-of-use tariff, and the turnover of the milling circuit product. The minimization is subject to various constraints such as the respect of the 4 × 4 model steady-state relationships and a required average throughput over the 7-day period.

The main results of the simulation study for a 2 MW mill were the following:

- controlled and manipulated variables comfortably remain within their constraints;
- the variability of the product particle size is low: standard deviation of 0.0031% for mean of 82.03% < 75 μm;
- when running at approximately 93% of achievable capacity, the mill power load shifting generates a cost saving of $9.90 per kg of unrefined platinum;
- when running at maximum capacity, mill power load shifting is not necessarily economically advantageous because the additional product may represent more money than what is saved in electricity costs.

13.4.4 Energy and Profitability in Flotation Processes

Flotation and centrifugal separation only accounts for 4% of the total energy consumption in mining and mineral processing [4], and thus the leverage for potential energy saving is very low. Even if energy efficiency and energy cost have the largest influence on the overall operating cost of the flotation process [39], the example described by Lelinski et al. [40] illustrates that it may not be profitable to minimize energy consumption in flotation. The energy cost over the 20-year lifespan of a single SuperCell™ was US$ 3.7 million for a specific power of 0.74 kW/m^3, and US$4.2 million and for a specific power of 0.84 kW/m^3. However, the additional specific power, and US$0.5 million energy cost, allowed generating a gain in recovery (2.6% for Cu and 3.1% for MoS_2), which would represent an additional US$160 million in total revenue.

This case study clearly illustrates that minimizing the energy consumption of a single processing stage may lead to poorer profitability.

13.4.5 Energy Efficiency of Concentration and Smelting

Through laboratory test work, mineralogical analyses, and modeling, Evans et al. [41] investigated the most energy-efficient circuit design and operating strategy for mineral concentration and smelting (copper–nickel sulfide). Instead of examining concentration and smelting independently, as it is usually the case, the overall energy consumption across the metal production chain was assessed. The overall energy consumption was estimated with and without a second grinding stage (re-grinding) while keeping constant the primary grinding energy. The results, summarized in Table 13.1, revealed that injecting more energy in the concentrator

Table 13.1 Energy consumption, with and without regrinding

	Without regrinding	With regrinding
Energy input to primary grinding (kWh)	1387	1387
Energy input to regrinding (kWh)	0	42
Energy input to Ni smelting (kWh)	3629	3123
Energy input to Cu smelting (kWh)	556	542
Total energy input (kWh)	5572	5094
Total energy consumed per ton of ore in feed (kWh/t)	55.7	50.9
Total energy consumed per ton of metal produced (kWh/t)	3910	3461

results, in some instances, in a reduction of the energy consumption for the integrated mill-smelter complex. Therefore, it is recommended that engineers design real-time energy optimization from a plant-wide point of view.

13.5 Conclusions

The mining industry faces tremendous challenges for improving its track record for energy management and carbon emissions. Analysis of the overall mineral/metal beneficiation chain reveals that mineral processing is one of the most energy-intensive stages, exhibiting an increasing share of the industry footprint. Among the various routes currently being examined to tackle this issue, automation stands out as a practical solution to achieve significant progresses within a relatively short-term timeframe.

The multiple functions of an automation system, namely process analysis, control, real-time optimization and management, all play a role in smart energy management of mineral processing operations. It is interesting to observe that reducing the variability of key variables using appropriate—often simple—process control strategies can lead to significant energy efficiency improvements resulting from improved process performance, better use of the circuit's capacity, higher throughput or reduced circulating loads. In some instances, this will be possible while maintaining or increasing throughput and quality even if the energy consumption is not explicitly considered in the design.

Moving to a higher level of the function hierarchical organization, real-time optimization is a powerful approach to explicitly manage energy consumption, throughput, and product quality, while handling operational constraints. It is more complex, but offers more capabilities than regulatory control. For instance, a real-time optimizer can decrease the footprint of not only individual units but also the overall circuit, and stir operating points to more favorable states minimizing the value of the economic criteria. It must be emphasized that the real-time optimization layer does not allow one to dampen the effect of fluctuating ore characteristics (hardness, composition, mineral associations, and size distribution) per se, so it cannot replace the control layer altogether. The ability to control key process variables is a necessary first step.

Future work needs to focus on developing more process control and real-time optimization industrial applications for individual units, sub-circuits and overall plants. As more case studies are developed, it will be possible to establish energy reduction benchmarks, define best practices and standards. Results could eventually influence design practices as higher energy efficiency could translate into leaner production units for both brown and green field projects.

Lastly, characterizing the effect of automation on other sustainability indicators should also be considered as maximizing energy efficiency could, for instance, influence the water consumption [41].

Acknowledgements D. Sbarbaro acknowledges the kind support of the support of Solar Energy Research Center (SERC-Chile) Fondap #15110019.

References

1. Tromans D (2008) Mineral comminution: energy efficiency considerations. Miner Eng 21:613–620
2. CEEC (2013) Innovation—mining more for less. http://www.ceecthefuture.org/comminution-2/innovation-mining-less/, 2013
3. Norgate T, Haque N (2010) Energy and greenhouse gas impacts of mining and mineral processing operations. J Clean Prod 18:266–274
4. U.S. Department of Energy (2007) Mining industry energy bandwidth study. http://www.energy.gov/sites/prod/files/2013/11/f4/mining_bandwidth.pdf, 2007
5. Wei D, Craig IK (2009) Grinding mill circuits—a survey of control and economic concerns. Int J Miner Process 90:56–66
6. Kanchibotla SS, Valery W, Morrell S (1999) Modelling fines in blast fragmentation and its impact on crushing and grinding. Presented at the Explo '99–A conference on rock breaking, Kalgoorlie, Australia, 1999
7. Torrealba-Vargas J, Esteban R, Roy D, Runnels D (2016) Mine to mill approach to optimize power consumption in a process plant operation by modelling and simulation. In: International mineral processing congress—IMPC 2016, Quebec City, 2016
8. Morrell S (2009) Predicting the overall specific energy requirement of crushing, high pressure grinding roll and tumbling mill circuits. Miner Eng 22:544–549
9. Van Der Meer FP, Gruendken A (2010) Flowsheet considerations for optimal use of high pressure grinding rolls. Miner Eng 23:663–669
10. Nordell LK, Porter B, Potapor A (2016) Comminution energy efficiency—understanding the next steps. In: IMPC 2016, Quebec City, 2016
11. Lessard J, De Bakker J, McHugh L (2014) Development of ore sorting and its impact on mineral processing economics. Miner Eng 65:88–97
12. Awatey B, Skinner W, Zanin M (2015) Incorporating fluidised-bed flotation into a conventional flotation flowsheet: a focus on energy implications of coarse particle recovery. In: Powder technology, vol 275, pp 85–93, May 2015
13. Tbaybi H (2015) Impact de la flottation éclair sur l'empreinte énergétique d'un circuit de broyage – Cas de la mine Nyrstar – Langlois. M.Sc., Département de génie des mines, de la métallurgie et des matériaux, Université Laval, 2015

14. Silva DO, Vieira LGM, Lobato FS, Barrozo MAS (2012) Optimization of the design and performance of hydrocyclones by differential evolution technique. Chem Eng Process 61:1–7
15. Radziszewski P (2013) Energy recovery potential in comminution processes. Miner Eng 46–47:83–88
16. Radziszewski P, Hewitt D (2015) Exploring the effect of energy recovery potential on comminution efficiency: the Glencore Raglan Mine case. In: *S*AG conference 2015, Vancouver, British Columbia, Canada, 2015
17. Bouchard J, LeBlanc G, Germain Y, Levesque M, Tremblay N, Légaré B, Dallaire B, Radziszewski P (2016) The CMIC/CanmetMINES comminution energy recovery potential initiative—The Agnico Eagle Goldex Division Case. In: International mineral processing congress—IMPC 2016, Quebec City, 2016
18. Levesque MY, Millar DL (2015) The link between operational practices and specific energy consumption in metal ore milling plants—Ontario experiences. In: Minerals engineering, vol 71, pp 146–158, Feb 2015
19. Nunez E, MacPherson G, Graffi D, Tuzun A (2009) Self-optimizing grinding control for maximising throughput while maintaining cyclone overflow specifications. In: 41st Annual meeting of the Canadian mineral processors, Ottawa, Canada, 2009, pp 541–555
20. Engell S, Harjunkoski I (2012) Optimal operation: scheduling, advanced control and their integration. Comput Chem Eng 47:121–133
21. Bonavita N (2013) Can process automation increase energy efficiency? Hydrocarbon Process 92:71–75
22. Sbarbaro D, del Villar R (2010) Introduction. In Sbarbaro D, del Villar R (eds) Advanced control and supervision of mineral processing plants. Springer, Berlin
23. Bascur O, Dunne R, Karageorgos J, Ruel M, Sbarbaro D (2016) Process control, optimization and performance management in mineral and metallurgical processing. In: Young C, Dunne R (eds) SME Process control handbook
24. Byron JHH, Landry J, Hart D (2003) Energy management information systems: achieving improved energy efficiency: a handbook for managers, engineers and operational staff. Office of Energy Efficiency of Natural Resources Canada, 2003
25. Efficiency New Brunswick (2010) Energy management information systems: planning manual and tool. Office of Energy Efficiency of Natural Resources Canada, 2010
26. Seshan A, Gorain BK (2016) An integrated mining and metallurgical enterprise enabling continuous process optimization. In: Lakshmanan VI, Roy R, Ramachandran V (eds) Innovative process development in metallurgical industry. Springer, Berlin
27. Friedmann PG (2006) Automation and control systems economics, 2nd edn. ISA Press
28. Tanaka K (2008) Assessment of energy efficiency performance measures in industry and their application for policy. *Energy Policy* 36:2877–2892
29. White DC (2003) The economics of advanced automation. In: AIChE 2003 spring meeting, New Orleans, 2003
30. Funk GL, Smith DE (1979) Estimating economic incentives for computer control systems: an applications approach. IEEE Trans Ind Appl IA-15:394–398
31. Hodouin D, Jamsa-Jounela SL, Carvalho MT, Bergh L (2001) State of the art and challenges in mineral processing control. Control Eng Pract 9:995–1005
32. Wei D, Craig IK (2009) Economic performance assessment of two ROM ore milling circuit controllers. Miner Eng 22:826–839
33. Desbiens A, Nunez E, del Villar R, Hodouin D, Poulin E (2008) Using process control to increase the energy efficiency of mineral and metal processing plants. Int J Power Energy Syst 28:145–152
34. Numbi BP, Xia X (2015) Systems optimization model for energy management of a parallel HPGR crushing process. Appl Energy 149:133–147
35. Numbi BP, Zhang J, Xia X (2014) Optimal energy management for a jaw crushing process in deep mines. Energy 68:337–348
36. Numbi BP, Xia X (2016) Optimal energy control of a crushing process based on vertical shaft impactor. Appl Energy 162:1653–1661

37. Nunez E (2002) Filosofía de control para molienda y clasificación en la planta concentradora de Toquepala y diseño de una estrategia de control utilizando DCS. Bachelor, Universidad Católica de Santa María, Arequipa, Perú, 2002
38. Matthews B, Craig IK (2013) Demand side management of a run-of-mine ore milling circuit. Control Eng Pract 21:759–768
39. Rinne A, Peltola A (2008) On lifetime costs of flotation operations. Miner Eng 21:846–850
40. Lelinski D, Govender D, Dabrowski B, Traczyk F, Mulligan M (2011) Effective use of energy in the flotation process. In: 6th Southern African base metals conference, Phalaborwa, North West Province, 2011, pp 137–148
41. Evans CL, Coulter BL, Wightman E, Burrows AS (2009) Improving energy efficiency across mineral processing and smelting operations; a new approach. In: Publication series—Australasian Institute of Mining and Metallurgy, vol 6/2009, pp 9–13, 2009

Chapter 14
Energy Management Systems in Copper Smelting: The Atlantic Copper Case Study

José Maria Tejera, Guillerno Rios, Tasio Martínez and Miguel Palacios

Abstract Copper smelters play an important role in the extractive metallurgy of copper, with 80% of mining output processed in primary smelters to produce copper cathodes, while the remaining 20% is refined at hydrometallurgical plants at the mines. Energy represents more than one-third of operating costs for copper smelting and refining, so good energy management is of vital importance to guarantee energy sustainability in an increasingly competitive economic setting, not to mention the ever-increasing demands for environmental protection. In 2009, Atlantic Copper launched a new energy management strategy for its copper smelter and refinery with the implementation of an Energy Management System, and it became the world's first copper smelter to receive the ISO 50001 (Int Organ Stand 16, 2011 [1]) certificate. This system can accurately monitor consumption, and shapes production planning in accordance with energy criteria in order to achieve continuous improvement in all company processes, based on the following:

- Integration of energy efficiency in the organization's scorecard.
- Assimilation of energy criteria in daily work decisions.
- Implementation of projects that can reduce specific energy consumption.
- Investment in projects that recover and utilize residual heat from the metallurgical processes.

This policy carried out between 2009 and 2014 enabled Atlantic Copper to become one of the copper smelters with the lowest specific energy consumption worldwide, achieving reductions of over 20% of the energy consumption, together with cuts in direct greenhouse gas emissions of more than 30%.

Keywords Copper · Smelter · Energy · Management

J. M. Tejera (✉) · G. Rios · T. Martínez · M. Palacios
Atlantic Copper SLU, Avda Francisco Montenegro S/N, 21001 Huelva, Spain
e-mail: jtejeraa@fmi.com

K. Awuah-Offei (ed.), *Energy Efficiency in the Minerals Industry*, Green Energy and Technology, https://doi.org/10.1007/978-3-319-54199-0_14

14.1 Introduction

Metal refining is an energy-intensive process. Consequently, many stakeholders have highlighted the need to improve the energy efficiency of this process. This chapter presents an industrial case study of a copper refinery's energy efficiency improvement program.

Atlantic Copper's business activity is based in the city of Huelva, in Spain, which is a member of the European Union. Its main shareholder is Freeport McMoRan, a world leader in metal mining with important operations across four continents, and one of the world's major producers of copper, gold, and molybdenum.

Atlantic Copper's smelter and refinery smelts more than one million tons of copper concentrate a year from the Freeport group as well as from other mining operators. Its main product is a high-grade copper cathode and it produces about 300,000 tons of that a year.

Briefly, the process consists of the following:

- The Smelter, equipped with a Flash Furnace licensed from Outokumpu (currently Outotec), four Peirce-Smith converters, three anode furnaces, and two casting wheels.
- Copper electrorefinery, using ISA permanent cathode technology.
- Three double absorption sulphuric acid production plants, designed by Lurgi (currently Outotec).
- A power plant, with an 11.5 MW nominal power steam turbine that produces around 20% of the electrical energy consumed at the Complex, that uses the residual heat from Flash Furnace off-gases and provides the steam power required by the other production plants.

All the technologies applied at the Huelva Metallurgical Complex (HMC) are state of the art. This enables Atlantic Copper to produce copper at a competitive price and, at the same time, match the highest quality standards and conduct its business activity in a manner compatible with respect for and conservation of the environment.

14.2 Energy Management: The First Step Is to Raise Expectations

The production of cathodes from copper concentrate is an energy-intensive process, which is a common feature of all copper smelters and the metallurgical industry, in general. Data from 2009 showed that energy costs at Atlantic Copper accounted for 30% of the total operating costs of smelting and refining at the HMC.

Like other similar installations, the HMC is an industrial plant that has been in operation for many years. The plant has been in operation for more than 40 years, since the commissioning of the Flash Furnace process. For years, operations of this type have monitored and optimized their energy performance based on mature cost calculation systems, fairly detailed reports on energy cost distribution per process unit or cost center, and projects to improve efficiency based on better technologies.

For this reason, it seems reasonable to think that the potential for reducing energy costs at a mature (in other words, totally optimized) metallurgical plant would be less than the average achievable in the industry in general. The first step must be to establish an attainable energy performance reference level, and use this figure to set the parameters for potential improvement by comparing it with the real level of energy consumption at the installations. To do this, we use three different approaches:

- Benchmarking;
- Reviewing theoretical studies; and
- Reviewing and updating the global assessment of the HMC's material and energy balances.

14.2.1 Benchmarking

According to the Wood McKenzie (formerly Brook Hunt) confidential report in 2007, based on figures for the year before, the Atlantic Copper smelter was already among the world's top five smelters in terms of energy efficiency. Although comparisons are not easy to make, and homogeneous data are required to standardize differences in terms of scope, technologies, process diagrams, etc., the logical conclusion from this report is that the potential for improvement in energy efficiency by benchmarking is slight (Fig. 14.1).

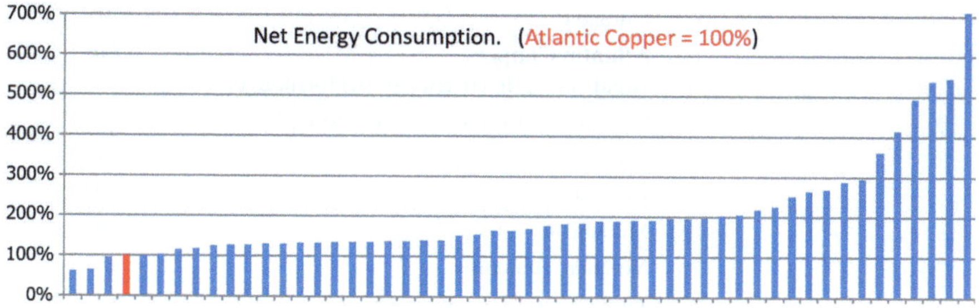

Fig. 14.1 Comparison of specific energy performance at various copper smelters around the world [6]

14.2.2 *Reviewing of Theoretical Studies*

In theory, each copper smelter has an attainable specific energy consumption baseline value that depends fundamentally on the technology used. Calculating this value requires assessing not only the material and energy balances of the processes but also taking into account numerous other aspects of the everyday operation of an installation, such as the efficiency of the equipment and machinery in use, the requirements of general services, like compressed air supply, cooling water, etc. This methodology for calculating a specific energy consumption value is complex. Atlantic Copper carried out an analysis of its energy performance comparing it to the results from a study made by Mackey et al. [2] which are shown in Table 14.1. The main conclusions were the following:

- With the process technologies of the 1970s, consumption baselines were significantly higher than those derived from technologies used this century. The Atlantic Copper plant was designed in the 1970s.
- In other words, we would expect to find better consumption ratios at those smelters designed and put into operation more recently.
- Atlantic Copper's energy performance in 2009 represented a significant improvement over the attainable baseline reference levels calculated for its technology (Outokumpu Flash).

In conclusion, we did not see much potential for improvement in Atlantic Copper's energy performance by applying this theoretical baseline.

Table 14.1 Theoretical baseline. Annual theoretical energy consumption in GWh for various technologies [2]

Designed in the 1970s [2]	
Reverberatory	1514
Outokumpu flash	**883**
Mitsubishi	1083
Noranda	795
Atlantic Copper (Real normalized energy performance)	**744**
Designed between 2000 and now [2]	
Flash-flash	531
Isasmelt	680
Mitsubishi	590
Noranda-Teniente	654

Outokumpu Flash Furnace and Atlantic Copper are bold for comparison purposes as Atlantic Copper is operating a Outokumpu Flash Furnace

14.2.3 *Global Assessment of Material and Energy Balances at HMC*

So, the only remaining alternative for us to avoid giving up on significantly improving energy performance at our Complex was to undertake our own exhaustive internal assessment of the HMC's material and energy balances, and establish a baseline attainable either by implementing changes in technology or in management processes. With this in mind, we carried out the following actions:

- Calculating and reporting energy consumption in terms of homogenous energy units. Cost and consumption analyses have normally been done by counting the number of physical units consumed, for example in kg of fuel oil, Nm^3 of natural gas, etc. This has the advantage of being able to measure, close-up, the operational variables in use (electrical power required, scheduled fuel costs, etc.), and provides a clear cost comparison between the various energy sources. However, this system does not allow us to compare, in energy terms, the quantities of energy consumed from each different source nor provide data on the cost per unit of each type of energy consumed. This is a first step toward detecting the opportunities for reducing costs in terms of energy price variations, or toward establishing priorities regarding what to analyze based on the quantities of energy used. Table 14.2 is a summary of the results from this type of energy performance analysis that was applied to normalized energy consumption figures at Atlantic Copper from 2009 onwards. The figures in grey represent the typical information provided by the cost and consumption control systems. The figures in black show the additional data derived from this new energy analysis on costs and consumption. Note that Oxygen is counted as a power

Table 14.2 Global energy consumption

	Consumption	Unit	Price (c€/unit)	Cost (MM€/a)	Consumption (GWh/a)	Price k€/GWh (% of 2009 average cost)	
Total Energy				32.7	744	44	(100%)
Electricity	311,285,552	kWh	5.3	16.4	311	53	(120%)
Natural gas	262,894,244	te	2.7	7.2	305	23	(53%)
Steam	202,352	t	2000	4.0	149	27	(62%)
Oxygen	153,478,541	Nm^3	5.9	9.1	84	108	(246%)
Fuel Oil	6,632,265	Kg	37.3	2.5	77	32	(73%)
Coke	2,028,693	Kg	16.3	0.33	18	19	(43%)
Diesel Oil	701,577	l	60.4	0.42	8	54	(123%)
Others	169,653	Kg	13.4	0.02	1	26	(58%)
Total energy consumed				40.0	954		
Energy recovered							
Steam	−202,352	t	2000.0	−4.0	−149	27	(62%)
Electricity	−60,052,239	kWh	5.4	−3.3	−60	55	(123%)
Total energy recovered				−7.3	−209		

supply because it is considered an over-the-fence supply. Hence, a calculation of the energy required for its production at external installations is included.

Unifying these data meant selecting a common energy measurement unit. We chose GWh for two reasons: (i) in Spain, our two main power supplies, natural gas, and electricity, are priced in kWh; and (ii) it is also a sufficiently broad measure for extracting accurate and easy-to-interpret figures, ranging from 0 to 1000, in terms of the HMC's annual consumption levels.

- The second action undertaken was to apply this new energy performance analysis to the various subprocesses in operation at the HMC, which would enable us to compare the amounts of energy involved in each subprocess and determine how much each energy supply contributed to the energy cost of the subprocess.
- In the third action, we added to our overall assessment, other factors such as the energy recovered from the individual process and the energy contributed by the raw materials. This required updating the energy and material balance of the whole plant. At long-established plants like ours, this type of information is not usually compiled, or if so, it is dispersed over various categories and documents; or the data might never have been updated since the plant has undergone so many expansions and modifications over the years. The data was not updated because each individual impact was not sufficiently significant to warrant a revaluation of the entire balance at the Complex. We should strive to keep the global energy and material balance updated at all times.

14.2.4 A Global Perspective

The results of the previous analysis shed new light on our energy costs and consumption. We now have a global perspective of energy consumption, a detailed breakdown according to each subprocess, and one which considers all the energy flows that occur at the HMC. Thus, each subprocess can now be evaluated by the organization in terms of its energy relevance. We can assess real energy flows, assess the amount of energy dissipated, and design maps that detail the origin and destination of the energy supplied to the process, among others. In addition, we can quickly and easily link the energy balance to other environmental issues such as the water consumption needed for cooling and, in particular, we can investigate the real possibilities of making significant reductions in energy consumption and boosting energy recovery in the process.

Figures 14.2 and 14.3 show the total global energy distribution, detailed by subprocess. It should be noted that the "energy size" of the various subprocesses when we only consider external energy supplies (energy purchases) is very different from the result that takes account of all real existing sources, including those

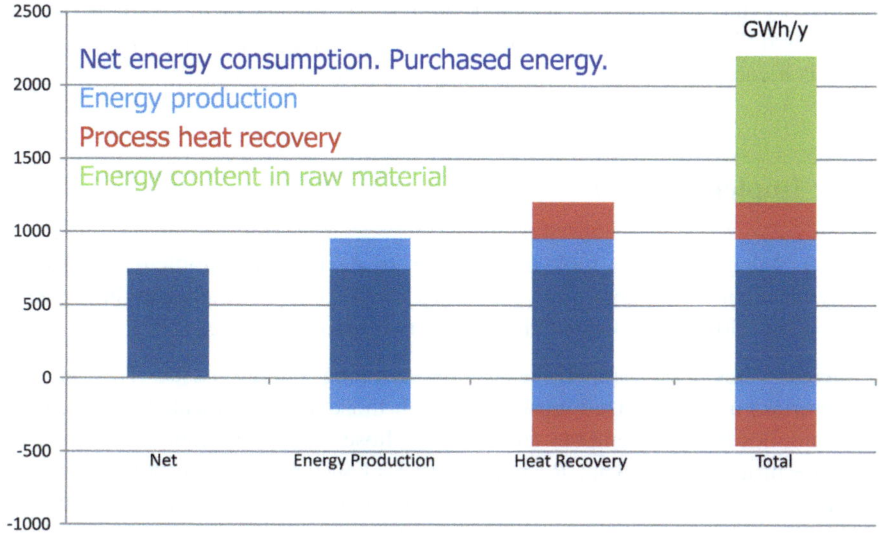

Fig. 14.2 Purchased energy versus total energy balance

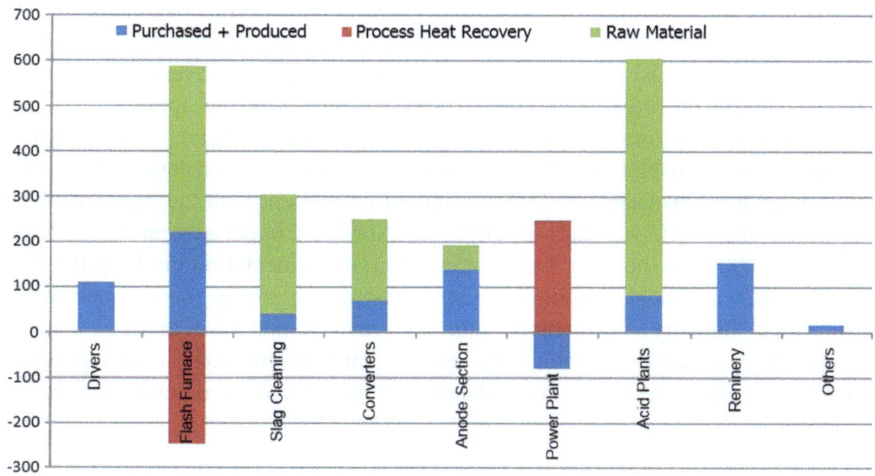

Fig. 14.3 Purchased energy versus energy balance by process

generated by the processes themselves. It is also significant that the total energy involved in the processes is almost triple the value of the energy purchased.

We can conclude then that, although, Atlantic Copper has an energy performance superior to the majority of copper smelters worldwide, and above the theoretically calculated baseline for its technology, there is significant potential for reducing consumption. This is the essential outcome of the "energy review" required by the ISO 50001 standard on Energy Management Systems, or the "energy audit" required by the European Union's Energy Efficiency Directive, and which is a current requirement for all major companies located in the EU.

14.3 Implementation

In 2010, Atlantic Copper took the decision to develop and implement a formal energy management system, thus becoming the first copper smelter in the world to gain the ISO 50001 certificate. There were two main objectives:

- Increase business competitiveness by reducing energy costs. This is very important in a globalized market in which those smelters located in developed economies have to compete on cost with those located in emerging markets.
- Make progress on Social Responsibility issues such as the responsible use of natural resources and the reduction of greenhouse gas emissions, in particular, the CO_2 associated with energy consumption.

As these objectives and associated methodology fit well with the ISO 50001 standard on Energy Management Systems, we decided to follow its guidelines for implementing the system. Moreover, we decided to pursue full certification for our energy management system. As a pioneer in this field, Atlantic Copper had to develop and take decisions on numerous details not directly covered by the standard, which can be broadly interpreted and can, therefore, be applied to a wide variety of organizations. In fact, our system was implemented before the ISO regulation was published, which is why our system was first certified at the start of 2011 taking the European EN-16001 standard as reference, in anticipation of the publication of the ISO standard. In 2012, the Atlantic Copper system was certified under ISO 50001. At the time, there were still no implementation guides published, and only a few specialist consultants with experience in energy management in the metallurgical industry. Fortunately, this has changed and now there is a lot of high-quality assistance available [3], which simplifies the job of setting up an energy management system, reduces the effort required and increases the possibility of success.

14.3.1 The Working Group or "The Energy Team"

In order to effectively implement the energy management system, there needs to be clear and decisive support from top management. It is also necessary to build a team capable of leading the change process. For successful implementation, this team must have the following features:

- Credibility in the organization, since this group is going to drive the changes forward.
- Maintain a global vision to ensure that the energy efficiency strategy is in line with the company strategy, and that individual actions are mutually compatible and coherent in terms of the strategy. Also, at a metallurgical complex such as ours, with several closely connected subprocesses and shared common service networks, it is clear that the global optimum is not equal to the sum of all local optimums. Hence, the key is to maintain an overall perspective and analyze the implications for the whole plant of all proposed modifications.
- Maintain a multidisciplinary approach. Energy management is not set apart from the management of other aspects of the business. In fact, its mission is to get the managers of the various units of the business to collaborate to improve energy performance. So, the "energy team" must have participants with experience not only in management systems but also in operations, maintenance, planning, process engineering, mechanical and electrical engineering, automation, energy procurement, project management, etc. The team must also have, what is known as, "energy knowledge".

14.3.2 Integration in the Organization

The successful integration of energy management in the organization is one of the key aspects of implementation. Hence, the strategy for implementation must adapt to the characteristics of the organization. Two factors must be taken into consideration:

- Integration in the hierarchy. At Atlantic Copper, the management of energy comes under the remit of the department for the Management of Planning and Control of Processes. We consider this to be the best option given the consumption characteristics of the HMC, where the factor that most influences consumption is the management of the processes. Monitoring consumption, developing, and tracking of energy indicators, or meetings to review the results, all form part of the existing systems and routines of process control. This simplifies management and eases the integration of energy management into the organization's daily work routine.
- Formal integration of the management system. Atlantic Copper has a lot of experience in Quality, Environmental and Safety Management Systems, so it makes no sense to design from scratch a whole new set of instructions and procedures required by an Energy Management System. We opted to "share" the already existing instructions of other systems, modifying them where necessary to conform to energy management requirements. In total, only six new documents specific to energy management were drawn up.

Simplicity is important, even more so when considering that, in the end, energy performance is the result of all the management actions in the organization. It is not enough to write up energy management requirements in documents that are then unrelated to or inconsistent with other documents that specify management requirements for other aspects of the business, such as quality or environment.

14.3.3 Measuring Energy Consumption

We do not need to discuss at length the obvious need for measurements in order to extract data. Both the ISO standards, Energy Management Systems implementation guides, and all the experts in this field, insist on this action. But some points are worth considering:

- It is not just a question of measuring energy consumption accurately, periodically, and in detail. We must also measure all those variables that can affect energy performance, which might include numerous operational parameters in production and raw material quality.
- Although the metallurgical industry typically requires a wide range of instrumentation, integrated within powerful real-time control systems for its operational needs, proper energy management sometimes requires additional measurement tools. Fortunately not many are needed, so investment is usually low. By carrying out or updating the material and energy balances mentioned previously, the additional instrumentation necessary for confirming these balances with real data will become evident, and these new measuring tools can provide further information to enable detection of deviations and to confirm increases in energy losses or reductions in equipment efficiency.
- All these field measurements must be processed and transformed into useful information by creating a system of indicators. When selecting a monitoring system from among the many that have been developed over the years, we must remember that the metallurgical industry energy performance is usually defined using specific consumption ratios that include at least one energy consumption metric and one nonenergy variable. The tracking system must be capable of integrating nonenergy measurements and carrying out the often complex calculations in order to estimate the required indicators. At Atlantic Copper, this work is done in-house using the organization's IT infrastructure to monitor the processes. Back in 2009, there was no alternative as there were no sufficiently developed energy-tracking systems on the market to deal with the complexities involved in energy management at a copper smelter.

14.3.4 Defining the Action Plan

Defining an action plan normally consists of deciding what we are going to change. In the case of implementing an energy management system, it is more important to decide what to investigate before defining the changes to be made. This is vital when considering that the first steps are always taken with limited resources. As a result, two additional tools were added to the global energy vision detailed previously.

The first one was an overall energy flowsheet. Given that improving efficiency actually consists of searching for and reducing dissipated energy, that is, energy losses, the flowsheet must show precisely that. Figure 14.4 presents an example of a flowsheet with figures estimated for this current case study.

There are three types of losses that happen at a metallurgical plant:

- Transformation: this is due to the efficiency of the machinery that transforms the energy supplies into the type of energy required in the process. For example, a boiler transforms the natural gas heat content as an energy increase of a water stream.
- Distribution: this is due to the efficiency of the plant′s utilities distribution networks. For example, losses in the electricity distribution network arising from voltage drops—heat losses—in transformers, cables, and connections.
- Process: the energy losses in the performance of process equipment. This is the core business and is due essentially to the technology used. For example, heat losses in an anode furnace.

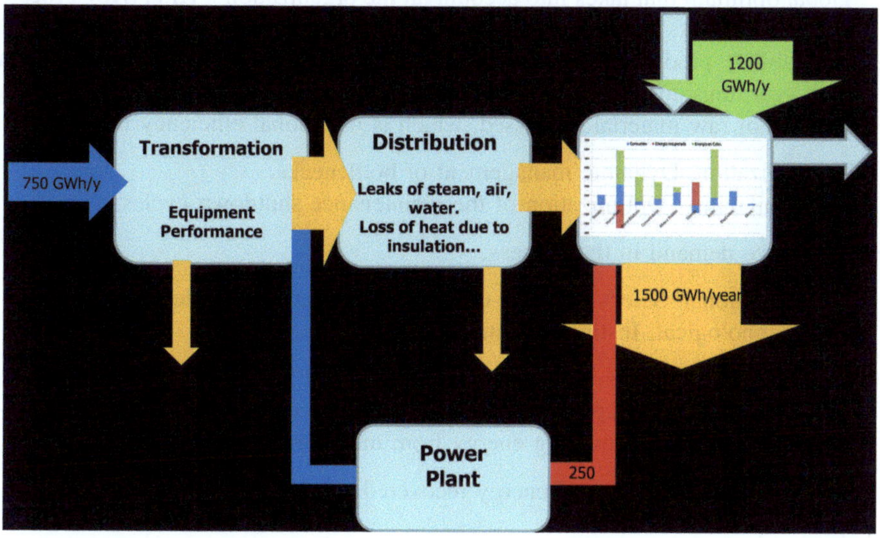

Fig. 14.4 HMC energy flow map

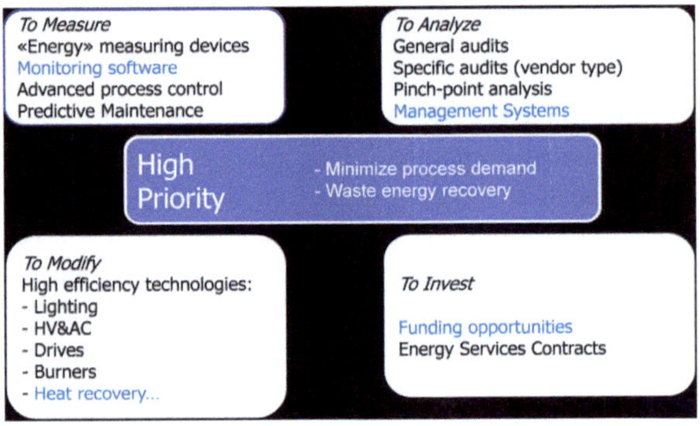

Fig. 14.5 Prioritizing the tools for the development of an energy management system

Of the three types, the process losses are the most pronounced at Atlantic Copper. In fact, they amount to double the energy purchased.

The second one was a classification of the tools and technologies on the market designed to improve energy efficiency. This involved investigating the functionalities of each one with the aim of selecting those which could improve aspects of energy management that were most in need of attention in our organization and discarding the rest with the minimum use of resources. It is important to sort the best from the array of proposed energy solutions on the market. However, we must be aware that assessments can become burdensome if too many resources are required to do them. Figure 14.5 shows some examples of these tools.

Based on this analysis we formulated an action plan, a strategy, whose main points were:

1. To integrate energy in the organizational scorecard. Determine an achievable global optimum that takes into account all the organization's KPIs (Key Process Indicators).
2. To stabilize production rate: avoid overconsumption and increase production each year, thus ensuring a double improvement in specific energy consumption per unit of raw material processed, which is our global efficiency indicator.

   2.1. Utilization. Dynamic management of bottlenecks.
   2.2. Availability. Optimization of the maintenance shutdown cycles.

3. To reduce demand in the process.

   3.1. Operational. Select the most economical operational mode.
   3.2. Technological. Instigate technological changes.

4. To increase energy recovery in the process. At our Plant, the most complex issue is to select the most appropriate uses for recovered energy and find the technology capable of recovering energy from metallurgical gases.

   4.1. To raise the amount of energy recovered.

4.2. To increase efficiency in the conversion of the residual energy in usable energy.

5. Other improvements in efficiency via automation, replacement of inefficient machinery, etc.

14.4 Results

In the five years we spent developing this strategy, we completed more than 30 projects involving management changes and investments. The global payback period was about 2 years, making it a highly profitable activity, partly because it coincided at the time with the highest prices for energy supplies in recent history.

14.4.1 Global Result

The total specific energy consumption per unit of raw material processed, which is the main indicator of global efficiency, fell by 21% in the 5 years from 2009 to 2014, and by 35% over 10 years, as can be seen in Fig. 14.6. Consumption for 2015 was slightly higher due to a major scheduled maintenance shutdown. In 2016 so far, energy consumption is at its lowest level ever recorded. These results compare favorably with those reported as best experiences in the Superior Energy

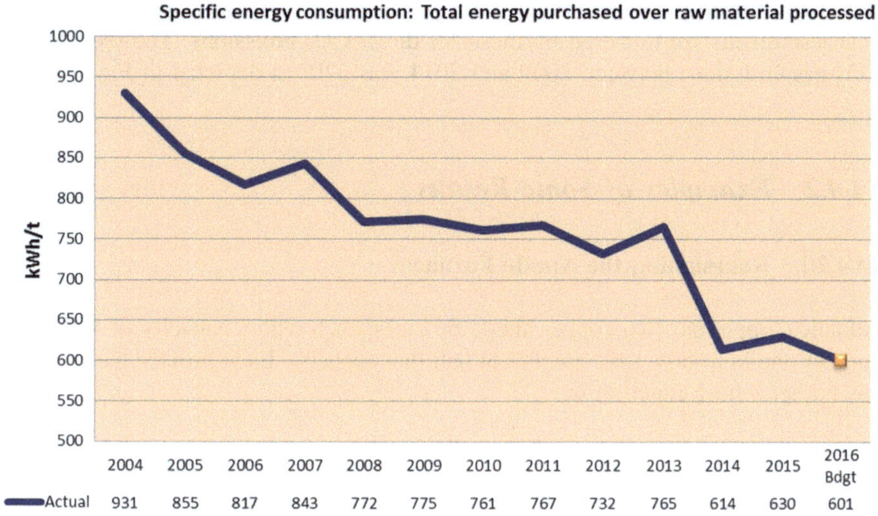

Fig. 14.6 Evolution of global specific energy consumption

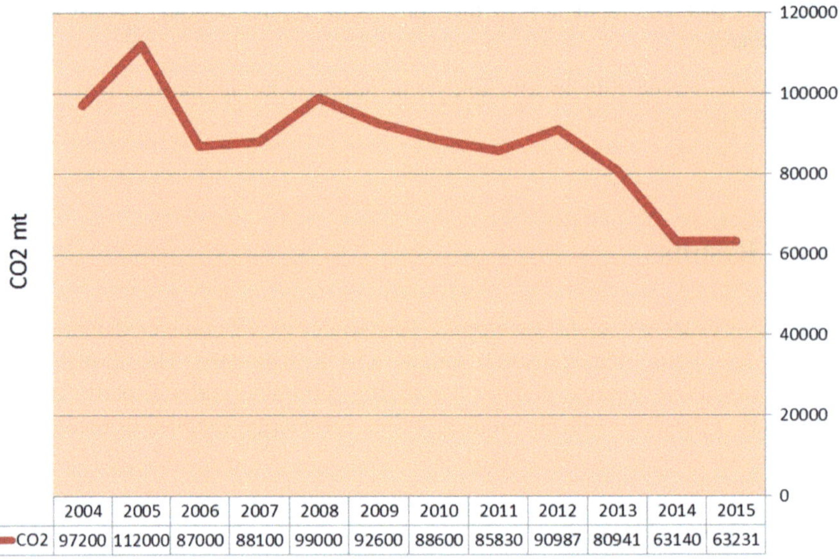

Fig. 14.7 Direct CO_2 emissions into the atmosphere

Performance program in the United States.[1] It is important to note that this indicator documents the overall evolution of the energy performance, which includes not only the modifications that improve energy efficiency but also the negative effects on energy consumption arising from the installation of new auxiliary plants, improvements in or extensions to installations to reduce the environmental impact, increases in the capacity of certain areas, etc., all of which mean significant increases in energy consumption.

These results are mirrored by those for direct CO_2 emissions. The reduction in CO_2 gas emissions between 2009 and 2014 was 32% as depicted in Fig. 14.7.

14.4.2 Examples of Some Results

14.4.2.1 Redesigning the Anode Furnaces

Atlantic Copper has two copper anode furnaces, each with a capacity of 400 t, and another one in reserve that operates at half that capacity. Each furnace has a natural gas burner whose function is:

[1]Superior Energy Performance: Platinum Certified Facilities available at http://energy.gov/eere/amo/certified-facilities.

Fig. 14.8 Refining furnaces (**a**) old manhole covers (**b**) new manhole covers

- To heat the furnace after being repaired.
- To keep the furnace hot.
- To keep the copper hot inside the furnace.
- To condition the interior of the furnace to a particular atmosphere, either oxidant or reductive, depending on the requirements of the stage of the process.

This investment project consisted of a series of modifications to the design of the furnaces and their auxiliary installations in order to reduce the quantity of natural gas consumed. The modifications were to minimize energy losses by:

- Installing new natural gas burners, incorporating Dilute Oxygen Combustion (DOC) technology by using pure oxygen as the oxidizing agent. This technology has been developed over decades through research into oxi-combustible combustion applied specifically to reheating copper [4, 5].
- Installing new furnace mouths with automatic pneumatically driven lids, replacing the manual closing system. This enables the furnace to remain airtight for longer and so reduces losses due to the presence of air. Figure 14.8 shows the old and new lids.
- Installing porous plugs and redesigning the furnace's refractory material, which led to an increase in the efficiency of the natural gas used as reactive agent and a reduction in furnace walls' heat losses. This reduced the quantity of natural gas needed to maintain the temperature of the molten bath.

The modifications to the refining furnaces were made during the general HMC shutdown in 2013 with an investment of 2.5 million Euros. This project has improved energy performance by more than 60 GWh/year, thanks to savings in natural gas consumption, and a reduction in direct CO_2 emissions of 11,000 tons per year as summarized in Fig. 14.9.

14.4.2.2 Reduction in Fuel Oil Consumption

Fuel oil is used in the Flash Furnace settler and provides the energy necessary for maintaining the furnace's energy balance. In this case, there were no modifications,

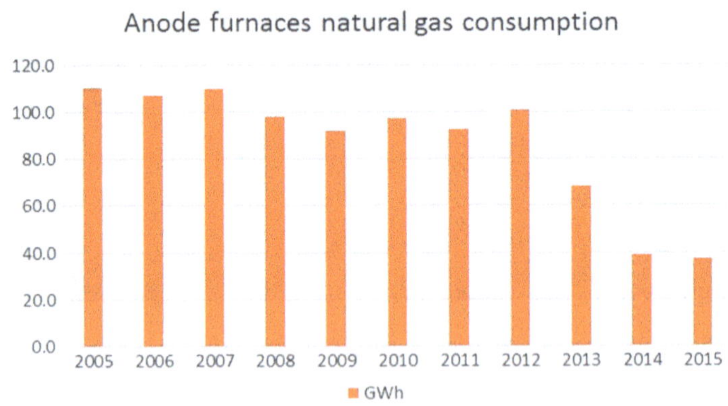

Fig. 14.9 Natural gas consumption in the refining furnace burners

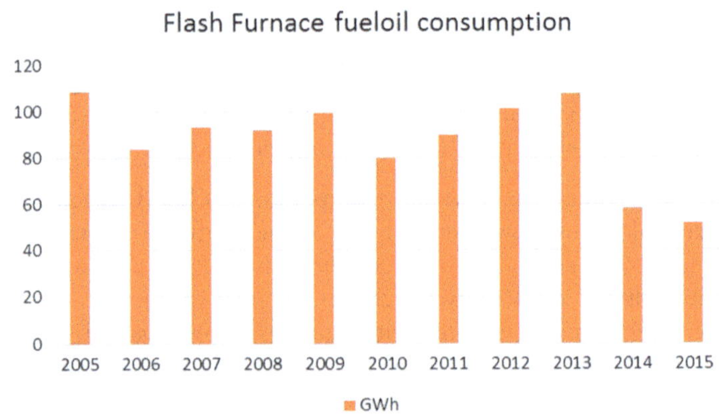

Fig. 14.10 Fuel oil consumption in the flash furnace

design changes or big investments, merely an example of saving by improving operations. The strategy was to tweak the operation of the furnace to improve its energy balance by selecting the point at which it functioned most economically, without affecting capacity, or the integrity and efficiency of the operation.

Fuel consumption has fallen by 45%, meaning a drop of 40 GWh/year, as shown in Fig. 14.10, and a reduction in direct CO_2 emissions of 11,000 tons per year.

14.4.2.3 Efficiency in the Power Plant

The area that we have worked on most is in improving the efficiency of converting the residual energy, recovered in the process, into useful energy. This work took place in the power plant. The installations there consist of the following:

- Boilers for the recovery of heat from the metallurgical gas, which produces steam.
- External natural gas-fired superheating boiler, which raises the temperature of the steam at the steam turbine inlet to design specifications.
- Steam-to-condensation turbine with intermediate extraction. From 11.5 MW of nominal power, it produces 40–70 GWh of electricity each year, which represents 20% of the electricity consumed in all fusion, refining, and auxiliary service operations. The steam extracted is used for services in the plant, in particular for conditioning the electrolyte in the copper refinery.
- Steam-generating auxiliary boiler to supply steam to the plants when there is no steam available from the residual heat.

This type of steam cycles based on residual heat from the industrial processes cannot yield high intrinsic efficiencies since they operate at low pressure and temperatures compared to the cycles of the conventional fuel-based power generation plants. Also, the turbines are of basic design, with no multiple extraction of steam or intermediate reheating taking place. The advantage is that part of the fuel is substituted by residual heat, thus improving its specific fuel consumption or apparent efficiency (electricity production/fuel consumption).

In order to improve efficiency at these facilities, we developed seven projects that led to a reduction of 73% in specific consumption (Fig. 14.11) without reducing electricity production. These projects were, in chronological order:

- Reduction of the minimum technical fuel consumption in the auxiliary boiler.
- Installation of a steam-driven mass heating system inside the Auxiliary Boiler, which guarantees that the machinery can start up securely in less than 30 min. This reaction time is sufficient for the processes that require steam in case of failure in the supply of steam generated in the recovery boilers. This enables the operations to function normally with the auxiliary boiler out of service (shutdown).

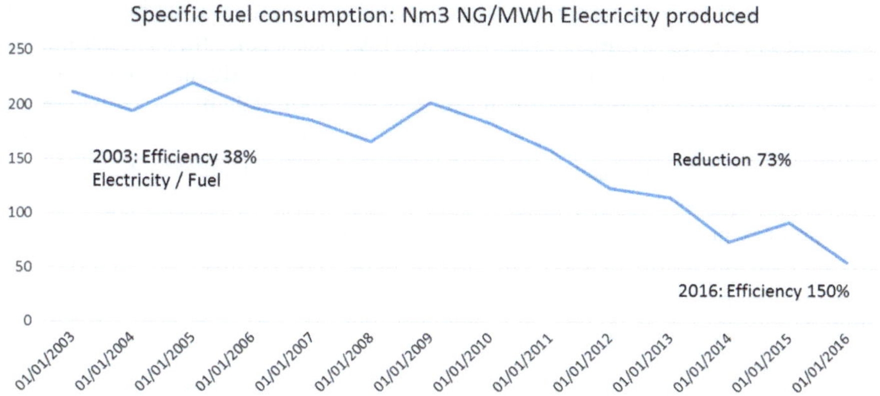

Fig. 14.11 Specific fuel consumption at the power plant

- Reduction of the live steam temperature from 500 °C (original steam turbine design temperature) to 400 °C. This enables us to reduce the natural gas used in the superheating boiler but at the cost of losing electricity production.
- Modification of the Flash Furnace Heat Recovery Boiler to produce superheated steam directly from residual heat. This means we can reduce fuel demand in the superheating boiler.
- Renovating the superheating boiler's fuel system to adapt it to the new fuel demands, which are now far lower than foreseen in the original design. This allows us to improve fuel efficiency in the superheating boiler.
- Installation of a heat recovery boiler in Acid Plant #3 which generates superheated steam directly from the residual heat coming from sulphuric acid production. This increases the residual heat used as well as electricity production.
- Other projects based on the selection of the most economical operational mode by using existing degrees of freedom to tune the energy balance of the various process units involved in the steam cycle (in the main, concentrate dryers, Flash Furnace, Acid Plant #3, power plant).

All these projects were consistent with the common objective, and followed a long-term strategy. These actions illustrate the concept of continuous improvement, which is the basis of all management systems, including those that manage energy. The result of all these projects is to maintain electricity production levels by substantially reducing fuel consumption and substituting it for residual heat from the metallurgical process. Thus, we simultaneously reduce costs, consumption and emissions of CO_2.

14.5 Conclusions

Good energy management is a process, not a single project or a series of projects. The opportunities for improvement are not always the result of investment in machinery or in developing more efficient processes. Improvements brought about by changes in management are usually the most profitable since they require little or no investment, and they can make a significant impact in reducing consumption.

The opportunities for improvement must be assessed globally by analyzing not only the effects caused in energy performance, but the implications for operations, maintenance and the environment. We cannot rule out the fact that these "collateral benefits" complement energy saving to such an extent that it encourages investment, and not just for the sake of cutting energy consumption. And the contrary is also true; in the sense that not all energy improvements that are, in principle, economically profitable are always viable when considering all the effects they would have on the organization.

To sum up, the key actions taken were:

- To change from managing supplies to managing energy.
- To change from looking at the installations to looking at the processes.
- To change from analyzing electricity bills to planning production with an energy consumption perspective.

And definitively, to change from "investing to make savings" to "operating to make savings".

ISO certification of our management system has helped us to:

- Explore all areas of energy management where savings could be made;
- Ensure continuity and monitoring of actions and results; and
- Grab the attention of managers and raise awareness of workers who, as always, are the key players.

References

1. International Organisation for Standardization (2011) ISO 50001—Energy management. Int Organ Stand 16
2. Coursol P, Mackey PJ, Diaz C (2010) Energy consumption in copper sulphide smelting. In: Proceedings of Copper 2010. Hamburg, Germany, pp 2543–2562
3. US Department of Energy Office of Energy Efficiency & Renewable Energy (2017) Software tools. In: Webpage. http://energy.gov/eere/amo/software-tools
4. Huang Y, Chen Y, Huang H et al (2011) Installation of oxygen combustion system at Jinlong Anode Furnace. Ind Heat
5. Department of Energy (2001) Kennecott Utah Copper Retrofits Smelting Applications from Air-Fuel to Oxy-Fuel Burners
6. Wood Mackenzie (2016) 3Q Report. Smelting and Refining 2014 Costs

Part IV
Renewable Energy and Miscellaneous Topics

Chapter 15
Solar Energy Applications in Mining: A Case Study

José Pablo Paredes Sánchez

Abstract In these times when sustainability is so crucial, clean energy resources have become increasingly important in the mining sector. Typically, about 30% of operational costs can be attributed to energy in mining activities. A mining company able to successfully embrace an integrated program that uses available renewable energy resources is often more successful. Renewable Energy Integration (REI) involves production, as well as managing the environmental and regulatory conditions. Renewable energy technologies are most attractive to mining projects in remote regions with little or no access to established electric grids. Inadequate energy supply has shifted the dynamic of solar energy development, as firms increasingly turn to renewable energies as one component of a basket of energy options used to maintain stable power at mining operations. The broad objective of the chapter is to foster a deeper understanding of solar technology and its integration in mines that enable them to address energy and sustainability issues more proactively and tactically. This chapter outlines recent developments in solar energy in the mining industry. It also discusses case studies where this framework has been applied and highlights the key emerging themes, such as energy management and environmental considerations, with benefits, weaknesses and future challenges.

Keywords Solar energy · Optimization · Mining · Energy management
Renewable energy integration

15.1 Introduction

The world of energy production is in a transition period, shifting from conventional to renewable energy sources. Moreover, the production of materials, especially raw minerals, is a major contributor to global energy use and corresponding greenhouse

J. P. Paredes Sánchez (✉)
Department of Energy, School of Mining, Energy and Materials Engineering of Oviedo, University of Oviedo, C/Independencia, 13, 33004 Oviedo, Asturias, Spain
e-mail: paredespablo@uniovi.es

K. Awuah-Offei (ed.), *Energy Efficiency in the Minerals Industry*, Green Energy and Technology, https://doi.org/10.1007/978-3-319-54199-0_15

gas (GHG) emissions [1, 2]. The global community committed to ambitious climate change mitigation efforts at the Paris UN Climate Change Conference held in 2015, which will only be achievable with a focused roll-out of sustainable energy generation and other sustainability initiatives in the world. In this context, a related priority is the need to reduce dependence on traditional energy sources through improvements in renewable energy use. Additionally, investments in this field contribute to the security of energy supply, while acting as a source of innovation, and are cost competitive, especially if energy prices remain high. Solar energy is going to be a key part of this new era, as it has become a cost effective and clean alternative to traditional energy sources. The technology permits us to collect thermal energy from the sun and use it for heating and cooling applications or power generation. The installed capacity of solar power has quickly reached a significant level in the world.

The mining sector is one of the biggest overall energy consumers. It is expanding into new and often remote locations in response to increasing demand from growing emerging markets. Mining operations require enormous amounts of energy to extract and process raw material, depending on the type of mineral resources and exploitation conditions, and even more on the amount of processing or beneficiation. For example, the production of cement and steel have been the focus of much attention in regards to GHG emissions, due to their significant contribution to industrial emissions [3, 4]. Energy is one of the most important cost factors for mining companies. The sector has traditionally relied on conventional fossil-based fuel sources to meet its energy demand. In the specific case of minerals, it is expected that ore grades will continue to decline and that, as a consequence, mining operations will require higher energy inputs and, therefore, higher costs, and GHG emissions [5, 6]. Energy management in mines has to respond to environmental sustainability and economic development goals as well.

Several case studies have already documented the importance of solar energy as a resource to be applied in industrial activities [7–9]. Solar energy is a decentralised energy source, which is an excellent means to deliver power to peripheral and remote consumers. In particular, it may be a key auxiliary energy source in remote areas, where solar energy use can be an essential source of energy. However, even some mines that are not so remote have adopted solar generation when the economics have been favourable [10].

Mining companies can improve their operating cost and environmental footprint by including solar power into their energy supply. However, mining companies still have to fully realize the potential to reduce their cost. This situation means that the mines that operate solar energy will have competitive advantages and outperform mines fully relying on traditional energy sources. There are already, in the literature, studies on renewable energy in the mining sector [11, 12]. This issue has been discussed in relation to specific countries [13, 14] as well as the influence of future global scenarios on mining practices [15]. Mining operations have studied solar in mining operations [16, 17]; but there is yet to be a top–down global study. The aim of this chapter is to fill this gap and analyse the industry baseline for solar energy operation and identify technological routes for a deeper understanding of solar integration in mines. The chapter defines current knowledge on the topic and the directions for the future research. This study will be a very useful tool for policy makers, researchers and energy users interested in promoting and developing renewable energy technology.

15.2 Methodology

This chapter focuses on solar energy use by the mining sector. It reviews the literature on solar energy applications in mines. Recent energy management initiatives are often discussed in technical reports and not in peer-review journal publications. However, the review is comprehensive and includes the latest literature in terms of reports, books, theses or dissertations, journal papers, conference proceedings, technical information, and a wide range of documents. The information presented for solar energy in the mining sector was collected using the available information from various sources.

The overview of the emerging solar energy technologies covered in this chapter is presented using a standard structure. First, the chapter briefly describes the technology. Next, by using key case studies, it shows the benefits, barriers, and challenges of solar energy. Finally, it discusses the relationship between energy and environmental management.

15.3 Solar Resources and Technology

The extraction and processing of mineral resources consume approximately 20% of the energy that the mining industry consumes and approximately 10% of world energy consumption [18]. Recent studies have examined the potential of renewable energy and concluded that we could use renewable energy to power all global needs in terms of the overall balance [19, 20]. Mining activities cannot exist without sufficient and secure supplies of both water and energy; without these, production is interrupted and costs increase [21]. We typically use renewable energy sources in the minerals industry, when economically feasible, to lower fossil fuel consumption, which in turn reduces GHG emissions [11]. Hence, solar energy used in the mining industry is part of the energy transition process toward a low-carbon economy.

From an energy management perspective, it is important that energy consumption in the mining industry is reduced efficiently. Hence, the main driver for changing to solar energy will be costs. From the solar industry perspective mining operations are a good fit, because:

- High energy consumption carries potential for large-scale solar power plants. Solar power can add value to mines for grid-connected and off-grid mines.

- Mining companies often have to deal with high energy costs due to remote locations. Moreover, mining companies in developing countries have to deal with unreliable electricity infrastructure, which makes it receptive for new solutions.
- Solar energy in the mining sector has potential to contribute to sustainable development efforts.

The energy industry can capture solar radiation and turn it into useful forms of energy, such as heat and electricity in a location. With the mining operations venturing into more complex areas, solar energy technologies are most attractive to mining projects in areas with little or no access to established electric grids. However, the technical feasibility and economical operation of the solar technologies at a specific location depends on the available solar resource. Measurements of solar energy are typically expressed as total radiation on a horizontal surface, or as total radiation on a surface tracking the sun. In terms of resource, the theoretical potential is the yearly radiation (kWh/year) to the Earth surface which is considered as an index that can be compared with other resources of fossil fuels [22]. The amount of sunlight falling on specific locations at different times of the year can be measured. Site-specific time series solar resource information, along with associated weather data, always has been essential to the successful design, deployment and utilization of solar energy systems.

There are many different ways to harvest the energy from the sun to provide energy. However, the technical feasibility and economical operation each method of harvesting the sun's energy, at a specific location, depends on the available solar resource. Some of these energy-converting systems are photovoltaic (PV) cells, the solar heating system (SHS), and concentrated solar power (CSP).

The use of photovoltaic effect is a method of generating electrical power by converting solar radiation into direct current (DC) electricity using semiconductors that exhibit this PV property (e.g., monocrystalline silicon or polycrystalline silicon). PV power generation operates solar panels composed of a number of solar cells. The solar cells are connected together in chains to form larger units known as modules and panels to boost the power output [23]. The solar industry can build PV systems to meet almost any electric power need, small or large, with components that take the DC electricity that cells produce and convert it to the alternating-current (AC) electricity. Solar heating systems harness solar energy to generate thermal energy and electricity. We can operate them for many applications, from water heating to air conditioning or even drying [24].

CSP systems use mirrors or lenses to concentrate a large area of sunlight, or solar thermal energy, onto a small area. CSP technology utilizes collected radiation from the sun to heat water in various ways (for instance, through reflection onto a central focal point or through individual solar panels). Electrical power is produced when the concentrated sunlight turns into heat, which drives a heat engine (usually by steam turbine) connected to an electrical power generator. CSP systems, also known as solar thermal power plants (STPPs), are flexible in the sense that we can use them with other renewable and conventional systems such as fired boilers, wind, and PV systems (hybridization) to extend plant life to a low-emission energy future [25]. Hybrid power plants combine at least two different energy types. In mining, the combination of diesel generator and solar power is rather common.

PV systems are the most common solar technology and have the most capacity for producing electricity, which is a key energy need in mining. However, PV cells are unable to efficiently store energy (as electricity is much more difficult to store than heat), meaning they are only effective when the sun is shining. Generally, we must install PV and SHS close to the load and, therefore, they require reliable solar data for the point of installation. CSP systems are capable of providing base load power to areas that have high direct solar insolation at costs similar to fossil fuel power generation. Moreover, power can be transported from insolation rich areas to remote areas. For example, power transportation is possible with only 11.5% losses for up to 3000 km [26]. A combination of PV and CSP technologies is the concentrator PV (CPV), which is a PV technology that generates electricity from sunlight by using lenses and curved mirrors to focus sunlight onto efficient solar cells.

15.4 Case Studies

Mining has a complicated relationship with renewable energy. Traditionally, the two might be viewed as rivals in the energy sector. Mining machinery consumes significant amounts of energy to perform extraction and mineral processing. Electricity and liquid fuels are the main resources used, both of which are based mainly on the use of fossil fuels. Fuels are consumed directly or indirectly as final energy depending on logistics conditions and energy supply. Solar for the mining operations is, at the moment, still a relatively small niche. This means that we should make significant efforts in order to increase the share of solar and clean energy sources of the entire energy supply structure. This section provides a brief insight into the nature of these issues by case studies in the context of the mining industry. The review indicates the additional benefits of solar energy supply systems for mining.

15.4.1 Power Energy

The common aim of mine management must be to ensure mine operations are environmentally sustainable, while diversifying energy sources to increase energy supply security. In grid-connected scenarios, mining companies often negotiate long-term power purchase agreements for renewable energy in regions with high electricity prices or unstable supply from the grid (e.g. South Africa). There are often good economic reasons for mining companies to source solar energy for their mining operations. Some mining companies have also invested in energy infrastructure and generated a part of the energy themselves. In some cases, remote mine sites are not grid-connected and generate their own energy using, typically, with large diesel generators. In no grid-connected scenarios, the remote locations of many mines lead to high transportation costs for the diesel fuel and losses during transportation. This situation considerably increases the total costs. There are already large-scale projects for grid-connected solar plants. Normally, solar power is less expensive than electricity from diesel in these locations. Any decrease in diesel consumption while maintaining production results in a reduction in fuel transportation and storage requirements because of diesel supply to the mine must be hauled and stored on site [27].

Cronimet Mining AG was the first mining company to adopt the hybrid solar application in the MW-class, integrating PV into one of their chrome mines in South Africa. The 1-MW solar plant with intelligent hybrid system is located in Thabazimbi in a chrome ore mine and produces up to 1.8 GWh of energy per year. The available diesel generators increase their performance automatically when the weather conditions are rather unfavourable and compensate for the brief deficits. The cost is frequently above $300/MWh. Given that the prices of PV plants have decreased considerably in recent years, solar energy is an attractive alternative. The mining sector expects to lower the consumption of mineral oil, used for the operation of the remaining diesel generators lowers, by half and remarkably cut the CO_2 emission. Furthermore, the cost advantage of solar solutions always depends, to some extent, on oil prices. Even with low oil prices, solar energy is often up to 50% less expensive than diesel-generated electricity. Usually, the existing power plant with diesel generators remains and a solar PV power plant is added [28].

The construction of a solar plant increases the total investment and capital costs in mining projects. Guaranteeing debt financing for the initial investment of energy infrastructure is a relevant economic factor. As in other renewable energy applications, switching from operating expenditure to capital expenditure can be a challenge without external financing capital. However, the investment increase can be mitigated by saving power supply during the operation years. After the pay-off period of the solar plant the mining project reaches additional cost savings. A high total cost of the power supply scenario is a driving force because mines have a finite life span and, therefore, represent a far-from-guaranteed long-term revenue stream. The case study of solar energy in copper mines shows additional benefits.

Solar projects are planned to supply electricity to Chilean copper mines [10] and the utilization of solar thermal energy in copper heap leaching processes has been investigated to increase solution temperatures, which leads to improved metal recovery [29, 30]. Mining group Minera El Tesoro (MET) installed the first solar thermal plant in South America. It installed a 10-MW plant at a copper mine in the Atacama Desert (Chile). The solar field has 1280 parabolic trough collector modules. The plant incorporates thermal energy storage that allows the delivery of heat for many hours after the sun goes down. The plant replaced more than 55% of the diesel fuel used in the mine heating process for electro-extraction of copper. The plant supplies heat in an efficient and sustainable manner. The solar plant delivers heat around the clock by using thermal energy storage, which it is ideal for the mining industry. The application of solar energy reduces CO_2 emissions by about 10,000 tonnes per year [31]. MET has built a CPV pilot plant in Sierra Gorda (Chile), installing four solar systems to provide renewable energy for a remote copper mining operation. The four CX-S420 systems installed for MET use CPV technology and two-axis tracking systems to generate a total installed capacity of 64 kWp [32]. The CPV systems feature very robust technology, particularly well suited for arid environments and do not need any water for cooling, making them perfectly suited for installation in desert regions.

15.4.2 Thermal Energy

The thermal energy usage from fossil fuels in the production of steel and non-metallic minerals contributes to 70% of energy consumption and 66% of emissions and is thus of high importance [11]. Thermal solar energy case studies in mines are a recent phenomenon. Goldcorp's Musselwhite Mine in northern Ontario (Canada) is a 4000-tonne-per day underground gold producer. It installed a solar air heater in the mechanical shop building and uses sunlight as fuel to preheat building ventilation air. As the air passes through the perforated metal solar collector the sun heats it, thus requiring less fossil fuel to heat the building; lowering fossil fuel usage, heating bills, and GHG emissions. The system is projected to offset 590 MWh of propane consumption, equaling $52,000 annual savings, and save over 120 tonnes of CO_2 emissions per year. This type of system may not be financially viable at every site. Mine managers and engineers should conduct an assessment prior to implementation to assess the available solar resource and the economics [33, 34].

15.5 Discussion of Key Considerations

The case studies presented in the previous section highlight the industry's challenges, related to meeting its energy needs, as it ventures into more remote locations. Based on the current knowledge on the topic, we need to consider three major groups in assessing the feasibility of solar energy: (1) mining operations, (2) energy management, and (3) environmental considerations.

15.5.1 Mining Operations

Mining companies focus on mineral resources and the sun is an additional resource to exploit. The changing landscape around carbon emissions and renewable energy creates a new challenge. In response, the mining sector has to deploy innovative renewable energy strategies and make substantial industry-wide direct investments into energy infrastructure. Third party developers can fund, build and operate solar energy systems as captive plants, with the mine committing to purchase the generated electricity at a fixed price over a certain time period. Mining operators expect full reliability and availability of their power systems, with very low tolerance for outages or shortfalls. Most mines operate at a 70–90% annual load factor relative to their peak load, but peak loads can occur at any time of day in any season. Due to the inherent variability of irradiation, PVs can only address a small fraction of energy needs in mines. However, CSP is as reliable and available as conventional technologies and capable of meeting a mining operations energy needs.

Incorporating solar power generation has many challenges. Improper positioning of the solar collectors, concentrators or panels relative to the sun can reduce the effectiveness of energy production. The sun only occupies half a degree of space in the sky, making it difficult to absorb all of its energy unless the solar system is perfectly positioned to face the sun directly. Moreover, the movement of the sun further reduces efficiency. The fact that power generation is limited to daytime hours is a huge obstacle that often makes solar power an infeasible option for many mining operations. This is especially troublesome in remote locations and uninterrupted 24-h mining production [35].

The development of mining projects is subject to the availability of resources and the capability to extract them. Most operations are unable to kick-start because of lack of capital to procure equipment and to fund operating costs. The economic viability is limited by the costs of the mining operations associated with transportation and infrastructure. As older ore bodies are being depleted, mining companies are increasingly looking to develop deposits in remote areas. The nature of these operations has marked the sector as a high risk one. In many cases in mining, electricity supply from the grid is either unreliable or nonexistent. The more remote the mine, the more likely off-grid power solutions are required because the nearest electricity grid for mining applications may be a long distance away. Mining operations are frequently vertically integrated and hold their own generation, transmission and distribution assets, as well as maintaining control of their own supply and demand in energy management practice [36]. Moreover, in most instances, electricity supply needs to be supplemented with on-site generation, typically large-scale diesel generation. In this sense, the use of solar diesel hybrids to supplying off-grid energy to mines via self-generation models faces several challenges. The technical requirements to develop solar energy in mining environments, particularly hybrid systems, are more complex than the common rooftop solar systems connected to the grid [37]. In future, the availability of technical expertise will be essential to scale up the deployment of distributed solar energy systems in the mining sector.

15.5.2 Energy Management and Renewable Energy Integration

A critical determinant of the growth in solar energy in mining will be the active support of national governments. Aside from direct financial incentives such as tax breaks and subsidies, establishing a policy framework, which allows mining companies to either sell energy independently or feed electricity back into the electricity grid, will increase capacity and help diversify the domestic energy supply. To address the barriers of the solar energy development in the mining sector, government policies such as feed-in tariffs or net-metering might be useful to overcome up-front capital costs. Building political will on green energy in mining could be particularly challenging, but any government support has the potential to see substantial dividends in increased electrical capacity.

The mining sector still faces numerous challenges that hamper its progress and growth such as scarce infrastructure in remote areas comprising poor roads, water systems, communication infrastructure and electricity. Solar power plants, with the right combination of renewable energy can operate with minimum reliance on fuels like natural gas or diesel. Natural gas supply can be very expensive, given the cost of building pipelines in remote areas. Diesel supply can have a heavy price since it is typically trucked to a remote mine over long distances. Solar energy can significantly reduce these costs and reduce energy supply risk.

15.5.3 Environmental Considerations

Life cycle assessment evaluates the energy consumed and emissions generated to provide electric energy in order to define the environmental performance of solar energy systems. Such analysis helps to answer two questions: (1) how much energy is needed to manufacture a solar generation system? and (2) how long is the system required to operate so as to recover the primary energy requirements? [38]. For most renewable energy technologies, including solar photovoltaic modules, the largest proportion of cumulative energy demand is the energy required for manufacturing.

Throughout the life cycle of solar energy, energy systems interact with the environment in a number of ways. The impacts are caused during their construction, operation and decommissioning [39, 40].

The production and consumption of energy places a wide range of pressures on the environment. Using solar energy produces no pollution or GHG emissions, but does have some indirect impacts on the environment. CSP may require water for regular cleaning of the concentrators and receivers and for cooling the turbine-generator. Using water from underground wells may affect the ecosystem. In addition, large solar thermal power plants can harm desert ecosystems if not properly managed. Birds can be killed if they fly into a concentrated beam of sunlight, such as that created by CSP systems. There are some toxic materials and chemicals that are used in the production of PV cells. Small amounts of these waste materials are produced. Some SHS utilizes potentially hazardous fluids (to transfer heat) that require proper handling and disposal.

Solar energy creates employment thereby contributing to economic development. Furthermore, it is an important input in mining. There is a need to diversify the energy mix in industry because of tightening environmental legislation and low-carbon business environment in order to meet the security of supply considering the risk of the growing energy demand.

15.6 Future Challenges and Research

The mining sector is facing growing requests from governments, customers and other key stakeholders to operate in a sustainable manner. Transformation of the sector's energy mix is driven by energy security concerns and resource-efficient [41]. The rising costs of traditional energy sources will certainly drive the transition to alternative energy sources. Additionally, given the long time horizon for "bridging" fuels such as nuclear or shale gas, it is uncertain whether these technologies will be attractive in the long run. This means that significant efforts should be made in order to increase the share of renewable and clean energy sources of the entire energy supply structure. Solar energy presents an excellent opportunity and holds great potential for mining companies that find themselves in isolated and energy-scarce regions. The future is all about potential: for mining companies to make steep power cost savings, for utilities to improve viability and for the population to benefit from faster expansion of the electricity grid. Growth of the solar sector, the development solar energy efficient technologies and the falling price of solar and storage solutions will be a main driver for further installation. The key research challenge for the future is the development of solar systems under economic conditions that are secure, reliable and efficient with due regard to the environment.

15.7 Conclusions

Solar energy used in mines is not only good as an action to mitigate climate change impacts, but may also meet the expectations and needs of people who live in the mining areas. With the mining industry venturing into more complex locations, new mining projects need to deliver both strong sustainability benefits while meeting the necessary technical requirements. The global mining industry has begun to embrace solar energy as a means of improving overall company performance, because solar energy helps companies to do business in a more sustainable and profitable way. As energy is one of the main cost drivers for mining companies, they can benefit from solar technology through considerable cost savings. It is obvious that economics remain a key driver in the decision to include solar energy projects in mine development plans. Moreover, there are already projects for grid-connected solar systems. Growth of the solar sector and the falling price of solar solutions will be a main driver for further installation. Solar energy could gain ground in countries with supportive legislative and fiscal framework. Solar energy programs in the mining sector should be initiated in order to improve the environmental awareness of all relevant stakeholders, so that they can grasp the advantages and disadvantages. Nevertheless, solar energy presents an excellent opportunity for mining companies in their energy management and business development. Technology transfer should also be supported so that advanced technologies can be transferred from the energy industry. Technological development and R&D on solar energy behaviour and transformation will be a key element in the future development of solar for mining applications. There will only be real solar sustainability by using sustainable installations.

References

1. Allwood JM et al (2010) Options for achieving a 50% cut in industrial carbon emissions by 2050. Environ Sci Technol 44(6):1888–1894
2. United Nations Statistical Division (UNSTATS) (2010) Greenhouse gas emissions by sector (absolute values). http://unstats.un.org/unsd/environment/air_greenhouse_emissions%20by%20sector.htm. Accessed 10 Aug 2016
3. Krausmann F et al (2009) Growth in global materials use, GDP and population during the 20th century. Ecol Econ 68(10):2696–2705
4. Worrell E et al (2009) Industrial energy efficiency and climate change mitigation. Energy Effic 2(2):109–123
5. Mudd GM (2010) The Environmental sustainability of mining in Australia: key mega-trends and looming constraints. Resour Policy 35(2):98–115
6. Mason L et al (2011) Availability, addiction and alternatives: three criteria for assessing the impact of peak minerals on society. J Clean Prod 19(9–10):958–966
7. Solangi KH et al (2011) A review on global solar energy policy. Renew Sustain Energy Rev 15(4):2149–2163
8. Halabi MA et al (2015) Application of solar energy in the oil industry—current status and future prospects. Renew Sustain Energy Rev 43:296–314
9. Varella FKOM et al (2009) A survey of the current photovoltaic equipment industry in Brazil. Renew Energy 34(7):1801–1805
10. Paraszczak J, Fytas K (2012) Renewable energy sources e a promising opportunity for remote mine sites? In: Proceedings of the international conference on renewable energies and power quality (ICREPQ'12), Santiago de Compostela (Spain), 28–30 Mar 2012
11. McLellan BC et al (2012) Renewable energy in the minerals industry: a review of global potential. J Clean Prod 32:32–44
12. Carvalho M et al (2014) Optimal synthesis of energy supply systems for remote open pit mines. Appl Therm Eng 64(1):315–330
13. Shao S et al (2016) Uncovering driving factors of carbon emissions from China's mining sector. Appl Energy 166:220–238
14. Baig MH et al (2015) The potential of concentrated solar power for remote mine sites in the Northern territory, Australia. J Sol Energy. doi:10.1155/2015/617356

15. Napp TA et al (2014) A review of the technologies, economics and policy instruments for decarbonising energy-intensive manufacturing industries. Renew Sustain Energy Rev 30:616–640
16. Paredes-Sánchez JP et al (2015) Solar water pumping system for water mining environmental control in a slate mine of Spain. J Clean Prod 87:501–504
17. Wang R, Taplin R (2014) The potential for carbon emission reduction using solar PV energy for the Mining Industry in China. In: Proceedings of MEA student conference 2014, University of Adelaide, Adelaide, Australia, 27 Oct 2014
18. Conti J et al (2016) International Energy Outlook (IEO). http://www.eia.gov/forecasts/ieo/pdf/0484(2013).pdf. Accessed 10 Aug 2016
19. Abbott D (2010) Keeping the energy debate clean: how do we supply the World's energy needs? Proc IEEE 98(1):42–66
20. Jacobson MZ, Delucchi MA (2011) Providing all global energy with wind, water, and solar power, Part I: technologies, energy resources, quantities and areas of infrastructure, and materials. Energy Policy 39(3):1154–1169
21. Nguyen MT et al (2014) Water and energy synergy and trade-off potentials in mine water management. J Clean Prod 84:629–638
22. Resch G et al (2008) Haas potentials and prospects for renewable energies at global scale. Energy Policy 36(11):4048–4056
23. Green MA (2000) Photovoltaics: technology overview. Energy Policy 28(14):989–998
24. Ekechukwu OV, Norton B (1999) Review of solar-energy drying systems II: an overview of solar drying technology. Energy Convers Manag 40(6):615–655
25. Lovegrove K et al (2012) Realising the potential of concentrating solar power in Australia. Australian Solar Institute, Canberra
26. Vogel W, Kalb H (2010) Large-scale solar thermal power: technologies, costs and development. Wiley-VCH, Weinheim
27. Rio Tinto (2010) Sustainable development report, Diavik diamond mine 2010. http://www.diavik.ca/ENG/resources/661.asp. Accessed 10 Aug 2016
28. Cromimet (2016) Technical report. http://www.crm-ps.com/energy. Accessed 10 Aug 2016
29. Piraino G et al (2012) 25+ MW solar heating commercially viable for solvent extraction electro winning (SX-EW) and heap leaching. In: Proceedings of 2nd international seminar on energy management in the mining industry, Salvador, Brazil, 10–12 Sept 2012
30. Metkemeyer L (2012) Improving copper recovery in heap leaching cycles in hydrometallurgical plants by solar heating. In: Proceedings of 2nd international seminar on energy management in the mining industry, Salvador, Brazil, 10–12 Sept 2012
31. Abengoa (2016) Chile plants. http://www.abengoasolar.com/web/en/plantas_solares/plantas_para_terceros/chile/index.html. Accessed 10 Aug 2016
32. Soitec (2016) Technical report. https://www.soitec.com/en. Accessed 10 Aug 2016
33. Goldcorp (2010) Energy initiatives at Musselwhite. http://csr.goldcorp.com/2010/6_energy.php. Accessed 10 Aug 2016
34. SolarWall (2010) Case study of SolarWall of Musselwhite. http://solarwall.com/media/download_gallery/cases/GoldCorpMussellMines_SolarWallCaseStudy_Y10.pdf. Accessed 10 Aug 2016
35. Solar Power Advice (2013) Solar panel technical report. http://www.solarpanelscostguide.com. Accessed 10 Aug 2016
36. Qureshi WA et al (2011) Impact of energy storage in buildings on electricity demand side management. Energy Convers Manag 52(5):2110–2120
37. Mohammed A et al (2015) A review of process and operational system control of hybrid photovoltaic/diesel generator systems. Renew Sustain Energy Rev 44:436–446
38. Akinyele DO et al (2017) Life cycle impact assessment of photovoltaic power generation from crystalline silicon-based solar modules in Nigeria. Renew Energy 101:537–549
39. Atilgan B, Azapagic A (2015) A Life cycle environmental impacts of electricity from fossil fuels in Turkey. J Clean Prod 106:555–564

40. Huang Y et al (2017) Life cycle assessment and net energy analysis of offshore wind power systems. Renew Energy 102:98–106
41. Awuah-Offei K (2016) Energy efficiency in mining: a review with emphasis on the role of operators in loading and hauling operations. J Clean Prod 117:89–97

Chapter 16
Energy-Efficient Mine Ventilation Practices

Nuray Demirel

Abstract Energy efficiency in mine ventilation, which is responsible for a substantial amount of total energy consumption, is of paramount concern in underground mining. Achieving energy-efficient mine ventilation practices is not only important for reducing total operating and energy costs but is also potentially the most effective means of reducing greenhouse gas emissions and environmental and occupational health and safety. This chapter presents a comprehensive review of the literature on energy-efficient mine ventilation practices and approaches to provide the current knowledge and research frontiers on energy efficiency in mine ventilation. Successful case studies, which resulted in efficiency increases, are also included to illustrate already existing energy efficiency alternatives and energy-saving opportunities. This review is expected to provide mining professionals a tool for improving current operations and achieving best practices.

Keywords Mine ventilation · Energy efficiency · Mine fans · Power consumption Ventilation optimization

16.1 Introduction

Energy efficiency can be defined as decreasing energy consumption without sacrificing an ultimate product and/or service. Increasing scarcity of energy resources and escalating global environmental problems necessitate adapting energy-efficient practices especially in energy-intensive industries such as underground mining, since energy efficiency is potentially the most effective means to reduce energy costs and greenhouse gas emissions.

In underground mining, mine ventilation is responsible for a substantial portion of total energy consumption. Costs incurred because of ventilation may range from

N. Demirel (✉)
Mining Engineering Department, Middle East Technical University, Universiteler Mah. Dumlupinar Bulvari, 06800 Ankara, Turkey
e-mail: ndemirel@metu.edu.tr

K. Awuah-Offei (ed.), *Energy Efficiency in the Minerals Industry*, Green Energy and Technology, https://doi.org/10.1007/978-3-319-54199-0_16

20 to 40% of the total energy costs and energy consumption in ventilation may reach 50% of the total energy consumption [1]. Papar et al. [2] estimated annual energy consumption for underground mine ventilation systems in the U.S. to be approximately 43.2 TJ. This amount is equal to about 3.5% of total energy used for lighting by the residential sector and the commercial sector in the United States in 2015 [3]. Du Plessis [4] claimed that every one percent saving on the ventilation and refrigeration energy costs amounts to US$80,000 per annum (at US ¢ 1.39×10^{-6} per J corresponding to US$0.05 per kWh). Considering this level of energy consumption, increasing efficiency of energy use associated with mine ventilation has become an emerging issue.

Achieving energy-efficient mine ventilation practices is not only important for reducing total operating and energy costs but also potentially the most effective means of reducing greenhouse gas emissions for underground mines. This is especially important given that recent policy initiatives dictated by several governments force industry to pay carbon taxes and other penalty costs for carbon emissions [5]. Unfortunately, experience has indicated that majority of existing ventilation systems have efficiencies of 65% or lower [1]. Increasing energy efficiency and adopting energy-efficient mine ventilation systems require systematic planning, optimized designs and their careful implementation to ensure maximized energy efficiency, minimized loss, and confirmed health and safety requirements.

There have been research-based case studies and industry practices for improving energy efficiency in mine ventilation. It is important to understand the current knowledge in order to achieve the best practices and propose pathways for future action. The goal of this chapter is to present a review of the literature on energy-efficient mine ventilation and propose future research directions. The chapter is organized into five main sections including this introductory section. Other sections include a discussion of the main elements of mine ventilation, mine ventilation efficiency improvement areas, some of the ventilation efficiency improvement strategies, and recommendations for future research and developments. This review is expected to serve as an essential tool for mining professionals who desire to improve their operations and achieve best practices with respect to energy-efficient mine ventilation.

16.2 Mine Ventilation Fundamentals

Mine ventilation, as part of total air conditioning, can be defined as creating an artificial atmosphere in underground mines to sustain workers' health and safety and to keep mining operations running continuously. The objective of mine ventilation is to provide airflows in adequate quantity and quality to dilute contaminants to acceptable levels in all parts of the mine where personnel are required to work or travel [6]. The significance of a carefully designed ventilation system is not limited to meeting regulatory and organizational health and safety requirements for workplaces. It is also required for operating diesel engines and blasting.

Mine ventilation air or fresh air is sent to the mine using mechanical ventilation fans. The magnitude and direction of air are controlled by regulators and fans through mine airways. The energy input to the fan should be high enough to overcome the mine resistances and to create sufficient pressure so that the mine air can reach the desired underground working areas. The basic fundamental formula to find the required pressure to be supplied by the fan is given by the square law as in Eq. 16.1 [7].

$$P = RQ^2 \tag{16.1}$$

In Eq. 16.1, P (Pa) is the required pressure difference to be supplied by the fan, R (gaul) is the mine resistance, and Q (m^3/sec) is the required amount of fresh air. Mine resistance is a function of friction (K), length of the airway (L), and size and shape of the airway as given by Eq. 16.2, in which C is the perimeter of the airway and A is the cross-sectional area of the airway. Mine resistance in SI unit is expressed as gaul [7].

$$R = \frac{KCL}{A^2} \tag{16.2}$$

Air power, which is the required power to create the desired pressure difference for a particular amount of air, is given by Eq. 16.3 [7].

$$\text{Air power (kW)} = \frac{P \times Q}{1000} \tag{16.3}$$

The fan input power, which is higher than the air power, is what should be supplied to the system in order to compensate for any losses in the system. The ratio of air power to fan input power is called as fan efficiency (Eq. 16.4) [7].

$$\text{Fan efficiency (\%)} = \frac{\text{Air power}}{\text{Input power}} \tag{16.4}$$

From above-mentioned principles, mine ventilation efficiency is related to mine design, fan and quantity-related factors. Design parameters such as, length of the airway, size of the shaft, and size of the airway are dictated usually by mine design features which are not quite flexible once operations begin. However, the planning and design of mine airways are critically important as they influence the pressure drop between inlet and exit, which further affect required air power and efficiency. Efficiency is directly proportional with pressure loss.

The air power is shown to be proportional to both the head loss and the air quantity. However, head loss in turn is proportional to the square of quantity (Q^2), resulting in the power being proportional to the cube of the air quantity.

This relationship between power and air quantity shows that quantity is the most important factor that influences air power. Therefore, in an effort toward improving the ventilation efficiency, air quantity should be cautiously considered during design and implementation of mine ventilation.

It is also important to recognize influence of fan laws, which dictate the relationship between fan power and ventilation parameters such as, flow rate, fan speed, and fan pressure, in studies targeting higher ventilation efficiency. Some of the relevant fan laws are [2]:

(i) Air flow rate is directly proportional to fan speed.
(ii) Fan pressure varies with the square of the fan speed.
(iii) Fan power is proportional to the cube of the fan speed.

A great emphasis should be placed on fundamental ventilation design principles, and fan laws to accomplish ventilation efficiency improvements. For example, Papar et al. [2] claimed that 10% reduction in fan speed will result in a 27% reduction in fan brake horsepower requirements.

16.3 Mine Ventilation Efficiency Improvement Areas

Mine ventilation is a complex optimization problem with various constraints and controllable and uncontrollable variables. Therefore, an optimized ventilation system requires careful planning and design. It is not that easy to alter designs once the implementation begins. For instance, once the shaft size is determined and driven based on the defined diameter, it will remain as it is. Therefore, increasing ventilation efficiency starts with the design and planning.

Knowing the fact that none of the designs can produce the best solution unless it is implemented properly, installing and operating main and auxiliary ventilation units and maintaining them with utmost care are of paramount concern for ventilation engineers. Therefore, in any attempt to improve ventilation efficiency, emphasis placed on the whole system starting from the design stage to the final output.

Various strategies have been proposed by researchers to improve mine ventilation efficiency. These improvement areas and techniques are grouped into two main groups: (i) efficiency improvement techniques related to modeling and design of ventilation system; and (ii) efficiency improvement techniques related to implementation of the designed ventilation system.

Importance of ventilation modeling and utilizing computer simulations to optimize ventilation systems are emphasized by the following advantages: [8]:

- *Economic sizing of main airways using ventilation modeling*: Engineers can simulate different airways of various sizes to estimate the associated ventilation economics and determine the optimum size of airways which yield the loweset overall cost.

- *Application of on-demand-based ventilation and on-demand-based heating and cooling control*: Ventilation models provide a fully dynamic simulation platform where airflow can continuously be changed and balanced based on the projected equipment moves, mining progress and air conditioning requirements. Mines can achieve optimized airflow splitting resulting in an minimized reduction in wasted air.
- *Intelligent active ventilation system control*: This system uses live modeling and online monitoring to allow for real-time evaluation of the underground environment and can be used to provide ventilation flow control to minimize fan power costs.
- *The application of main fan energy management, with reduced air flows during selected peak and off peak periods, resulting in substantial reduction in peak power demand*: Optimization models such as genetic algorithms can be utilized to determine the most effective combination of the fan operational duties, periods and and locations to minimize the operational fan power costs.

Moreover, a number of areas in improvement of energy efficiency of ventilation systems have been addressed. These are:

- Retrofit of main fan installations
- The use of variable pitch axial fans that can be adjusted down during periods of low activity
- The use of variable speed drives to provide speed control, reduce mechanical stress on the fan and motor and reduce energy consumption
- The upgrading or replacement of an impeller with a design impeller to suit the actual ventilation requirements, resulting in increased fan efficiencies
- The use of composite materials (lighter than steel, with higher resistance to fatigue) limiting fan impeller and blade failure.

16.4 Some Ventilation Efficiency Improvement Strategies

16.4.1 Mine Ventilation Overall System Optimization

In mine ventilation, mine air is supplied by ventilation hardware system which is mainly composed of power supply, motor, coupling, fan, flow control devices, ducts, and passageways [2]. Usually, fans are considered as the most important component of the system to improve the efficiency, however, as suggested by Papar [2], overall system approach should be adopted and mines should seek overall efficiency [2]. It has been reported that the power consumption of ventilation fans may account for up to 25% of total power consumption in a coal mine, but the average efficiency is only 57% [9].

It is also important to recognize that dampers, vanes, elbows, and other directional changes in the ducting systems have significant impact on fan performance

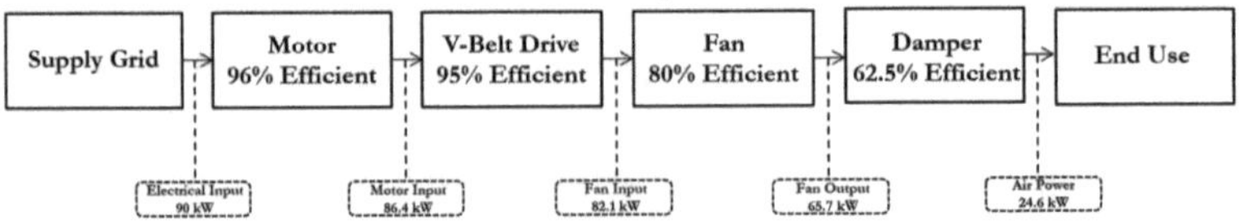

Fig. 16.1 Ventilation hardware system components [2]

and the energy efficiency of the mine ventilation system [2]. Du Plessis also claimed that reducing the operating costs significantly requires overall system optimization in addition to appropriately designed, energy-efficient equipment single ventilation and cooling components [10].

Focusing on individual efficiencies may lead to incorrect conclusions about the system efficiency. Similarly, life cycle of the system should be assessed rather than focusing on the short-term needs and solutions associated with them. Typical ventilation hardware system is presented in Fig. 16.1 [2].

Figure 16.1 shows that although the individual component efficiencies for the motor, belt drive, and fan are relatively high, the overall system efficiency is just 27% due to the fact that the pressure losses associated with the outlet damper control contribute significantly to the overall system performance [2]. This example clearly validates that using a systems approach instead of a component-based approach is critically important in considering potential efficiency improvement alternatives [2].

Papar and colleauges [2] conducted a case study in a gold mine requiring 141,000 cfm of air with the two existing axial-flow fans. However, they were not providing enough supply air into the mine because of the following reasons:

- The two fans had been mounted only inches apart from each other directly on the shaft bulkhead. No provision had been made to install outlet cones on the fans which would have allowed the proper development of flow and pressure before the air was dumped swirling into the mine shaft.
- The fans did not have inlet cones. This omission increases inlet losses and results in a poor flow profile into the fans.
- The fans were installed inside a heater house. Due to the configuration of the heater and the entry into the house, the air had to make a 90 turn just before the fan inlets.
- System performance tests indicated that the mine resistance was approximately 20% higher than design.
- A gap in the shaft collar was allowing approximately 4000 cfm to escape at the bulkhead.

Based on the field study, it was established that the two existing fans are able to meet the requirement of 141,000 cfm provided that a proper aerodynamically designed duct system was installed. This was completed and once the new inlet cones were installed, it was found that the reconfigured system met the 141,000 cfm airflow requirements. The cost of the new ductwork, inlet cones and relocation work was approximately $60,000. The streamlined arrangement resulted in an avoided horsepower increase of over 300 hp. The savings on this avoided energy use are approximately $112,000/year [2].

Minimizing various losses involved in fan system can improve efficiency, which in turn facilitate savings in energy consumption. Fan efficiency is greatly dependent on the profile of the blade [11]. Panigrahi and Mishra [11] conducted Computational Fluid Dynamics (CFD) simulations of drag and lift coefficients of six different airfoils using the ANSYS Fluent software to help with the selection of an energy-efficient blade profile for mine ventilation fans.

Another method of efficiency control is variable pitch angle in motion fans. The blades on axial-flow fans will adjust to meet the changing system requirements while in motion which assists in achieving higher efficiencies [2]. The fan manufacturers may be able to modify the fan by deblading and removing fan blades [12]. Power intake could be minimized by selecting angle of blades and rotation velocity of the fan at the pre-set minimum depression levels [13].

De Souza [4] presented case studies conducted to reduce power consumption, to lower operating cost, and to increase ventilation efficiency. In one of the case studies, two surface exhaust fans operating in parallel were inspected and surveyed. The fans have a diameter of 2.1 m and hub diameter of 0.8 m and they had 261 kW motors operating at 1170 rpm. De Souza [4] claimed that although fan assemblage was well designed with acceptable resistance pressure losses, the fans were fitted with very inefficient cones which were 1.5 m long and 2.4 m in outlet diameter. The cone losses were estimated at 0.161 kPa, the fan velocity pressure including losses was 0.41 kPa and the total pressure was estimated at 1.9 kPa with operating power per fan was calculated to be 230 kW.

The existing system was modified by replacing existing exhaust cones with more efficient cones of 4.3 m long and 3.05 m in diameter. With this new configuration the cone losses were reduced to 0.149 kPa. The new fan velocity pressure including losses and the total fan pressure were estimated as 0.25 and 1.74 kPa, respectively. The operating power per fan was reduced to 211 kW. Therefore, the total operating power savings are 38 kW and the annual savings in operating cost is $37,330 based on a power cost of $0.112 per kWh [1].

In the same underground mine, another modification on fan operation yielded a decrease of 22% in annual operating cost. The previously utilized booster fans of 1.67 m in diameter and 0.66 m hub diameter had 112 kW motors installed operating at 1200 rpm. The blade settings were 20°. The fan was a higher pressure fan, however, because of the low system resistance they were deemed to be operating inefficiently [1]. In order to improve the efficiency without incurring any additional investment cost, the fans were operated in half-blade and with a blade setting of 22°. In doing so, the fans supplied the same required flow but with an increased efficiency of 59.5% and decreased annual operating cost of $117,470 which used to be $150,000. These two case studies showed that modifying fan configuration may result in significant improvement in the ventilation efficiency.

In a deep underground mine in South Africa, energy savings of 10,400 MWh per annum were achieved, resulting in an energy cost saving of US$2 million per annum, by changing the inlet guide vane setting of the main surface fans. It would be possible for the mine to save an additional 3300 MWh and effect a further energy cost saving of US$0.5 million per annum by stopping an underground booster fan. The main penalty attached to these changes is that the underground airflow would be reduced by nominally 7% [10].

16.4.2 Auxiliary Ventilation

In order to increase the efficiency of mine ventilation, auxiliary ventilation, and auxiliary equipment stations could be utilized where necessary. There are various devices and auxiliary ventilation arrangements to ventilate dead-end openings such as check curtains, line brattice, jet fans, injectors, diffusers, scrubbers, booster fans, spray fans and vent tubing [7]. The well-designed auxiliary ventilation system not only provided supplemental flow to assist main ventilation system but also increase the effectiveness of dust and gas control. A case study conducted in a long wall mine revealed that effective utilization of auxiliary facilities could save significant amount of air [14]. In the study conducted by Kazakov et al. [13], transferring a part of the main fan load onto additional draft sources located nearby difficult-to-air sites that partially reuse the site return air allowed energy saving in ventilation [13].

In an attempt to improve mine ventilation efficiency, it needs to be ensured that installation, inspection, and maintenance of auxiliary ventilation system and its components are done properly and any cause of inefficiency in the system should be investigated. A case study conducted in an underground hard rock mine showed that improving an improper auxiliary ventilation system with improperly hung fans, leaky duct-to-fan connections, and damaged ducts, the ventilation efficiency yielded a reduction in cost by 51% [1].

16.4.3 Ventilation on Demand

Since air is a costly commodity in mines, ventilation on demand [15] principles, although it still is not commonly implemented [4], should be applied. Ventilation on demand can be described as supplying sufficient amount of air where necessary and when necessary rather than keeping constant airflow quantities at all times. Keith et al. [16] described ventilation on demand as supplying airflow to only working areas of the mine while minimizing airflow to remaining areas. It is applied to metal/nonmetal mines and not coal mines. While applying ventilation on demand principles, all the risks should be assessed in advance and no health and safety-related issue could be jeopardized.

One of the other strategies for energy saving is redistribution of air flows by backward control devices given their optimized location and selected values to fit the minimum main fan depression sufficient to ensure the required air flow [13].

System energy efficiency is achieved by operating mines at optimum airflow quantities and cooling capacity, and by cyclical and on-demand operation [10]. De Souza claimed that application of ventilation on demand may result in reduced energy usage by 20–40%. The results of some of the field studies have shown that a fully automated ventilation on demand system can have an electric power savings of up to 50% over the conventional mine ventilation systems [16].

16.4.4 Leakage Control

Another important issue that needs to be addressed for efficient mine ventilation is leakage; that is, unintended losses of mine air from intake to the return. Quantity of air, even if it may not increase, could be maintained by preventing leakage. In coal mines, leakage may reach up to 80% of the total volume of air circulated. Leakage may occur through stoppings, overcasts, and doors and also through crushed pillars and improperly packed gob. The leakage should be estimated precisely to determine the required air quantity at the fan inlet and should be kept minimal as it is the most common cause of inefficient distribution of air in mines.

In order to project the route of leakage, Zang and colleagues [17] used coal and rock permeability and pressure gradient between nodes of a ventilation network and developed pressure gradient matrix for a hypothetical ventilation system [17].

NIOSH [18] claimed that leakage could be best minimized by reducing the number of ventilation structures. This can be accomplished by lengthening the distance between crosscuts to minimize the number of stoppings per unit of development [19].

In a case study conducted in an underground hard rock mine, where push system with primary surface fresh air fans installed, 42% leakage amount was reduced to 9.9% by sealing off and shotcreting all bulk headed raise connections and installing appropriate doors. The reduction in leakage by 30.1% resulted in 64.4% savings in an annual fan operating cost of $370,160 [1]. Figure 16.2 shows the fan operation characteristics before and after leakage control. A 10% reduction in leakage may yield 30% reduction in the overall operating cost [1].

16.4.5 Load Clipping

Du Plessis [10] discussed energy-saving methods and strategies and as one of the options the authors suggested load clipping (energy use is reduced for certain parts of the day). If it is applicable for the mine schedules, the implementation of the ventilation and cooling system might be done during the time of the day when there are no personnel in the production zones. However, since the conventional approach to mine ventilation is to ventilate and cool the entire mine all of the time, this approach may not be suitable for every mine. In South Africa where power

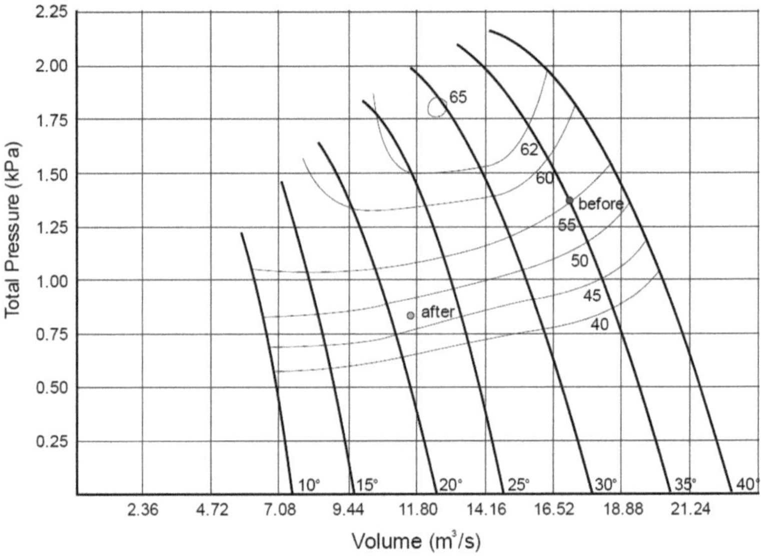

Fig. 16.2 Fan operation before and after leakage control [1]

tariff structures vary throughout the day and week, in most of the deeper underground mines, load clipping is implemented using inlet guide vane control to reduce the load during periods of peak power demand [10].

Mines utilize ice as an energy-saving strategy to reduce the amount of water circulated and to save on pumping energy. Ice is produced as either hard ice or ice slurry. Both are difficult to transport to where it is used and needed. This problem has largely been solved with the development and use of pipe conveyer belts [20]. In some South African mines, significant power savings are attainable using previously produced ice blocks at refrigeration plants during the low electrical demand periods and at a lower price and melting these ice during high electrical demand periods [16]. Impact of hard ice use for refrigeration and cooling was investigated by Mackay et al. [21] and concluded that it is a more attractive approach at lesser depths.

16.4.6 Underground Shop Air

If the quality of shop air is good enough to be reused in the mine, then routing of the underground shop air can be utilized to increase the mine airflow quantity [18]. In order to reuse the shop air in underground mines, the content of the shop air should be analyzed and contaminants should be determined. Mine ventilation engineers need to ensure that the risks are managed and all regulatory and organizational requirements are met prior to employing shop air. A case study conducted by NIOSH [14] revealed that use of mine shop air has a potential to improve the efficiency of mine ventilation by increasing the mine airflow. In this study, mine production areas of an underground room and pillar mine in which mine level airflows were 236 m^3/s and shop airflows were 24 m^3/s were provided with additional 19 m^3/s air saved out of 24 m^3/s and total mine airflow and subsequently efficiency is increased.

16.5 Recommendations for Future Work

Future research initiatives and case studies toward improving ventilation efficiency are critically important for two reasons. One is that increasing energy and electricity cost is one of the largest cost component in underground mining and any energy-saving option increases the economic success of mining projects. Further, it contributes to sustainability of the mineral industry even when commodity prices are low. Second, and more important, the benefit of such studies is to decrease the environmental burden of energy-intensive mining operations as the pressure to lower CO_2 and other greenhouse gas emissions is increasing. In addition to these benefits, efficient ventilation systems result in reduced maintenance costs and downtime, increased reliability and productivity, and improved work conditions [2]. Therefore, mine ventilation engineers should seek for energy-saving alternatives and more efficient ventilation systems.

Existing mine ventilation systems should be assessed and any inefficiency caused by improper design and implementation should be removed from the system. Optimum ventilation systems should be designed using computerized simulations and implemented with the view of overall system efficiency and life cycle costing principles. For the last decade, CFD analysis has been in use for large and complex ventilation systems in three dimensions. While it is important to recognize the benefits of using computational tools, the representativeness of models, the limits of programs, and accuracy of the results should be well analyzed through proper assessments, which requires experience and expertise.

Moreover, the general belief that "supplying more than adequate air does not harm" [2] should change. Ventilation on demand principles should be adopted where applicable and safe. Continuous monitoring of concentrations and locations of contaminants, harmful gases and particulates, is essential. Methane drainage systems might be utilized to remove the explosive methane gas in advance of operations and even converting it to an energy should be implemented where technically and economically feasible. Methane drainage also decreases the essence of ventilation requirements while increasing the potential energy sources.

In the future, the industry is expected to use more powerful information technologies as Wi-Fi, Bluetooth, tablet, smart phone and smart watch applications that allow engineers to communicate in real time to monitor air quantity and quality, exact location of contamination, etc. Virtual reality can also be utilized to simulate certain ventilation problems or recreate mine ventilation related accidents to manage the risks if available.

16.6 Summary

Energy efficiency in mine ventilation, which is responsible for a substantial amount of total energy consumption, is of paramount concern in underground mining. Achieving energy-efficient mine ventilation practices are not only important for reducing total operating and energy costs but also potentially the most effective means of reducing greenhouse gas emissions.

Significant energy savings are attainable by adopting integrated ventilation efficiency strategies. Ventilation engineers should be motivated, in the design and implementation of ventilation and cooling systems, to ensure efficient and optimized systems.

This chapter presents an overview the current ventilation efficiency improvement studies and successful case studies to help mining professionals toward improving the current operations and achieving best practices.

References

1. De Souza E (2015) Cost saving strategies in mine ventilation. In: Abstracts of the Canadian Institute of Mining, Metallurgy and Petroleum (CIM) 2015 Convention, Montreal, 10–12 May 2015
2. Papar R, Szady A, Huffer WD, Martin V, McKane A (1999) Increasing energy efficiency of mine ventilation systems. Lawrence Berkeley National Laboratory, University of California, Tech. Rep. Available via http://citeseerx.ist.psu.edu/viewdoc/download?doi=10.1.1.608.6290&rep=rep1&type=pdf. Accessed 12 July 2016
3. US Energy Information Administration (2016) https://www.eia.gov/tools/faqs/faq.cfm?id=99&t=3. Accessed 28 July 2016
4. Du Plessis JJL, Marx WM, Nell C (2007) Main fan power control. In: Proceedings of the mine ventilation society of South Africa conference, Johannesburg, South Africa
5. Awuah-Offei K (2016) Energy efficiency in mining: a review with emphasis on the role of operators in loading and hauling operations. J Cleaner Productions 117:89–97
6. McPearson MJ (2009) Subsurface ventilation engineering. Mine Ventilation Services Inc, Fresno, California
7. Hartman HL, Mutmansky JM, Ramani RV, Wang YJ (1997) Mine ventilation and air conditioning. Wiley, USA
8. De Souza E (2015) Improving energy efficiency of mine fan assemblages. Appl Therm Eng 90:1092–1097
9. Man LI, Wang X (2009) Performance evaluation methods and instrumentation for mine ventilation fans. Min Sci Tech 19:819–823
10. Du Plessis JJL, Marx WM, Nell C (2014) Efficient use of energy in the ventilation and cooling of mines. J South Inst Min Metall 114:1033–1037
11. Panigrahi DC, Mishra DP (2014) CFD Simulations for the selection of an appropriate blade profile for improving energy efficiency in axial flow mine ventilation fans. J Sustain Min 13 (1):15–21
12. Loring D (2007) Development of a tool to predict performance of debladed mine fans at Henderson Mine. SME Annual Meeting, Denver, CO
13. Kazakov BP, Shalimov AV, Kiryakow AS (2013) Energy-saving mine ventilation. J Min Sci 49(3):475–481. doi:10.1134/S1062739149030155

14. NIOSH (2014) Methods to improve mine ventilation system efficiency. Available via http://www.cdc.gov/niosh/mining/userfiles/works/pdfs/mtimv.pdf. Accessed 10 July 2016
15. Bartsch E, Laine M, Andersen M (2010) The Application and implementation of optimized mine ventilation on demand (OMVOD) at the Xstrata Nickel Rim South Mine, Sudbury. In: Proceedings of 13th United States North American mine ventilation symposium
16. Keith W, Brian P, Daniel SJ (2015) The practice of mine ventilation engineering. Int J Min Sci Tech 25:165–169
17. Zhang J, Wang H, Wu F, Chang X (2008) Analysis of mine's air leakage based on pressure gradient matrix between nodes. J Coal Sci Eng 14(3):443–446
18. NIOSH (2014) Methods to improve efficiency of mine ventilation systems. Available via https://www.cdc.gov/niosh/mining/UserFiles/works/pdfs/mtieom.pdf. Accessed 10 July 2016
19. Grau R, Krog R (2008) Using mine planning and other techniques to improve ventilation in large opening mines. SME Annual Meeting, Salt Lake City, UT
20. Schutte A J (2014) An integrated energy efficiency strategy for deep mine ventilation and refrigeration. Dissertation the North-West University
21. Mackay L, Bluhm S, Walter K, De Wet J (2014) Refrigeration and ventilation systems for ultra-deep platinum mining in the Bushveld ignous complex. In: Proceedings of the 10th international mine ventilation congress, Sun City, South Africa, 2–8 August 2014

Chapter 17
Technology Selection and Sizing of On-Board Energy Recovery Systems to Reduce Fuel Consumption of Diesel-Electric Mine Haul Trucks

Petrus J. Terblanche, Michael P. Kearney, Clay S. Hearn and Peter F. Knights

Abstract Installing an energy recovery system (ERS) on a mining haul truck has the potential to save a significant amount of fuel by recovering energy while descending into the pit and reinjecting this energy to reduce fuel usage for acceleration and ascent out of the pit. This chapter presents an initial investigation into the technical and economic feasibility of such an ERS for diesel-electric drive mine haul trucks. A simulation model incorporating the haul route, the truck and drive system characteristics, and the ERS is employed to evaluate the changes to fuel used and impact on payload for an ERS of a specific technology and size on a given pit depth, from which cost savings and fuel savings per tonne of material moved are inferred. Lithium-ion batteries and electrolytic double-layer capacitors were found to be generally infeasible due to, respectively, poor charging rate and cycle life, and low energy density. Both lithium-ion capacitors and electromechanical flywheels promise fuel efficiency improvements of greater than 10% for a large range of pit depths. Electromechanical flywheels are judged the most cost-effective option, with an expected payback period of less than 1.2 years.

Keywords Diesel-electric mine haul trucks · Hybrid vehicles · Energy recovery
Electromechanical flywheel · Lithium-ion capacitor

P. J. Terblanche (✉) · M. P. Kearney · P. F. Knights
School of Mechanical & Mining Engineering, University of Queensland,
St Lucia, Brisbane, QLD 4072, Australia
e-mail: p.j.terblanche@uq.net.au

C. S. Hearn
Center for Electromechanics, University of Texas at Austin,
10100 Burnet Bldg 133, Austin, TX 78758, USA

K. Awuah-Offei (ed.), *Energy Efficiency in the Minerals Industry*, Green Energy and Technology, https://doi.org/10.1007/978-3-319-54199-0_17

Abbreviations and Acronyms

AC	Alternating current
DC	Direct current
DEMHT	Diesel-electric mine haul truck
DOD	Depth of discharge
EDLC	Electrolytic double-layer capacitor
EM-FW	Electromechanical flywheel
ERS	Energy recovery system
EVM	Empty vehicle mass
Li-S	Lithium sulphur
$LiFePO_4$	Lithium iron phosphate
LIC	Lithium-ion capacitor
$NaNiCl_2$	Sodium nickel chloride
NCA and $LiNiCoAlO_2$	Lithium nickel cobalt aluminium oxide
SE	Specific energy
SP	Specific power

17.1 Introduction

Diesel-electric drive mine haul trucks (DEMHT) used in surface mining have an empty vehicle mass (EVM) of between 85 t [1] and 250 t [2, 3]. These trucks (powered by engines of up to 3000 kW [4]) haul loads of between 108 t [1] and 363 t [4] of broken ore, coal or overburden out from surface mining pits. Mining pits vary significantly in depth, with a few attaining depths greater than 430 m and the deepest currently at 1200 m deep [5]. For trucks operating at an open pit mine, each cycle involves descending into the pit, being loaded by a mining shovel, ascending and off-loading. A truck may execute as many as 60 cycles each day depending on the haul distance (which depends on the depth of the pit, the height of the waste dumps, distance of crushers or stockpiles from the exit of the in-pit ramp, etc.) with the larger trucks consuming more than 1.5 ML of fuel each year. When a 250 t EVM truck descends into a pit 100 m deep (typical of open-cut coal operations), it loses approximately 245 MJ [68 kWh] of potential energy in 2 min. The amount of energy lost is equivalent to the energy needed to power a typical four-person household in a first world country for three days. While some of this energy is used to overcome rolling resistance most is dispelled as heat via a resistor bank. The ability of an energy recovery system (ERS) to capture as much of this lost energy as possible and reinject it on acceleration and pit ascent appears to hold significant potential to improve the fuel efficiency of haul operations and reduce harmful emissions. This chapter presents an initial investigation into the technical and economic feasibility of a DEMHT ERS using state of the art energy storage technologies. It is a significant extension of work done by Esfahanian [6] including investigation of an appropriate selection of energy recovery technologies and using a simulation model with appropriate modelling of the dynamics and performance of each technology if used in an on-board ERS.

There has been one attempt by General Electric from 2002 to 2007 to develop a mine haul truck ERS for a Komatsu DEMHT using sodium nickel chloride batteries ($NaNiCl_2$) as the energy storage technology [7–10]. There has been no known commercial outcome to date, and the overall outcomes were not reported, except that the batteries did not perform up to expectations [10]. The absence of a commercially available mine haul truck ERS suggests that there are fundamental limitations with respect to the application of the $NaNiCl_2$ batteries or other chemical batteries for the application.

The unloaded-descent followed by loaded-ascent nature of the DEMHT working cycle make the energy flows over the cycle similar to that of loading equipment such as excavators and cranes, where ERSs have been installed to achieve significant fuel savings. These ERSs operate by recovering and storing energy when the load is decelerating or lowering, and reinjecting the stored energy to assist with accelerating and raising the load. Caterpillar's 6120B H FS shovel makes use of electrolytic double-layer capacitors (EDLC) [11] to reduce fuel cost by 25% [12]. Ricardo's hybrid excavator uses a hydraulic drive flywheel storage system to reduce fuel consumption by 10–30% [13], while an electromechanical flywheel (EM-FW) storage system, used in a mobile gantry crane, potentially reduces fuel consumption by 20% [14]. Komatsu Mining Corp. claims the flywheel-based storage system, integrated into the drive system, improves the fuel consumption of their newly developed 18 HD wheel loader by 20% [15]. These existing applications suggest that significant fuel savings can result from installing an ERS, and that non-chemical-battery energy storage technologies may be the more appropriate technology for this application.

This chapter presents and follows a framework for identifying and evaluating an energy storage technology that could be beneficially used for the recovery and reuse of recoverable potential or kinetic energy on-board DEMHT. The required steps can be summarised as: (i) Identify a set of operating parameters for a representative DEMHT and haul route, (ii) Use these parameters to characterise the requirements for an ERS in terms of energy, power and cycle requirements, (iii) Evaluate the technical feasibility of different energy storage technologies using the developed requirements and the impact on payload, (iv) Use simulation to verify the approach taken and (v) Identify the most promising solution(s) and provide an initial assessment of economic viability.

The structure of the chapter is as follows: Sect. 17.2 outlines the problem, its implications and technical requirements; Sect. 17.3 considers candidate storage technologies with a particular focus on EM-FW energy storage; Sect. 17.4 describes the working of the simulation model and inputs; Sect. 17.4.3 presents the results obtained from the simulation model incorporating the use of different ERS technologies in a mine a hundred metres deep and the fuel savings potential of the various ERS in mines of different depths. The section also highlights the implications of having an under- and oversized ERS based on EM-FW installed and presents results of an initial investigation of cost implications and service life.

Section 17.5 provides a summary of the different technologies' potential to satisfy DEMHT requirements and improve fuel efficiency. Section 17.6 contains concluding remarks.

17.2 Problem Characteristics and Requirements

An ERS, consisting of an energy storage device and some appropriate power conditioning and control equipment, can be directly integrated into the existing DEMHT electrical system. The existing electric drive system (solid lines in Fig. 17.1) has a similar structure to a series-topology hybrid electric vehicle [17, 18], though it lacks the energy storage and has a resistor grid (rated at up to 4750 kW [2]) to dissipate power generated while braking. The dashed lines in Fig. 17.1 shows how the proposed ERS could be installed across the DC link, in parallel to the rectifier, and the braking resistor grid including the braking chopper. Here, it will be able to both extract power from the DC link and inject power into the DC link.

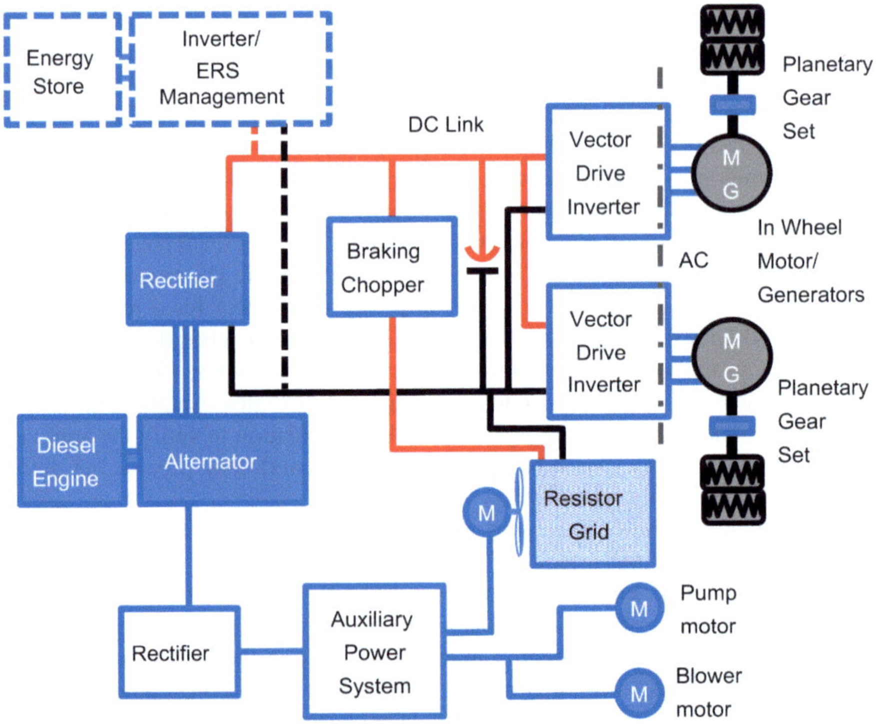

Fig. 17.1 Typical AC DEMHT drive arrangement (based on concept reflected by Siemens [16]). Solid lines show typical configuration, dashed lines show envisaged ERS interface

The key technical requirements for the ERS are the required charge/discharge rate, the required energy storage capacity and the cycle life. Satisfying these demands may not be sufficient to guarantee the success of the ERS. It is also important to consider how the ERS will impact mining operations preferably with no impact on productivity and acceptable performance in the often harsh mining environment.

17.2.1 Fundamentals of Calculating Energy Flow Rates and Quantities for Mine Haul Trucks

To determine energy flows and power requirements, we use a single dimension vehicle dynamics method (Fig. 17.2), as discussed by Guzzella and Sciarretta [19]. The truck's longitudinal motion is described by Eq. 17.1,

$$m_{\mathrm{v}} \frac{\mathrm{d}}{\mathrm{d}t} v_{\mathrm{v}}(t) = F_{\mathrm{t}}(t) - \left(F_{\mathrm{r}}(t) + F_{\mathrm{g}}(t) + F_{\mathrm{a}}(t) + F_{\mathrm{b}}(t)\right), \tag{17.1}$$

where m_{v} is the mass of the vehicle and load, $v_{\mathrm{v}}(t)$ is the vehicle velocity as a function of time (t), $F_{\mathrm{r}}(t)$ is the rolling resistance, $F_{\mathrm{g}}(t)$ is the force caused by the grade and $F_{\mathrm{a}}(t)$ is the aerodynamic resistance. $F_{\mathrm{b}}(t)$, the braking force and, $F_{\mathrm{t}}(t)$ the traction force do not act simultaneously. For the diesel-electric drive system, $F_{\mathrm{t}}(t)$, is provided by the in-wheel electric motors powered by an on-board diesel generator. The power produced by the diesel generator is reduced by energy conversion losses (in the AC-generator, rectifiers and vector drive inverters), transmission losses (in electrical cables) and friction losses in the power train (such as in-wheel motors and planetary gearing), occurring between the generator and the wheels. This reduces the maximum traction force available at the wheels to accelerate or drive the truck up ramp while fully loaded. At lower speeds, the traction force is limited by system parameters. The traction force is expressed as a function of truck speed and engine power as described by Eq. 17.2 and shown in Fig. 17.3.

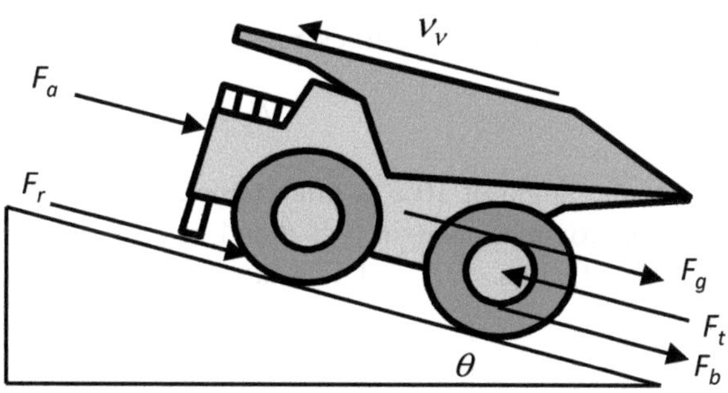

Fig. 17.2 Positive force directions with truck on uphill grade

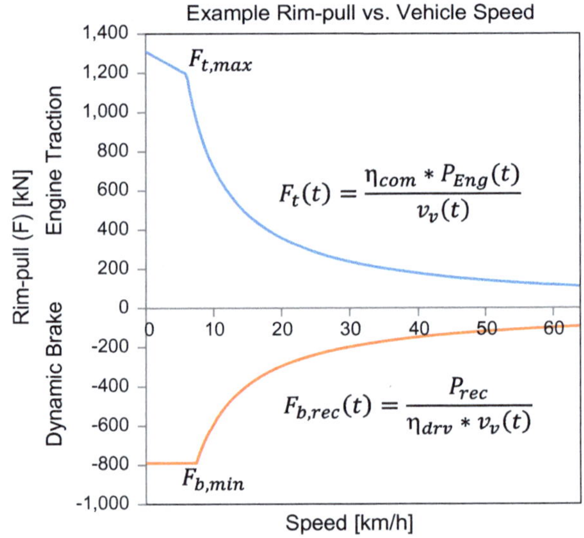

Fig. 17.3 Traction available at the wheel resulting from engine power (P_{eng}) assumed to be 2390 kW and brake force resulting from reverse torque of traction motors working as generators with P_{rec} = 1390 kW as explained in Sect. 17.2.3. $F_{t,max}$ and $F_{b,min}$ were determined using data from Caterpillar [2]

$$F_t(v_v(t)) = \min\left[\left(F_{t,max}\right), \left(\frac{\eta_{com} * P_{eng}(t)}{v_v(t)}\right)\right] \tag{17.2}$$

Efficiency between the engine and DC link and between DC link and wheels, respectively, denoted η_{gen} and η_{drv}, are not fixed in practice and vary depending on the operating point. In this work, it is assumed to be constant with η_{gen} and η_{drv}, respectively, assumed to be 95 and 85% from parameters given in Mazumdar et al. [20]. The combined drive system efficiency (η_{com}) is the product of η_{gen} and η_{drv} and thus equals 80.75%.

For a DEMHT, the rolling resistance is typically expressed as a percentage of the weight (normal force) of the truck and a value of 2% assumed [21–24]. For a given loading condition, the term $F_r(t)$ becomes independent of time. Thus,

$$F_r = 0.02\, m_v g \cos\theta, \tag{17.3}$$

where 'g' is the gravitational constant, assumed to be 9.81 m/s^2 and θ is the ramp angle.

The force acting on a truck resulting from driving on a ramp is given by

$$F_g = m_v g \sin\theta. \tag{17.4}$$

A ramp angle of 10% is often targeted on mine sites [21–24]. Using the small angle approximations ($\sin\theta \approx \theta$ and $\cos\theta \approx 1$) for this ramp angle (descending into the pit), Eqs. 17.3 and 17.4 are, respectively, simplified to

$$F_{\rm r} = 0.02\, m_{\rm v} g, \text{ and} \tag{17.5}$$

$$F_{\rm g} = 0.10\, m_{\rm v} g. \tag{17.6}$$

Equations 17.1, 17.5 and 17.6 form the basis of calculations in the following sections that aim to establish the basic requirements for a DEMHT ERS. For all these calculations, except where otherwise specified or clear from the context, typical rolling resistance force is assumed to be 2% of the truck weight and ramp angle is assumed to be 10% (θ = 5.71°). The aerodynamic drag $F_a(t)$ was found to be relatively insignificant due to low operational speeds. At 30 km/h it is approximately 5% of the value of the rolling resistance. To avoid the complication of a nonlinear speed-dependent variable that has very little effect, $F_a(t)$ is neglected in the discussion in the remainder of Sect. 17.2 and in Sect. 17.3. It is appropriately incorporated in the simulation model discussed in Sect. 17.4 and the Results of Sect. 17.4.3 and is determined using Eq. 17.7 with variables Cd, ρ and A, respectively, assumed to be 0.9, 1.2 kg/m^3 and 70 m^2;

$$F_{\rm a} = 0.5\, Cd\rho A (v_{\rm v}(t))^2. \tag{17.7}$$

17.2.2 Recovering Potential Energy from Descent

The available potential energy that can be recovered to the DC link is given by

$$E_{\rm pot,rec} = \eta_{\rm drv} \int_0^s \max\left[F_{\rm g}(x) - F_{\rm r}(x), 0\right] {\rm d}x, \tag{17.8}$$

where x is the distance along the descent, s is the total descent distance. The recovery efficiency that captures mechanical and electrical losses between the wheels and the DC link and is assumed to be the same as the drive system efficiency ($\eta_{\rm drv}$) which was assumed to be 85%. For this characterisation, it is assumed that the truck mass, ramp gradient and rolling resistance are consistent across the descent, which results in a simplification of Eq. 17.8 to

$$E_{\rm pot,rec} = \eta_{\rm drv} (F_{\rm g} - F_{\rm r})\, s. \tag{17.9}$$

Equation 17.9 assumes that the ramp angle is sufficiently steep that $F_g > F_r$. Note that Eqs. 17.8 and 17.9 are independent of descent speed.

Figure 17.4 shows the amount of energy recoverable to the DC link per metre of vertical descent from an empty haul truck of 250 t EVM, as influenced by different combinations of rolling resistance and gradient of the pit access ramp. Higher rolling resistance substantially reduces recoverable energy while greater ramp

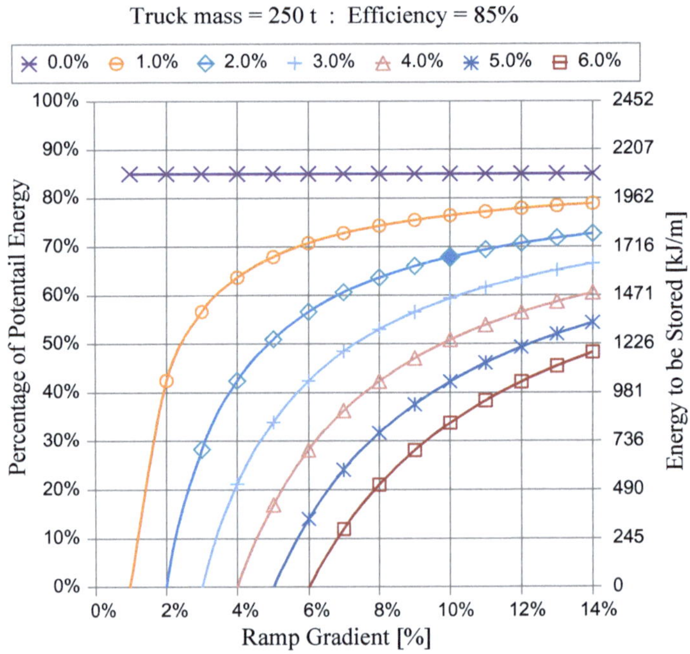

Fig. 17.4 Recoverable potential energy expressed as percentage of theoretical maximum (left hand vertical axis) and physical amount per m vertical descent (right hand vertical axis) as it varies with rolling resistance (in range 0–6%) and ramp gradient (0–14%)

angles favour recoverable energy. On a ramp of 10% gradient and a rolling resistance of 2% (accentuated on the 2% curve in Fig. 17.4), energy expected to be returned to the DC link is calculated as 1667 kJ/m vertical descent [463 Wh/m]. This equates to a recovery of 68% of the theoretical potential energy of 2452 kJ/m [681 Wh/m].

17.2.3 Required Energy Absorption Rate (kW)

DEMHTs primarily use their dynamic brake systems to slow down or restrain the truck on-ramp descents. It does this by operating the wheel motors as generators and the generated power is dissipated as heat using fan-cooled resistor banks. To maximise energy recovery, the ERS must be able to sink the recovered energy at the produced rate. Any excess power would be lost as heat through the resistor banks. Assuming a constant down-ramp speed, Eq. 17.1 can be appropriately rewritten and P_{rec} (the rate at which energy is recovered to the DC link through the dynamic brake system), can be determined as

$$P_{\mathrm{rec}} = \eta_{\mathrm{drv}} F_{\mathrm{b}} v_{\mathrm{v}} = \eta_{\mathrm{drv}} \left(F_{\mathrm{g}} - F_{\mathrm{r}} \right) v_{\mathrm{v}}. \tag{17.10}$$

Equation 17.10 shows that P_{rec} increases with the steepness of the gradient (through F_{g}), and is a linear function of the descent speed. For a descent with 10% gradient, 2% rolling resistance and a typical permissible downhill speed of 30 km/h the required energy absorption rate of the ERS, which equals P_{rec}, is 1390 kW.

17.2.4 Recovering Kinetic Energy

The ERS would also be able to recover kinetic energy when the truck uses the dynamic braking system to slow the truck down. The rate at which energy returns to the DC link will depend on the efficiency of the system, the deceleration rate, the current vehicle mass and the speed of the truck at any instant, and is given by

$$P_{\mathrm{rec}}(t) = \eta_{\mathrm{drv}} F_{\mathrm{b}}(t) v_{\mathrm{v}}(t) \tag{17.11}$$

To maximise energy storage, the recovery rate needs to be limited to the ERS recovery power. The maximum useable brake force $F_{\mathrm{b}}(t)$ then becomes $F_{\mathrm{b,rec}}(t)$, which, at lower speeds, is still limited by drive system parameters (evident from the brake rim-pull curves in Caterpillar [2] and Liebherr [4]) but at higher speeds should be limited by the maximum energy absorption rate, P_{rec}, of the ERS. The useable brake force as influenced by truck speed is thus given by

$$F_{\mathrm{b,rec}}(t) = \max\left[\left(F_{\mathrm{b,min}} \right), \left(\frac{P_{\mathrm{rec}}}{\eta_{\mathrm{drv}} * v_{\mathrm{v}}(t)} \right) \right]. \tag{17.12}$$

Figure 17.3 shows how $F_{\mathrm{b,rec}}$ changes as a function of speed for P_{rec} = 1390 kW. However, the dynamic braking power for a 250 t EVM truck could be as high as 4750 kW [2], and any braking power higher than the maximum recovery power will cause energy to be dumped into the resistor banks. Assuming the driving style and braking rate is adapted such that no energy is diverted to the resistor banks, the amount of energy that can be recovered can be determined using

$$E_{\mathrm{kin,rec}} = \eta_{\mathrm{drv}} \int_{0}^{T} F_{\mathrm{b,rec}}(t) v_{\mathrm{v}}(t) \mathrm{d}t. \tag{17.13}$$

Table 17.1 compares the amount of recoverable kinetic energy that can be returned to the DC link by stopping a loaded truck of 570 t mass from an initial speed of 45 km/h for different ERS recovery power ratings. Doubling the power rating from that calculated for the ramp decent in Sect. 17.2.3 only results in a small increase in captured kinetic energy, suggesting that the ERS should be sized for

Table 17.1 Values for kinetic energy recovery to the DC link

P_{rec} (kW)	E_{kin} [MJ (kWh)]	$E_{kin,rec}$ [MJ (kWh)]	$E_{kin,rec}/E_{kin}$ (%)	Equivalent descent (m)
1390	44.5 (12.4)	24.5 (6.81)	55.0	14.7
2780	44.5 (12.4)	29.5 (8.19)	66.2	17.7

Values reflect slowing the loaded truck from a speed of 45 km/h on level ground. Equivalent descent is based on 1667 kJ/m [463 Wh/m] (see Sect. 17.2.2)

potential energy not kinetic energy recovery. The energy captured is equivalent to the amount of energy recoverable from descent into a 14.7 m and a 17.6 m deep pit, respectively, meaning that, particularly for shallow pits, the capture of kinetic energy is significant.

17.2.5 Required Cycle Life

Depending on assumptions about truck cycle time, pit depth and truck utilisation, a DEMHT could perform several thousand (Lang [25] uses six thousand) descend-load-ascend-dump cycles per annum. For an ERS with storage capacity matched to a specific mine depth, each of the cycles would involve the full charge of the ERS on descent, and the full discharge of the ERS on ascent. The storage device will therefore be required to perform an equal number of charge–discharge cycles per annum. This does not include additional charge–discharge micro-cycles, which would occur several times during each haul cycle due to slowing the truck down using regenerative braking. The storage device in the ERS system should, therefore, be capable of a high number of cycles or it should be inexpensive and easy to replace.

17.2.6 Implications of Mining Operations on ERS Sizing and Selection

DEMHTs are designed to remove as large as possible payloads from mining pits as quickly and efficiently as possible [26], and any increase in mass of the empty truck (such as the addition of an ERS) will result in a corresponding reduction of the payload. The mass of the storage system is primarily determined either by the energy release rate of the application and the specific power (W/kg) of the chosen technology or, the required energy storage capacity of the application and the specific energy (J/kg or Wh/kg) of the chosen technology. A technology leading to an excessively heavy storage system, due to its low specific power or specific energy, can potentially negate the fuel savings potential. Since the ERS repeatedly captures and reuses energy, it should also reduce the amount of fuel that needs to be carried, providing a small reduction in the additional mass. Although the necessity for the resistor bank may be significantly reduced, its removal to provide a further mass reduction is not envisaged as it would provide the back-up braking capacity if the ERS should fail or reach full storage capacity during operation. The sizing of an ERS is further complicated by continual changes to the haul cycle over the truck's operational life. These changes lead to the effective depth of the haul route taken by the truck varying significantly over its life. A mismatch between haul route and ERS does not provide the highest level of energy recovery, and thus fuel savings, if the ERS is too small, or unnecessarily impacts payload, hence productivity, if the ERS is oversized. Ideally, the storage capacity of the ERS would be able to be modified to match the recoverable energy for the present haul route. The ability to easily and affordably decrease or increase the ERS storage capacity to better match current requirements, using a modular design, for instance, could be highly beneficial.

17.3 Candidate Energy Storage Technologies and Their Potential for Beneficial Use in Mine Trucks

The predominant energy storage technology used in hybrid electric and all-electric mobile applications are chemical batteries [27–29]. However, alternatives such as super or ultra-capacitors [11, 30, 31], and electromechanical flywheels [32, 33] are gaining favour in certain applications. This section of the chapter focuses on determining the most suitable ERS technology that can provide the required recovery power, P_{rec} of 1390 kW determined and storage capacity requirement E_{rec} of 166.7 MJ [46.3 kWh] (as determined for a 100 m descent) determined in Sect. 17.2.2, while offering an appropriate cycle life. The key metric in determining the most suitable technology (given satisfaction of the cycle life requirements) is the mass of the ERS, as each kilogram of ERS corresponds to a kilogram reduction in maximum payload. In Table 17.2, we compare a selection of chemical batteries, EDLC's and Li-ion capacitors (LIC's) for the 100 m pit case study. In Tables 17.3 and 17.4, we provide information on the characteristics and expected sizes of EM-FWs.

The effective mass of an ERS is higher than what would be expected from the cell-based specific energy or specific power of the storage device typically stated in the literature. This is primarily due to the omission of the mass of essential ancillary equipment such as casing and power convertors. For example, the 'DryLyte' battery has a cell specific energy ($\mathrm{SE}_{\mathrm{cell}}$) of 220 Wh/kg [34]; however, the effective specific energy, once installed in a module ($\mathrm{SE}_{\mathrm{module}}$) is 130 Wh/kg [35]. For the Maxwell BCAP3000 P270 ultra-capacitors $\mathrm{SE}_{\mathrm{cell}}$ is 6 Wh/kg [36] but $\mathrm{SE}_{\mathrm{module}}$ is 2.3 Wh/kg [36]. Equations 17.14 and 17.15 estimate two useful ratios based on this information. Where needed, due to a lack of published data, these ratios are used to provide estimates of the specific energy (indicated in Row 5 of Table 17.2) and

Table 17.2 A selection of energy storage technologies' characteristics and implications for use in mine haul truck ERS

	a.	b.	c.	d.	e.	f.	g.	h.
1.		Units	$NaNiCl_2$	$LiFePO_4$	NCA	Li-S	EDLC	LIC
2.	Example of use or test application	–	Railway, GE Scoop [39]	BYD All Electric Bus, Fisker Karma	Tesla	Portable/Military	Cat 6120 B FS Shovel	Railway
3.	Example of prominent or sole research organisation or manufacturer	–	FIAMM [40], General Electric [39] (Durathon)	a123 [41] (Nano Phosphate)	Panasonic [42] (NCR)	Oxis [43], Sion	Maxwell [36]	JSR Micro [30]
4.	Specific energy (cell)	Wh/kg	120 [18]	131 [41]	225 [42]	300 [43]	6.0 [44]	12.1 [30]
5.	Specific energy (module)	Wh/kg	80 [40]	76.9 [41] (Eq. 17.13)	132 [28]	176 [43] (Eq. 17.13)	2.3 [36]	4.64 [30] (Eq. 17.14)
6.	Energy storage required (see Sect. 17.2.2)	kWh	46.3	46.3	46.3	46.3	46.3	46.3
7.	Mass (energy storage based)	kg	579	602	351	263	***20,131***	***9978***
8.	E-rate (charge rate/E-storage)	h^{-1}	≈0.50* [45]	1.0 [41]	0.50 [42]	0.50 [46]	95.2 [30]*	40.9 [30]*
9.	Specific power (module—charging)	W/kg	40.0	76.9	66.0	88.0	219.0	189.7
10.	Charge rate required (see Sect. 17.2.3)	kW	1390	1390	1390	1390	1390	1390
11.	Mass (charge rate based)	kg	***34,750***	***18,087***	***21,061***	***15,795***	5890	7326
12.	Minimum energy storage capacity	kWh	2780	1390	2780	2780	46.3	46.3
13.	Service life	Cycles	1000 [47]–3500 [48]	3000 [41] to 83% EOL	500 [42] to 80% EOL	500–1500 [46] to 42% EOL	10^6 [36]	10^6 [49]
14.	Cycle life test conditions (if available)	DOD Charge/Discharge	3000 to 80% DOD [40]	100% DOD [41] +1C/−2C @23 °C	100% DOD [42] +0.5C/−1C @25 ° C	100% DOD [46] +0.5C/−1C @? ° C	–	–

Note $NaNiCl_2$—Sodium nickel chloride; $LiFePO_4$—Lithium iron phosphate; NCA (or $LiNiCoAlO_2$)—Lithium nickel cobalt aluminium oxide; Li-S—Lithium sulphur; EDLC—Electrolytic double-layer capacitor; LIC—Lithium-ion capacitors. Although batteries can be charged at higher rates than indicated in Row 9, that tends to reduce cycle life [37, 38], and leads to notable decreases in efficiency as is evident from test data in Burke and Miller [30]. Information marked * are best estimates based on available/accessible information. Values in Row 9 are the products of values in Rows 5 and 8 of corresponding columns. Values in italics in Rows 7 and 11 reflect minimum mass of a system satisfying the more demanding of storage capacity and charge rate required while Row 12 shows the storage capacity of such a system. Values reflected in Table 17.2 are meant to be indicative of performance only. Where data from specific supplier brochures are used this is done with the intent of providing useful information and is not implied as a guarantee of performance from such suppliers specific power (indicated in Row 9 of Table 17.2) of some chemical battery (Bat) and ultra-capacitor (Cap) modules based on available cell-based specifications

$$SE_{ratio(Bat)} = \frac{SE_{module(Bat)}}{SE_{cell(Bat)}} = 0.591 \tag{17.14}$$

$$SE_{ratio(Cap)} = \frac{SE_{module(Cap)}}{SE_{cell(Cap)}} = 0.383. \tag{17.15}$$

17.3.1 Chemical Batteries, EDLCs and LICs

Table 17.2 shows the characteristics of a selection of energy storage technologies and their implications in terms of mass required for use in a mine haul truck ERS. Values included in Table 17.2 reflect a combination of specific energy and specific power for each technology applicable during charging. Charge power capacity of storage devices is generally lower than the discharge power capacity [30, 41, 42]. In practice small changes in cell chemistry or structure, charge and discharge rates, temperatures and depth of discharge (DOD) have a distinct impact on the practically achievable values of efficiency, specific energy, specific power, energy density, power density and cycle life [37, 50, 51]. However, this preliminary study uses indicative values and does not consider these variations.

When the required minimum mass of the energy storage system is based on the required energy storage capacity, chemical batteries appear to provide a solution that is surprisingly light as indicated in Table 17.2 Row 7. Such a light battery system will not, however, be able to absorb energy at the rate required without overheating and giving a very short service life. Salasoo [8] points out that in the context of chemical batteries, it is the charge rate that determines the storage system capacity for mine haul trucks. When the required mass is calculated as determined by the required energy absorption rate of 1390 kW, the chemical battery-based solutions become excessively heavy at 15,795 kg or more (Table 17.2 Row 11). For $LiFePO_4$ technology (which provides a good combination of cycle life and charge rate for chemical batteries) the solution is likely to weigh about 18,100 kg. While this mass will have a significant impact on the payload, being oversized by a factor of about 30, it will inherently provide a large storage capacity which, at 1667 kJ/m [463 Wh/m], should in theory be sufficient for any depth of mine down to 3000 m deep. The depth of charge–discharge will inevitably be affected by the depth of mine but even considering depths to 420 m will imply a DOD of 14% and to 100 m a mere 3.3%. These values will naturally vary slightly as a result of recovering kinetic energy in addition to potential energy and variation of truck EVM due to changes in the ERS mass. Using an approximation of the results from Wang et al. [37] and the cycle life of 3000 at 100% DOD (See Table 17.2 Rows 13 and 14) these equate to approximately 21,400 and 90,000 charge–discharge cycles at DOD of 14 and 3.3%, respectively.

Considering the low DOD values, the implications on battery service life will be reconsidered once the cycle duration and number of cycles as influenced by the depth of mine is determined as part of the simulation in Sect. 17.4. For EDLC and for LIC solutions, where the energy storage capacity is the limiting factor, solutions for a 100 m deep descent is expected to weigh 20,131 and 9978 kg, respectively. Considering all the technologies presented in Table 17.2 and their expected impact on payload, LIC's seem to be the more likely to provide a beneficial ERS also offering cycle life significantly better than all but the EDLCs.

17.3.2 Electromechanical Flywheels (EM-FWs)

EM-FWs store energy as the kinetic energy of a rotating flywheel. A motor-generator speeds up and slows down the rotating flywheel to store and extract energy [52, 53]. The motor-generator is connected to the DC link through a bidirectional inverter. Flywheels are capable of exceptionally high cycle life [52] (at least an order of magnitude greater than that of the best chemical batteries) and can be designed for high power delivery and recovery rates. However, their specific energy values do not compare favourably with high-quality chemical batteries, as they are limited by the maximum allowable rotor speeds. Specific energy values of between 10 and 30 Wh/kg are given in [47] and can be calculated from information in Werst [54]. Unfortunately, it is not clear whether the mass of needed auxiliary equipment is reflected in the quoted mass values. Compared to other energy storage technologies, flywheels tend to lose stored energy more quickly. However, high-performance flywheels have spin down times between 34 and 93 h [55]. These times are dependent on the quality of the vacuum and the bearings used, with magnetic bearings performing better than roller bearings. Flywheels using roller bearings have an energy loss rate of approximately 21% per hour [33]. With the cycle times of the haul trucks being in tens of minutes, the self-discharge loss of high-performance flywheels is expected to be small over the duration of a cycle but it may not be negligible when the design uses roller bearings.

Flywheel energy return efficiency values have been quoted as 85% in Beacon Power [56], which includes two power conversion processes on each of the storage and return stages [57]. On DEMHT, one of these power conversions will be avoided from the DC link to the flywheel storage so we expect a higher return efficiency.

There is a perception of danger associated with flywheels aggravated by news reports of flywheel-related failures [58, 59]. What appears to be less well known is that a significant number of flywheel storage applications have been safely in use for many years [60]. To minimises risk and ensure responsible and safe development of the technology, a number of approaches discussed in some detail in Hansen

Table 17.3 A selection of flywheel designs and the parameters of a concept design for use in a 100 m deep mine

	a.	b.	c.	d.	e.	f.
1.		Units	Existing designs			Conceptual design
2.	Example of use or test application		NASA [62]	ALPS [32, 63]	ATB [64, 65]	100 m mine
3.	Example of prominent or sole research organisation or manufacturer		University of Texas at Austen Center for Electromechanics			–
4.	Design mass	kg	113.4 [54]	8618 [54]	204 [54]	4574
5.	Peak power	kW	5 [54]	2000 [54]	150 [54]	–
6.	Rated power (50% of peak)	kW	2.5*	1000*	75*	1390
7.	Energy storage capacity	kWh	3.6 [54]	133 [54]	1.9 [54]	46.3
8.	Specific energy (system)	Wh/kg	31.7*	15.4*	9.31*	10.12
9.	Specific power (system—charging or discharging)	W/kg	22.1*	116.0*	367.6*	303.9
10.	E-rate (charge rate/E-storage capacity)	h^{-1}	0.69*	7.52*	78.9*	30.0*
11.	Service life	Cycles	–	–	–	100,000+

Information marked with * are best estimates based on accessible information

and O'Kain [60] including "absolute containment" should be considered, as is done by research organisations such as Ricardo [61].

Flywheels are attractive for applications where a long service life with frequent charge–discharge cycles at high power are some of the key drivers of the design solution. A mine haul truck is such an application.

In contrast to chemical batteries and ultra-capacitors, the ratio between specific power and specific energy for EM-FW can be varied during the design stage (Table 17.3 Row 10). Specifically, varying the mass moment of inertia or maximum rotational speed of the rotating mass enables manipulation of the storage capacity, while changing the design of the motor-generator and power conditioning hardware changes the charge and discharge power rating. For an ERS on a mine haul truck, the required charge/discharge rates do not change with pit depth, essentially fixing the motor-generator and inverter mass. However, the required storage to recover all available potential energy increases proportionately with depth. Hence, the ideal ratio between specific energy and specific power increases as the depth of the mine increases. To size a minimum mass flywheel for given depth, it is necessary to find an appropriate match of the specific energy and specific power of the flywheel ERS. We evaluated existing designs from the University of Texas at Austin's Center for Electromechanics (UTA CEM) summarised in Werst

Table 17.4 Flywheel ERS characteristics as determined by pit depth for a 250 t EVM truck. Bold value reflects expected mass for 100 m deep pit

P_{rec} = 1390 (kW)		Required energy storage (kWh)	Mass of FW storage (kg)	Specific power (SP) (W/kg)	Specific energy (SE) (Wh/kg)	E-rate (SP/SE) (h^{-1})
Height of descent (m)	10	4.63	920	1510.6	5.03	300.0
	20	9.26	1491	932.2	6.21	150.0
	30	13.89	1978	702.9	7.02	100.0
	40	18.52	2416	575.3	7.66	75.0
	50	23.15	2823	492.5	8.20	60.0
	100	46.30	**4574**	303.9	10.12	30.0
	200	92.60	7412	187.5	12.49	15.0
	300	138.90	9830	141.4	14.13	10.0
	400	185.20	12,010	115.7	15.42	7.5
	500	231.50	14,029	99.1	16.50	6.0
	1000	463.00	22,734	61.1	20.37	3.0
	1200	555.60	25,812	53.9	21.53	2.5

[54] to suggest some trends which could be used to provide approximate minimum mass values. The flywheels are the NASA—[62], ALPS—[32, 63] and ATB—[64, 65] flywheel designs. Table 17.3 shows the specific energy and specific power calculated from information in Werst [54]. Assuming that the torque characteristic of the motor-generator is relatively constant across the operating range, the torque-limited power would vary between 50 and 100%. As a result, the rated power reflected in Table 17.3 was assumed to be only 50% of the peak power values. In the Ragone plot in Fig. 17.5 the specific energy and specific power characteristics of the three UTA CEM flywheel designs are plotted with those of other energy storage technologies.

On the Ragone plot two trends can be observed for the three flywheel designs reflected in Table 17.3. The specific energy decreases marginally as the specific power increases, and the three flywheels almost sit on a straight line on the log versus log scale plot which implies approximating an exponential relationship. Fitting a curve to these data points gives specific energy $E_{\mathrm{Wh/kg}}$ of a flywheel-based energy storage system as a function of its specific power $P_{\mathrm{W/kg}}$. This function is given by

$$E_{\mathrm{Wh/kg}} = 122.369\, P_{\mathrm{W/kg}}^{-0.43595569}. \tag{17.16}$$

In this chapter, it is assumed that the power rating for charge and discharge for a particular flywheel design is the same. The validity of this assumption is supported by information from Beacon Power [57]. Using Eq. 17.16, it is possible to determine an approximate mass of the lightest flywheel-based ERS capable of providing the energy storage power and energy storage capacity requirements as it varies for

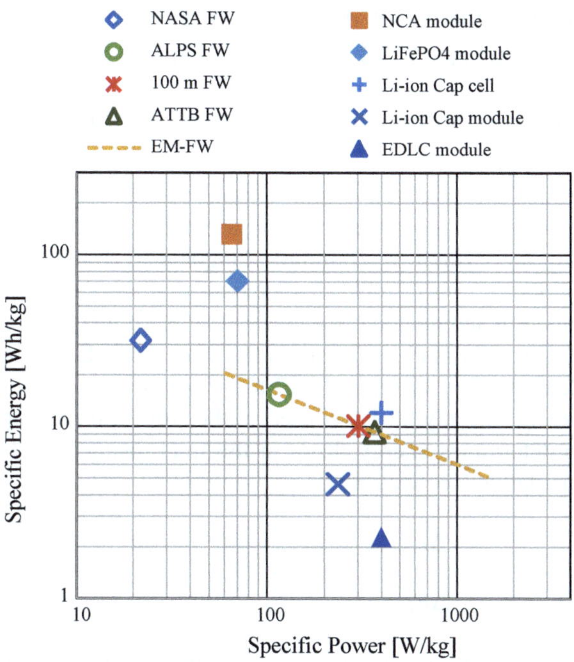

Fig. 17.5 Specific energy and specific power of a selection of energy storage technologies when charging. EM-FW (produced plotting the data from Table 17.4) reflects the mass dependent characteristics of a series of flywheel designs each designed to provide the best combination of specific power and specific energy for different depths of mine. High specific power and lower specific energy (to the lower right of the curve) is required for shallow mines while the lower specific power and higher specific energy is required for deeper mines

different depths of mine. Values appropriate for the generic 250 t EVM truck are reflected in Table 17.4. In Fig. 17.5 the position of a flywheel-based ERS design providing the 46.3 kWh as needed for a 100 m pit is indicated as '100 m FW' and its mass is given as 4574 kg in Table 17.4. We used the other data points in Table 17.4 to generate the dashed line denoted 'EM-FW' in Fig. 17.5. The line suggests the range of flywheel designs that would satisfy the spectrum of surface mine depths currently mined across the world.

At 4574 kg the EM-FW-based ERS is substantially lighter than that based on $LiFePO_4$, EDLC and LIC technologies for a 100 m pit application. It is expected that, with appropriate design or through a service life extension involving appropriate replacement of low-cost elements such as bearings, it would be possible to match the required cycle life.

17.4 Simulation Study of Mine Haul Truck with ERS Fitted to Assess Fuel Savings Benefit

17.4.1 Modelling, Simulation and Input Parameters

Simulation provides a cost-effective and flexible way to evaluate the effect of an ERS on truck fuel consumption. We, therefore, developed an appropriate simulation programme using the Python programming language. The simulation programme allows one to input and vary truck characteristics, storage device characteristics, haul route information and related parameters. It is based on the approach presented in Chap. 2 of Guzzella and Sciarretta [19] and uses Euler integration with Eqs. 17.1, 17.2, 17.5, 17.6, 17.7 and 17.12 to model the linear motion of the truck. A small step size, dt, of only 0.01 s is used in the simulation to improve accuracy. The model also determines energy flows in the truck using Eqs. 17.17 and 17.18 with variables as defined in Sect. 17.2,

$$\Delta E_{\text{pot,rec}} = \eta_{\text{drv}} F_{\text{b,rec}}(t) v_{\text{v}}(t) \text{d}t \tag{17.17}$$

$$\Delta E_{\text{kin,rec}} = \eta_{\text{drv}} F_{\text{b,rec}}(t)\, v_{\text{v}}(t) \text{d}t \tag{17.18}$$

While executing a haul cycle, a truck can be in one of six primary operational states or modes. The states are: (a) being stationary with the engine running (such as at the loading device or dump); (b) accelerating to attain target speed on level ground; (c) maintaining a constant speed on level ground; (d) decelerating on level ground; (e) maintaining a constant target speed while descending the ramp; or (f) ascending the ramp under full power. For a real truck operating on a real haul cycle, these states can change often and quickly. For the simulation, it is assumed that each of these states lasts for some time.

In addition to providing propulsion power, the engine also powers a number of auxiliary systems. In this research, the power draw for auxiliary services is denoted P_{Aux} and it is assumed to demand a constant power draw at the flywheel of the engine of 146 kW while the engine is running. In the simulation programme, it is assumed that the maximum traction force, determined using Eq. 17.2, is used when accelerating or ascending the ramp, with P_{eng} = 2390 kW at the engine's flywheel. Some of the energy produced is lost due to generator (η_{gen}) and drive system (η_{drv}) inefficiencies (as per Eq. 17.2), in overcoming rolling resistance (Eq. 17.5) and air resistance (Eq. 17.7). The remainder of the energy is stored in the vehicle by virtue of its mass, either as kinetic energy through increased speed of the truck ($\Delta E_{\text{kin}} = 0.5\ m_{\text{v}}(v_2^2 - v_1^2)$) or as potential energy due to an increase in elevation (h) of the truck ($\Delta E_{\text{pot}} = m_{\text{v}} g(h_2 - h_1)$), or a combination of both.

When the truck is decelerating or descending into the pit, an opportunity arises to recover some of the energy stored in the truck by using the dynamic brake system to apply a braking force to decelerate or restrain the vehicle. The amount of energy returned to the DC link is reduced by drive system (η_{drv}) inefficiencies, overcoming rolling resistance (as defined by Eq. 17.5) and some air resistance (Eq. 17.7).

The rate at which an ERS can store energy from the DC link is the product of the particular technology's specific power (as reflected in Table 17.2 Row 9 for $LiFePO_4$, EDLC and LIC-based ERSs) and the system mass. However, for an EM-FW-based ERS, the specific power is assumed to be 1390 kW/ERS mass (for a mass of 4574 kg the specific power would thus be 303.9 W/kg). When energy is returned from the in-wheel generators to the DC link at a rate that exceeds the ERS charging rate, the excess energy would be diverted to the resistor grid and would be lost. In the simulation it is assumed that the maximum braking force, determined using Eq. 17.12 and P_{rec} = 1390 kW, is used when decelerating or descending the ramp.

The storage capacity of the ERS is determined by its mass and specific energy. The specific energy is reflected in Table 17.2 Row 5 for $LiFePO_4$, EDLC and LIC-based ERS and determined from Eq. 17.16 for EM-FW. Once the ERS has been recharged to capacity all excess energy will be diverted to the resistor grid and lost. Due to a lack of access to reliable efficiency information, appropriate for different technologies over a range of power levels, we assumed a fixed storage efficiency of 90% for charging and for discharging giving a return cycle efficiency (DC link-storage-DC link) of 81% for all the technologies investigated.

The recovered energy can be used in several ways including productivity improvement [15], fuel consumption reduction [14] and emissions reduction [66]. In the research for this chapter, stored energy was used to reduce overall fuel consumption and was, therefore, only used in lieu of required engine power. The simulation model is thus set-up to use stored energy (when available) in lieu of engine power as soon as an opportunity arises. In practice, this approach would minimise the size of storage required and minimise losses of energy while in storage. As the truck travelling along the haul route is simulated, the code determines at each instant whether an increase or decrease in speed is required based on the truck speed and the sector target speed (see Table 17.5). If a speed increase is required, the amount of energy required for the next time step is calculated; if it is more than what can be provided by the engine, it is capped within the engine's capacity. If energy is available from storage it is used to replace an amount of engine power. The code does this in such a way that the power at the wheels is the same whether provided by engine power alone or partially provided from storage. The storage energy release rate is limited to the lower value of either the max ERS discharge rate (based on the technology used and mass) or 410 kW. If an ERS is capable of capturing all recoverable energy during the ramp descent, we found that an energy release rate of about 410 kW maximises reuse of stored energy, assisting acceleration of the loaded truck as well as providing an even release of energy along the ramp ascent.

The simulation model also determines and keeps a record of the amount of fuel consumed continuously powering auxiliary services and propelling the truck as needed. The fuel consumption is calculated by determining the product of the amount of power delivered by the engine ($P_{eng}(t)$ + P_{Aux}) at the flywheel and the

Table 17.5 Haul route definition

Sector number	Sector length (m)	Grade (%)	Speed limit (m/s)	Truck mass (kg)	Time delay (s)	
0	25	0	0.625	250,000	40	Unladen
1	125	0	4.167	250,000	0	
2	175	0	12.500	250,000	0	
3	50	0	8.333	250,000	0	
4	Calculated	−10	8.333	250,000	0	
5	250	0	12.500	250,000	0	
6	100	0	4.167	250,000	0	
7	20	0	2.000	250,000	0	
8	1	0	4.167	570,000	120	Fully loaded
9	100	0	4.167	570,000	0	
10	250	0	8.333	570,000	0	
11	Calculated	10	8.333	570,000	0	
12	50	0	8.333	570,000	0	
13	175	0	12.500	570,000	0	
14	125	0	4.167	570,000	0	
15	25	0	0.48	570,000	0	

duration (dt), using an assumed brake specific fuel consumption (BSFC) of 210 g/kWh. The use of recovered energy reduces the required P_{eng}, thereby reducing the amount of fuel consumed.

Table 17.5 shows the haul route definition used in the simulation. It starts with the truck waiting to unload and unloading outside the pit, the truck driving towards the entry ramp whilst adhering to different speed limits, driving down the ramp (assumed to have a 10% grade) while limited to 30 km/h (8.333 m/s), driving to and being loaded at the loading point, and returning in a loaded condition. The haul cycle includes 1471 m of flat haul and manoeuvring in addition to the on-ramp distance, which is calculated from the pit depth and ramp grade.

17.4.2 Using the Simulation Model

To generate results, the code is used to repeatedly simulate the truck operation along the haul cycle defined in Table 17.5 for a given pit depth. In the first cycle, the parameters of the basic truck are used with no ERS. This establishes a baseline for comparing fuel consumption when an ERS is added. In subsequent cycles, the ERS mass increases incrementally by 250 kg and repeats simulation of the haul cycle until the maximum ERS mass (typically 5 or 10% of the truck EVM) is reached. The ERS mass is added to the EVM of the truck, and consequently reduces payload by the same amount. For each cycle, the model determines the fuel consumption and records the payload to enable calculation of the fuel savings per tonne mined. The same procedure is used for each of the technologies reflected in Fig. 17.8 and for each of the range of depths in Fig. 17.9.

17.4.3 Results

17.4.3.1 Energy Recovered and Fuel Savings Expected in a 100 m Deep Mine

Figure 17.6 shows results obtained from the simulation model, reflecting the use of a 4500 kg flywheel-based energy storage system. This mass is very close to the mass required for a flywheel system that captures all energy from a 100 m deep pit. This implies that potential energy can be captured at the rate at which it becomes available with nothing lost to the resistor grid. Examination of the 'Ramp Descent' stage in Fig. 17.6 shows that the recovery of potential energy (as evident from the steep rise in the 'Energy in store') makes a much more significant contribution to the amount of recovered energy than recovering kinetic energy. The ratio between potential and kinetic energy will naturally depend on the depth of the mine.

The graphs reflected in Fig. 17.7 are the results of gradually increasing the mass of the energy storage system while adjusting its specific energy and specific power characteristics using Eq. 17.16. Based on the assumed mass, the flywheel parameters are recalculated to capture the energy at the rate of 1390 kW and store the

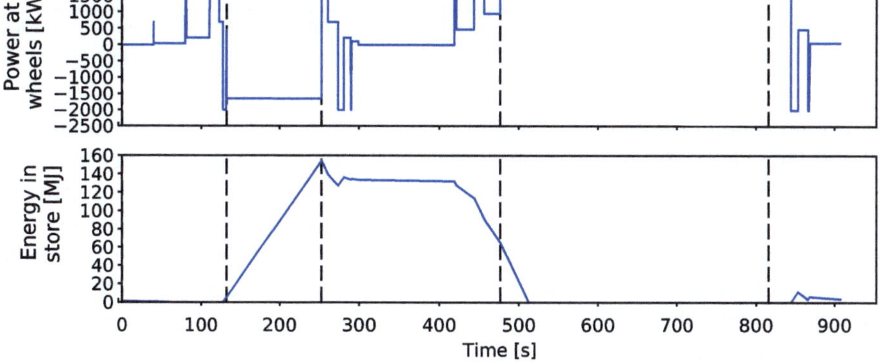

Fig. 17.6 Speed and power trajectories of DEMHT on haul route with 4500 kg flywheel-based ERS and resulting energy in storage as it varies over the haul cycle

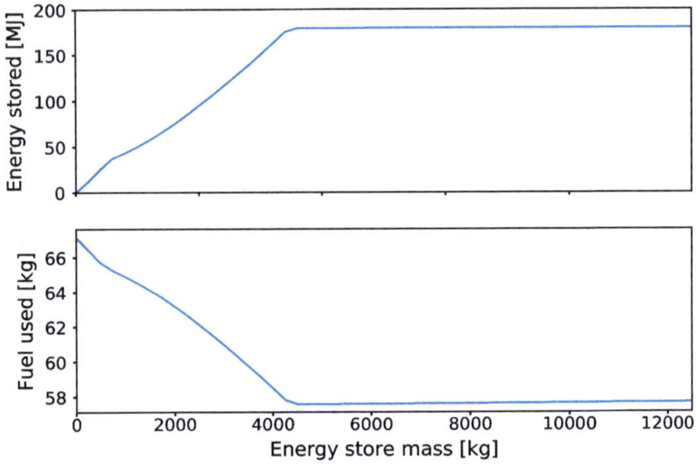

Fig. 17.7 Total amount of energy stored to flywheel-based ERS and fuel used to complete 100 m deep haul cycle as influenced by mass (hence capacity) of ERS

Fig. 17.8 Expected fuel savings per tonne hauled as it varies with ERS mass for indicated technologies in a 100 m deep mine 'EM-FW'—electromechanical flywheel; 'LIC'—lithium-ion capacitor; 'EDLC'—electrolytic double-layer capacitor; 'LiFePO$_4$'—lithium iron phosphate battery

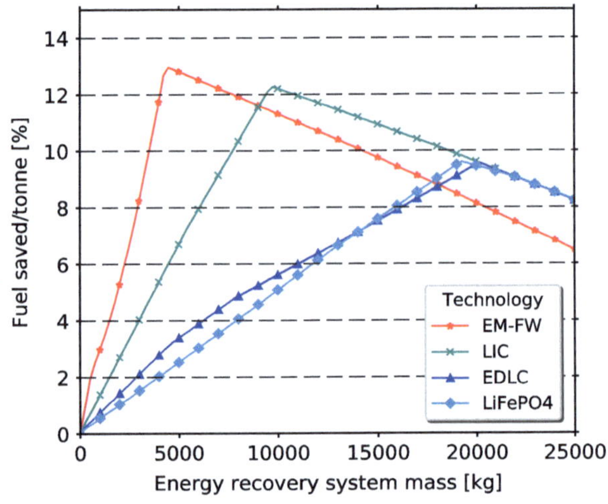

maximum amount of energy possible. Mass is increased up to a maximum of 12,500 kg which is 5% of the empty vehicle mass and slightly more than needed for a 420 m deep descent, as indicated in Table 17.4. In Fig. 17.7 the 'Energy stored' and 'Fuel used' curves flatten out beyond approximately 4500 kg because the additional storage capacity provides no additional benefit as all the energy available from the 100 m deep descent can already be captured by the system. Although the amount of fuel used remains essentially constant after this point, the additional mass becomes a hindrance as it reduces payload and hence increases the fuel used per tonne hauled. This is clearly visible in Figs. 17.8 and 17.9 in the steep decline in curves after the peak point.

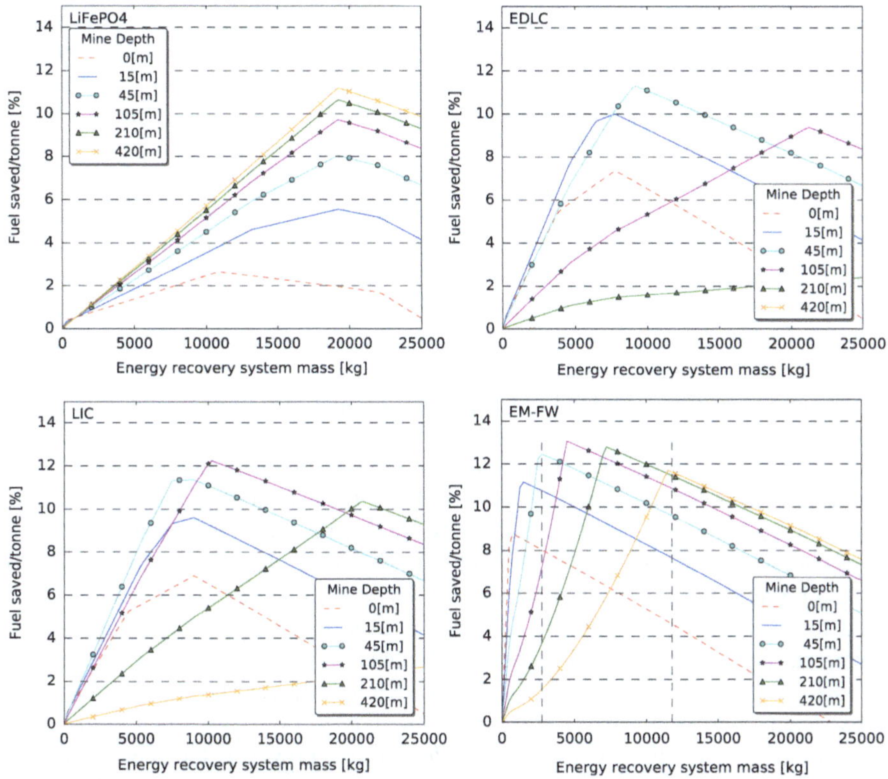

Fig. 17.9 Expected fuel savings per tonne hauled for a selection of mine depths and mass for indicated ERS technologies

Figure 17.8 compares the fuel savings benefit of the different ERS technologies as it changes with mass on-board the DEMHT in a 100 m deep mine. With sufficient mass all the technologies are able to provide a noteworthy fuel savings benefit. However, the EM-FW shows the most promising results achieving much greater improvement in fuel consumption per tonne hauled at a significantly lower ERS mass. The LIC-based ERS also achieves good results but uses a system with more than twice the mass to get comparable results.

17.4.3.2 Fuel Savings Expected Over a Range of Mine Depths and Fuel Saving Benefit of Oversizing Versus Under-Sizing ERS

The collection of graphs in Fig. 17.9 shows the fuel savings benefit expected for each of the ERS technologies for a range of mine depths as determined by the ERS mass. The '0[m]' route implies the recovery of kinetic energy only. As the depth of the mine increases and potential energy recovery becomes the predominant recoverable energy source, a larger and heavier ERS tends to provide greater benefit for that depth of mine by capturing more of the recoverable energy leading to greater fuel efficiency up to the peak point. The peak point at the top of each curve indicates the point where the storage capacity matches the amount of energy recoverable for the specific depth of descent using the ERS technology under consideration. The best peak point for each technology is closely related to the point where the E-rate (see Table 17.2 Row 8) of the technology and the power to energy ratio of the particular depth is the more closely matched. For EDLC technology the matched depth would be between 15 and 45 m and for LIC it would be between 45 and 105 m. The EM-FW's ability to be 'designed to requirement' enables improved performance at a greater number of depths.

The graphs in Fig. 17.9 tend to have a steeper approach towards the peak points with a gentler departure. This suggests that a proportionately undersized ERS performs worse in terms of fuel savings than a proportionately oversized ERS. The 45 and 420 m curves for 'EM-FW' clearly shows the benefit of an oversized ERS in terms of its ability to provide improved fuel efficiency. The best mass of a flywheel-based ERS sized for the 420 m depth of descent is approximately 11,750 kg and it is expected to provide 12.4% fuel saving per tonne hauled when used for a mine at this depth. If this ERS is used in a 45 m deep mine, the fuel saving per tonne hauled is still expected to be close to 9.8%. In contrast, the mass of an ERS sized to the requirements of a 45 m depth of descent is about 2750 kg and is expected to provide a fuel saving per tonne hauled of 12.6% when used with a 45 m deep descent. If used in a 420 m deep mine, however, the fuel saving per tonne hauled is less than 1.6%. In spite of the excess mass, the larger ERS still performs well even in shallow operations by providing a significant fuel consumption reduction. Admittedly, the larger ERS will inevitably be more expensive to acquire and to maintain and harder to accommodate on-board. Due to its impact on productivity, it is not expected to be viewed favourably by mining companies.

17.4.3.3 Cost Implications and Service Life Expectations

In this section, the cost of procuring and installing an ERS is compared to the reduction in the cost of fuel over the lesser of the truck or ERS's service life, for a 100 m deep mine. This analysis is not comprehensive, as other factors such as capital costs, maintenance, labour and effects on productivity, would need to be considered before adopting the technology. However, it provides a preliminary estimate of the benefit of installing an ERS.

Finding reliable cost data for the various technologies is challenging and cannot be done with great accuracy without full appreciation of the system design and costing requirements. Ricardo [67] provides some guiding values for lithium-ion batteries, EDLC and EM-FW while the cost per kWh for LIC is quoted in [34]. In Table 17.6, we use the various costs in combination with the minimum installed energy storage capacity that will satisfy the more demanding of the power and the storage capacity requirement (refer Table 17.2 Row 12), to provide a cost estimate for each ERS type.

Although the cost per kWh of lithium-ion batteries is significantly lower than that of the other technologies, the initial cost of the installation, expected to be approximately US$417,000, is significantly greater than that of the EM-FW and LIC technologies and almost the same as the EDLC technology. While the other technologies are expected to last as long as the truck, batteries may need to be replaced after a few years of service. The service life is closely linked to the DOD and would, therefore, be very dependent on the pit depth. In Table 17.7, the cycle time as well as the DOD (both obtained from the simulation model) is shown against different depths. Assuming 6000 h of operation per year the cycle time can be used to calculate the number of haul cycles per year. Using an approximation of the results from Wang et al. [37] and the cycle life of 3000 at 100% DOD for $LiFePO_4$ from Table 17.2 Row 13 and 14, the corresponding service life can be calculated. Table 17.7 clearly shows that the service life in all but the shallowest of mines (less than 45 m deep) would be less than half of the truck's 10 year life. If used in a 100 m deep mine, it will be about 3.5 years suggesting that it would need to be replaced twice during the life of the DEMHT.

Table 17.8 provides an indication of the potential of different technologies to provide payback if used in a 100 m deep mine. The simulation model was used to provide an estimate of the fuel cost to move a given amount of material using the following assumed values: 6000 operating hours per year, fuel cost is US$0.623/l (US$2.358 per US gallon), and fuel density is 0.840 kg/l. Cycle time is determined to be 908 s giving 23,784 cycles per year. Because the payload is reduced by the ERS but the number of cycles fixed, using a payback period in terms of years would skew the comparison with some of the technologies getting advantage from moving a smaller amount of material. To avoid this, the model was used to determine the fuel cost of moving 7,610,880 tonnes (the amount of material moved by an unmodified truck in one year) of material using each of the technologies. From Table 17.8, it is clear that the service life of lithium-ion batteries is similar to its expected payback period therefore holding very little if any potential for use on-board DEMHT. Due to the high initial cost and weaker performance in a 100 m deep mine the EDLC also does not perform as well as the LIC and EM-FW. (To enable comparison, the impact on productivity of the different technologies is also shown in Table 17.8.)

Engine wear and the amount of fuel consumed are closely related [70] (p. 155). If an improvement in fuel consumption better than 12% can additionally achieve a proportional reduction in engine hours and proportionate increase in engine time between overalls (TBO), the presence of an ERS is expected to leverage economic benefit that is greater than just the fuel savings.

Table 17.6 ERS cost estimates for 100 m deep mine

Technology	Cost $[(kWh)^{-1}]$	Assumed cost $[(kWh)^{-1}]$	Energy storage required to satisfy power requirement (kWh)	System cost
Li-ion batteries	US$200–US$360 [67] US$300 [68]	US$300	1390	US $417,000
EDLC	US$9000 [67]	US$9000 [67]	46.3	US $416,700
LIC	US$5000 [34]	US$5000 [34]	46.3	US $231,500
EM-FW	US$4000 [67] US$3872[a] [69]	US$4000 [67]	46.3	US $185,200

System cost is the product of 'Assumed cost' and required installed energy storage capacity
[a]This value is calculated using only equipment cost reflected in [69]

Table 17.7 Influence of mine depth on cycle time and implications for service life for li-ion battery-based ERS

Depth of mine (m)	Cycle time (s)	Number of haul cycles per year	DOD (%)	Service life (years)
0	444	48,697	0.64	9.63
15	511	42,309	1.06	6.72
45	651	33,185	1.97	4.58
105	932	23,185	3.75	3.45
210	1423	15,181	6.92	2.86
420	2405	8980	13.2	2.53

17.5 Review of Technologies' Potential to Satisfy DEMHT ERS Requirements and Improve Fuel Efficiency

Comparing the mass of an EM-FW-based ERS suitable for a 100 m deep mine with the mass of ERS's based on other technologies, the flywheel is expected to provide a solution which is substantially lighter than any of the other solutions and at a lower cost (see Table 17.6). When considering the use of EM-FW over a range of mine depths a single size ERS design is not expected to be optimal and a number of designs would be necessary. A range of designs optimised with due consideration of the mine development plan is expected to provide the best solution. Considering the high cycle rates reflected in Table 17.7 care would be required in the design to ensure a suitable cycle life. The major downsides of the flywheel for the DEMHT application are the difficult operational environment (shock, vibration, dirt), and the

Table 17.8 Productivity and payback considerations

Technology	1 year's operation			To haul 7,610,880 tonnes				
	Payload average (kg)	Tonnes hauled per year	Productivity (%)	Fuel cost to haul 7,610,880 t (US$)	Fuel cost savings (US $)	System cost (US $)	System cost/Fuel cost savings	Service life of ERS (years)
No ERS	320,000	7,610,880	100.00	1,185,320	0	0	–	–
Li-ion bat	301,913	7,180,699	94.35	1,077,335	107,985	417,000	3.86	≈3.5
EDLC	299,869	7,132,084	93.71	1,071,354	113,967	416,700	3.66	10
LIC	310,022	7,373,563	96.88	1,039,881	145,440	231,500	1.59	10
EM-FW	315,426	7,502,092	98.57	1,028,100	157,220	185,200	1.18	10

Due to impact on productivity the **System cost: Fuel cost savings** ratio approximates but is not the same as payback period in years likely need for a gimbal in installation. With these issues resolved it shows great potential to improve fuel efficiency in the mine haul process.

LIC provides a good combination of specific energy and specific power potentially satisfying a useable range of mine depths. Although heavier and somewhat more expensive than the EM-FW-based design, the LIC solution may, in practice, have a number of other advantages over that of flywheel designs. Some of the benefits expected include the ability to segregate the storage system and distribute modules to suitable locations on-board the truck easing integration. A modular design based on LIC may also provide a comparatively simpler way of increasing storage capacity as the depth of a mine increases, with the growth of the storage system being an extension of an existing design, rather than a new design as is expected for EM-FW. LIC's avoid the need for a bulky gimbal and the element of risk associated with flywheels. Although LIC's are still expensive compared to lithium-ion batteries ($5000/kWh [34] vs. $300/kWh [68] in 2015), it may be that their cost could reduce to the same extent as has happened for lithium-ion type batteries, the cost of which has reduced from over $1000/kWh in 2007 to around $300/kWh in 2015 [68]. It may become even more cost-effective in coming years. A significant number of capacitors will be necessary to provide the required storage capacity and would require appropriate control to ensure even distribution of temperature and charge amongst cells. This is expected to be a challenging aspect of design. LIC's may also not be as tolerant to temperature extremes sometimes encountered at mine sites.

A beneficial attribute of the battery-based system is that an installation that is sufficient for a 100 m deep mine would also be sufficient for mines that are much deeper, requiring no change to the installed system due to the power capacity being the tight design requirement. However, despite this beneficial attribute, the results of Sect. 17.4.3.3 show that lithium-ion batteries are not expected to provide economically beneficial use in the ERS of DEMHT.

The EDLC is expected to perform reasonably well in shallow mines with best performance around 45 m deep. However, as a result of its low specific energy, deeper mines are shown to demand a dramatic increase in mass to accommodate the additional energy.

17.6 Conclusions

This investigation demonstrates that there is significant potential for an energy recovery system (ERS) to reduce fuel consumption per tonne hauled for diesel-electric mine haul trucks (DEMHTs). However, this benefit is dependent on the technology used in the ERS and the depth of the mine. For typical mine and truck characteristics, up to 68% of the potential energy lost during the decent can be recovered to be reinjected, reducing fuel consumption per tonne hauled by as much as 10–12% for a mine of an appropriate depth.

Designing an ERS for DEMHTs presents a challenge to all modern energy storage technologies because it requires high power, large storage capacity and high cycle life while limiting impact on payload. Chemical batteries do not have a sufficient specific power or cycle life and prove to be too expensive for this application. Electrolytic double-layer capacitors (EDLCs) show some potential for shallower applications but the expected installation cost and the mass required, due to low specific energy of EDLC, make it an unattractive proposition. In contrast, the design of an electromechanical flywheel (EM-FW)-based ERS using available technology offers the potential to be sized for a given depth of mine. Although it does not offer a high specific energy, flywheel-based storage technology allows the ERS to be customised to the power and storage requirement of the application during design. This technology, therefore, promises a lighter, more balanced storage solution that can be designed to provide a cycle life to match the design life of DEMHTs. Based on available cost information, it also promises a comparatively short estimated payback period of 1.2 years. Although adding mass on-board mine haul trucks is contrary to the impetus put on reducing empty vehicle mass (EVM), the results presented in this chapter show that even if the flywheel-based ERS is relatively oversized, implying a greater than needed impact on the EVM of the truck, it could still provide an improvement in fuel consumption per tonne hauled greater than 10%. Although it is recognised that there are significant technical challenges associated with the use of flywheels on-board mobile applications, these have been dealt with in a number of existing designs [13–15, 57, 60], which includes a recently developed underground mining loader.

Lithium-ion capacitors (LICs) are not as flexible in optimising the power and energy storage capacity to the specific demands of the application as flywheel designs potentially are. However, LICs can facilitate a modular design and allow segregated installation while avoiding some of the negative aspects of EM-FW. They also have the potential to provide a more desirable solution, especially, for mining applications around 45–105 m.

Further work is required to more accurately determine the costs and benefits of an ERS on DEMHT over its operating life within mining operations, which typically requires the truck to operate over a range of depths. Specifically, the technical feasibility of modular ERSs, of which the capacity can be changed with the depth of the mine, and the effect of a fixed-size ERS over multiple depths need to be investigated. The installation of the ERS will also influence other factors that affect the operational cost and/or productivity of the DEMHT, such as the return on capital, the cost per tonne of maintenance, labour, engine wear and tyre wear. The negative impact of the ERS on productivity may be compensated for, or productivity may even be improved if using stored energy to increase the power and hence speed on the ramp ascent rather than only aiming to increase fuel efficiency. Investigating the trade-off between using the ERS for increasing power and for increasing fuel efficiency is another important future research avenue.

References

1. XEMC (2011) Heavy equipment SF31904 Motorized Wheel Dump Truck (Xiangtan Electric Manufacturing Corporation Ltd). EveryChina. http://www.everychina.com/products/heavy_equipment_sf31904_motorized_wheel_dump_truck_welcome_to_xemc-32845373-za96a6d-detail.html. Accessed 29 Mar 2016
2. Caterpillar (2013) 795F AC Mining Truck (Doc. No.: AEHQ6882-01). Caterpillar, USA
3. Komatsu (2008) Komatsu 960E-1 Electric Drive Truck (Doc. No.: AESS647-00). Komatsu America Corp., USA
4. Liebherr (2014) Mining Truck T 284 (Doc. No.: LME 11481630-0.85-02.14_enGB). Liebherr, USA
5. Mining_Technology (2013) Top 10 deep open-pit mines. http://www.mining-technology.com/features/feature-top-ten-deepest-open-pit-mines-world/. Accessed 13 July 2015
6. Esfahanian E (2014) Hybrid electric haulage trucks for open pit mining. The University of British Columbia, Vancouver
7. 21CTP (2007) 21st Century Truck Partnership—Project Quad Sheets (trans: DOE/DOT/EPA). 21CTP-004 edn
8. Salasoo L (2004) Advanced hybrid propulsion and energy management system for high-efficiency, off-highway, 320-ton class, diesel electric haul truck (trans: Energy USDoEEEaR). Annual Progress Report for Heavy Vehicle Systems Optimisation, FY 2003 edn., Washington, D.C
9. Richter T (2006) Advanced hybrid propulsion and energy management system for high-efficiency, off-highway, 320-ton-class, diesel electric haul trucks (trans: Energy USDoEEEaR). Annual Progress Report for Heavy Vehicle Systems Optimization Program, FY 2006 edn., Washington, D.C
10. Richter T (2007) Advanced hybrid propulsion and energy management system for high efficiency, off highway, 240 ton class, diesel electric haul trucks (trans: Energy USDoEEEaR). Annual Progress Report for Heavy Vehicle Systems Optimization Program, FY 2007 edn., Washington, D.C
11. Maxwell_Technologies (2013) Maxwell Technologies Supplying Ultracapacitors to Caterpillar for Energy Recuperation, Power Assist in Fuel-Efficient Hybrid Mining Shovel. PRNewswire. http://www.prnewswire.com/news-releases/maxwell-technologies-supplying-ultracapacitors-to-caterpillar-for-energy-recuperation-power-assist-in-fuel-efficient-hybrid-mining-shovel-223115091.html. Accessed 13 Feb 2015
12. Caterpillar (2012a) Caterpillar announces development of largest hydraulic shovel. Caterpillar. http://www.cat.com/en_GB/news/machine-press-releases/caterpillar-announcesdevelopmentoflargesthydraulicshovel.html. Accessed 28 May 2015
13. Morey B (2013) Ricardo sees a future in flywheel hybrid excavators. SAE International. http://articles.sae.org/12282/. Accessed 05 June 2014
14. Flynn MM, McMullen P, Solis O (2008) Saving energy using flywheels. Ind Appl Mag IEEE 14(6):69–76. doi:10.1109/MIAS.2008.929351
15. Komatsu_Mining_Corp (2017) Hybrid Loaders—Joy 18 HD. https://mining.komatsu/product-details/joy-18hd. Accessed 08 July 2017
16. Siemens (2009) SIMINE CIS TR. http://www.industry.usa.siemens.com/verticals/us/en/mining/mobile-mining-solutions/Documents/Brochure_SIMINE_TR_en.pdf. Accessed 12 Mar 2015
17. Sciarretta A, Guzzella L (2007) Control of hybrid electric vehicles. Control Syst IEEE 27 (2):60–70
18. Tie SF, Tan CW (2013) A review of energy sources and energy management system in electric vehicles. Renew Sustain Energy Rev 20:82–102
19. Guzzella L, Sciarretta A (2013) Vehicle propulsion systems: introduction to modeling and optimization, vol Book, Whole, 3rd edn. Springer, New York, Heidelberg

20. Mazumdar J, Köllner W, Moghe R (2010) Interface issues of mining haul trucks operating on trolley systems. In: Applied power electronics conference and exposition (APEC), 2010 Twenty-fifth annual IEEE, Palm Springs, CA, 21–25 Feb 2010. IEEE, pp 1158–1165. doi:10.1109/APEC.2010.5433357
21. Caterpillar (2015) Caterpillar performance handbook, 45 edn. Caterpillar
22. Fiscor S (2013) Developing the drive system for the world's largest haul truck. Eng Min J 214 (10):2
23. Ghojel J (1993) Haul truck performance prediction in open mining operations. Paper presented at the national conference on bulk materials handling, Yeppoon QLD
24. Parreira J (2013) An interactive simulation model to compare an autonomous haulage truck system with a manually-operated system. The University of British Columbia, Vancouver
25. Lang R (2010) Increasing mine productivity with an appropriate mine truck body
26. Moore P (2014) Celebrating the workhorses. International Mining
27. BYD (2014) BYD Unveils world's largest battery electric vehicle. http://www.byd.com/news/news-256.html. Accessed 06 Nov 2015
28. Tesla_Motors (2015) Increasing energy density means increasing range. http://my.teslamotors.com/roadster/technology/battery. Accessed 20 Mar 2015
29. Voelcker J (2014) Toyota racks up 7 million hybrids sold since 1997. http://www.greencarreports.com/news/1094753_toyota-racks-up-7-million-hybrids-sold-since-1997. Accessed 28 May 2015
30. Burke A, Miller M (2011) The power capability of ultracapacitors and lithium batteries for electric and hybrid vehicle applications. J Power Sour 196(1):514–522. doi:10.1016/j.jpowsour.2010.06.092
31. Millikin M (2007) Toyota hybrid race car wins Tokachi 24-hour race; in-wheel motors and supercapacitors. Arlotto M. http://www.greencarcongress.com/2007/07/toyota-hybrid-r.html. Accessed 12 Mar 2015
32. Caprio MT, Murphy BT, Herbst JD (2004) Spin Commissioning and drop tests of a 130 kW-hr composite flywheel. In: 9th International symposium on magnetic bearings, Lexington, KY, Aug 2004, pp 3–6
33. Hearn CS, Flynn MM, Lewis MC, Thompson RC, Murphy BT, Longoria RG (2007) Low cost flywheel energy storage for a fuel cell powered transit bus. In: Vehicle power and propulsion conference (VPPC), Arlington, TX, 09–12 Sept 2007. IEEE, pp 829–836
34. Mueller M (2015) Life after lithium—power rangers. Electric and hybrid vehicle technology international, January 2015 edn. UKiP Media & Events, UK
35. SEEO (2014) DryLite (TM) Automotive Pack. http://seeo.com/drylyte-automotive-pack. Accessed 20 Mar 2015
36. Maxwell_Technologies (2014b) Datasheet 125 V heavy transportation module. Doc. No.: 1014696.7 edn
37. Wang J, Liu P, Hicks-Garner J, Sherman E, Soukiazian S, Verbrugge M, Tataria H, Musser J, Finamore P (2011) Cycle-life model for graphite-$LiFePO_4$ cells. J Power Sour 196(8):3942–3948. doi:10.1016/j.jpowsour.2010.11.134
38. Broussely M (2010) Battery requirements for HEVs, PHEVs, and EVs: an overview. In: Pistoia G (ed) Electric and hybrid vehicles—power sources, models, sustainability, infrastructure and the market. Elsevier, pp 305–345
39. Donahue RW (2013) Breakthrough energy technology successfully powers GE mining scoop. GE. http://www.getransportation.com/news/breakthrough-energy-technology-successfully-powers-ge-mining-scoop. Accessed 13 June 2015
40. FIAMM (2016) 110RW80. FIAMM. http://www.fiamm.com/en/oceania/industrial-batteries/products/110rw80-global-asia.aspx. Accessed 12 Feb 2016
41. A123 (2012b) Nanophosphate lithium ion prismatic pouch cell AMP20M1HD-A. http://www.a123systems.com/prismatic-cell-amp20.htm Accessed 11 Mar 2015
42. Panasonic (2012a) Panasonic lithium ion NCR18650A. http://industrial.panasonic.com/lecs/www-data/pdf2/ACA4000/ACA4000CE254.pdf. Accessed 19 Mar 2015
43. OXIS (2014) It's safer with OXIS lithium sulfur rechargeable batteries. OXIS

44. Maxwell_Technologies (2014a) Datasheet K2 ultracapacitors—2.7 V series. Doc. No.: 1015370.5 edn
45. Veneri O, Capasso C, Patalano S (2017) Experimental study on the performance of a ZEBRA battery based propulsion system for urban commercial vehicles. Appl Energy 185, Part 2:2005–2018. doi:http://dx.doi.org/10.1016/j.apenergy.2016.01.124
46. Song M, Zhang Y, Cairns EJ (2013) A long-life, high-rate lithium/sulfur cell: a multifaceted approach to enhancing cell performance. Nano Lett 13(12):5891–5899
47. IEC (2011) Electrical energy storage: white paper; Dec 2011. International Electrotechnical Commission, Geneva Switzerland
48. General_Electric (2015) Durathon batteries: battery power, Reimagined vol Brochure #20194-A. General Electric
49. Banas J, Peterson M (2012) Advances in ULTIMO™ lithium ion capacitor (LIC) technology. In: NCCAVS "Technology for Clean Air", San Jose, 22 Feb 2012
50. de Guibert A (2013) Advances of Li-ion use in industrial applications. Paper presented at the IBA2013 (International Battery Association Meeting), Barcelona, Spain, 12 Mar
51. Masih-Tehrani M, Ha'iri-Yazdi MR, Esfahanian V, Safaei A (2013) Optimum sizing and optimum energy management of a hybrid energy storage system for lithium battery life improvement. J Power Sour 244:2–10. doi:10.1016/j.jpowsour.2013.04.154
52. Onar O, Khaligh A (2015) Hybrid energy storage systems. In: Emadi A (ed) Advanced electric drive vehicles. Energy, power electronics, and machines. CRC Press, pp 283–316. doi:10.1201/b17506-9
53. Aneke M, Wang M (2016) Energy storage technologies and real life applications—a state of the art review. Appl Energy 179:350–377. doi:http://dx.doi.org/10.1016/j.apenergy.2016.06.097
54. Werst M (2010) Rotating machinery as energy storage and power management systems
55. Hearn CS (2013) Design methodologies for advanced flywheel energy storage
56. Beacon_Power (2014b) Beacon Power 450 XP Performance Specifications. Beacon_Power
57. Beacon_Power (2014a) Beacon power flywheel energy storage systems. Beacon_Power
58. Nearing B (2011) Flywheels fail at energy project. Timesunion
59. 10News_Digital_Team (2015) 11 K Pound flywheel caused poway explosion
60. Hansen JGR, O'Kain DU (2011) An assessment of flywheel high power energy storage technology for hybrid vehicles. Oak Ridge National Laboratory, USA
61. Wheals JC, To W, Dalby J, Vigar M, Hodgson J, Buchanan J, Robertson A, Macpherson J, Taylor J, Lanoe W, Heaton M (2015) Viable flywheel system for rail 1–22
62. Hebner R, Beno J, Walls A (2002) Flywheel batteries come around again. Spectr IEEE 39 (4):46–51. doi:10.1109/6.993788
63. Herbst JD, Caprio MT, Gattozzi AL, Graf C (2005) Challenges and solutions for the use of flywheel energy storage in high power applications. Paper presented at the EESAT 2005, Electrical Energy Storage and Technologies Conference, San Francisco, California, U.S.A., 17 Oct 2015
64. Hayes RJ, Kajs JP, Thompson RC, Beno JH (1999) Design and testing of a flywheel battery for a transit bus. SAE Technical Paper
65. Hayes RJ, Weeks DA, Flynn MM, Beno JH, Guenin AM, Zierer JJ, Stifflemire T (2003) Design and performance testing of an advanced integrated power system with flywheel energy storage. SAE Paper (01):2302
66. Grondin O, Thibault L, Querel C (2015) Energy management strategies for diesel hybrid electric vehicle. Oil Gas Sci Technol 70(1):125–141. doi:10.2516/ogst/2013215
67. Ricardo (2016) Energy storage systems. http://www.ricardo.com/en-GB/Our-Markets/Clean-Energy-and-Power-Generation/Energy-Storage-Systems/. Accessed 18 July 2016

68. Nykvist B, Nilsson M (2015) Rapidly falling costs of battery packs for electric vehicles. Nat Clim Change 5(4):4. doi:10.1038/nclimate2564
69. Akhil AA, Huff G, Currier AB, Kaun BC, Rastler DM, Chen SB, Gauntlett WD (2013) DOE/EPRI 2013 electricity storage handbook in collaboration with NRECA. Sandia Report: SAND2013-5131. Sandia National Laboratories, Albuquerque, USA
70. Caterpillar (2012b) Operation and maintenance manual 795F AC off-highway truck—excerpt (Doc. No.: SEBU8349–08). Operation and Maintenance Manual, USA

MIX
Papier aus verantwortungsvollen Quellen
Paper from responsible sources
FSC® C105338

If you have any concerns about our products,
you can contact us on
ProductSafety@springernature.com

In case Publisher is established outside the EU,
the EU authorized representative is:
Springer Nature Customer Service Center GmbH
Europaplatz 3, 69115 Heidelberg, Germany

Printed by Libri Plureos GmbH
in Hamburg, Germany